Natural Water

RETAINS ALL NATURAL MINERALS

20 FL. OZ.

SODIUM FREE
NO ADDITIONAL PROCESSING OR
MODIFICATIONS ARE REQUIRED FOR THIS
NATURAL WELL WATER FROM

Nutrition Facts
Serving Size 6 oz. (180ml)
About 2.5 Servings Per Container

Amount per serving

Calories 0   0%
Total Carbohydrate 0 g   0%
Protein 0 g   0%
                               Calcium 2%

*Percent Daily Values are based on
a 2,000 calorie diet.

♲ PLEASE RECYCLE

Bottled by RIMROCK WATER CO., RIMROCK ARIZONA 86335

ELEMENTARY ALGEBRA FOR COLLEGE STUDENTS

ELEMENTARY ALGEBRA FOR COLLEGE STUDENTS

Seventh Edition

IRVING DROOYAN
Los Angeles Pierce College, Emeritus

KATHERINE FRANKLIN
Los Angeles Pierce College

PRENTICE HALL, Englewood Cliffs, New Jersey 07632

Library of Congress Cataloging in Publication Data:

Drooyan, Irving.
  Elementary algebra for college students.

  Includes index.
  1. Algebra. I. Franklin, Katherine.
II. Title.
QA152.2.D748 1988   512.9   87-6180

TO THE STUDENT: A student's solutions manual for the textbook is available through your college bookstore to accompany this textbook. The solutions manual can help you with course material by acting as a tutorial, review, and study aid. If the solutions manual is not in stock, ask the bookstore manager to order a copy for you.

 © 1988, 1984, 1979 by Prentice-Hall, Inc.
A Division of Simon & Schuster
Englewood Cliffs, New Jersey 07632

Printed in the United States of America

10 9 8 7 6 5 4 3 2

ISBN 0-13-261439-1

Prentice-Hall International (UK) Limited, *London*
Prentice-Hall of Australia Pty. Limited, *Sydney*
Prentice-Hall Canada Inc., *Toronto*
Prentice-Hall Hispanoamericana, S.A., *Mexico*
Prentice-Hall of India Private Limited, *New Delhi*
Prentice-Hall of Japan, Inc., *Tokyo*
Simon & Schuster Asia Pte. Ltd., *Singapore*
Editora Prentice-Hall do Brasil, Ltda., *Rio de Janeiro*

PREFACE

This edition of *Elementary Algebra for College Students* retains the basic subject matter and pedagogy of the previous editions but it differs from them in several significant ways.

CHANGES IN THIS EDITION

Chapters 1, 2, and 3 have been extensively reorganized to introduce equations and word problems in a spiral approach. Problem solving is developed gradually and integrated with the introduction of standard algebraic topics. Simple equations that can be solved with one or two algebraic operations are treated in Sections 1.4 and 1.5, respectively. Word problems that can be modeled by these equations follow in Section 1.6. Here the emphasis is on first identifying the unknown quantity and then writing an equation for the conditions stated in the problem. This careful approach to constructing models is reinforced in Chapter 2 as students develop their skills using integers. Section 3.3, which is concerned only with constructing models for word problems, has been added to prepare the student for word problems in Sections 3.4 and 3.5, which require the use of the distributive law.

The careful step-by-step approach used to solve word problems in Chapters 1 to 3 is continued in Section 4.6, which involves formulas from geometry; in Sections 7.8 and 7.9, which involve algebraic fractions; in Section 9.5, where systems of equations are used as models; and in Section 10.4, which involves quadratic equations.

Instructions for those exercise sets that involve word problems specify that students follow the step-by-step method emphasized in the text. An appropriate model for each odd-numbered word problem is given in the answer section.

The text has also been strengthened by broadening the scope of many topics and by incorporating greater variety and more challenging problems into the exercise sets. Several new sections have been added, including Section 4.4 on the laws of exponents, Sections 11.7 and 11.8 on Nth roots and fractional exponents, and Section 12.6 on radical equations. In addition, existing material on scientific notation, use of formulas, operations with fractions, and graphing of quadratic equations has been expanded and strengthened. Several of the new sections are labeled optional and may be omitted without loss of continuity.

EXCERPTS FROM THE PREFACE OF THE PREVIOUS EDITION THAT ARE APPLICABLE TO THIS EDITION

This textbook has been written for students who are beginning their study of algebra at the college level and who are scheduled to complete two semesters of high-school work in one semester.

The general organization of the material is traditional. Algebra is developed as a generalized arithmetic, and the assumptions underlying the operations of both arithmetic and algebra are stressed. The textual material is brief. However, a large number of sample problems is included.

REVIEW MATERIAL

Subject matter is continually reviewed through the use of chapter and cumulative reviews at the end of each chapter. Ten additional cumulative reviews are included at the end of the book.

REFERENCE MATERIAL

Chapter summaries appear at the end of each chapter. A glossary of new terms introduced in the text is placed at the end of the book. In addition, a list of symbols introduced in the text appears on the inside front cover and a summary of the properties and operations appears on the inside of the back cover.

FEATURES

Annotations are used to highlight parts of examples.

Subheadings are used to highlight different topics of each section.

Common student errors are emphasized in the text, exercise sets, and cumulative reviews.

ANSWERS

Answers are provided for the odd-numbered exercises and for all exercises in the chapter reviews and cumulative reviews. A solutions manual containing completely worked-out solutions to all even-numbered exercises is available as a student supplement.

ACKNOWLEDGMENTS

We gratefully acknowledge the helpful comments from the reviewers of this edition: Vincent R. Aleksey, Grays Harbor College; Arthur P. Dull, Diablo Valley College; Deborah Cochener, Angelo State University; Mary Coughlin, University of Toledo; Charles F. Lindblade, Loop College; Robert Muksian, Bryant College.

We wish to thank William Wooton, a coauthor on previous editions, for giving us permission to prepare this edition.

We also extend a special thanks to our production supervisor Lucille Buonocore for her professional help in the preparation of this edition.

Irving Drooyan
Katherine Franklin

CONTENTS

ELEMENTARY ALGEBRA FOR COLLEGE STUDENTS

1 ELEMENTARY NOTIONS OF ALGEBRA USING WHOLE NUMBERS

In this book, we focus our attention on numbers. We will use the same procedures and symbols we used in arithmetic, together with certain new symbols. The vocabulary used in arithmetic will apply in algebra. In short, we will be studying arithmetic, but from a more general point of view.

1.1 ALGEBRAIC EXPRESSIONS: SUMS AND PRODUCTS

The numbers we use to count things are called **natural numbers.** The numbers

$$1, 2, 3, 4, 5, 6, 7, \underbrace{\ldots}$$

Read "and so on."

are natural numbers, whereas $\frac{2}{3}$, 3.141, and $\sqrt{2}$ are not.

When the number 0 is included with the natural numbers, the numbers in the enlarged collection

$$0, 1, 2, 3, 4, 5, \ldots$$

are called **whole numbers.** Thus, we can refer to numbers such as 2, 3, and 6 as natural numbers or whole numbers.

The natural numbers 2, 4, 6, . . . are called **even numbers,** and the natural numbers 1, 3, 5, . . . are called **odd numbers.** The even numbers are always multiples of 2. (The whole number 0 is even, since 0 is a multiple of 2.) A number is odd if it leaves a remainder of 1 when it is divided by 2.

Mathematics is a language. As such, it shares a number of characteristics with any other language. For instance, it has verbs, nouns, pronouns, phrases, sentences, and many other concepts that are normally associated with a language. They have different names in mathematics, but the ideas are similar.

In language, we use pronouns such as *he, she,* or *it* to stand in the place of nouns. In mathematics, we use symbols such as *x, y, z, a, b, c,* and the like, to stand in the place of numbers. Letters used in this way are called **variables.** In this chapter, variables will always represent whole numbers.

In a language, the verbs are action words, expressing what happens to nouns. In mathematics, operations such as addition, multiplication, subtraction, or division express an action involving numbers. The symbols we use for these operations, and the properties these operations have, are the same in algebra as in arithmetic.

SUMS

When we add two numbers *a* and *b* the result is called the **sum** of *a* and *b*. We call the numbers *a* and *b* the **terms** of the sum.

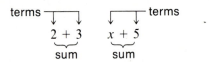

A number of different English phrases are used to indicate the operation of addition, including "the sum of," "increased by," "plus," "more than," "add," "exceeds," and "exceeded by."

PRODUCTS

When we multiply two numbers *a* and *b*, the result is called the **product** of *a* and *b*. We call numbers *a* and *b* the **factors** of the product. In arithmetic, we used the symbol × to represent multiplication. But in algebra, the symbol × may sometimes be confused with the variable *x*, which we use so frequently. So in algebra we usually indicate multiplication either by a dot between the numbers or by parentheses around one or both of the numbers.

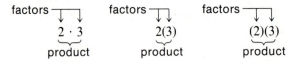

Multiplication of variables may be written the same way or may be written with the symbols side by side. For example,

factors ⇃⇂
ab
product   factors ⇃⇂
3*x*
product

where

ab means "the number *a* times the number *b*," and
3*x* means "the number 3 times the number *x*."

The operation of multiplication is indicated in English by such phrases as "times," "multiply," "twice (two times)," "of," and "product of."

PRIME FACTORS

A **prime number** is a natural number greater than 1 that is exactly divisible only by itself and 1—that is, a multiple of no natural number other than itself and 1. For example,

$$2, 3, 5, 7, 11, \text{ and } 13$$

are prime numbers, whereas 4 and 21 are not, since 4 is divisible by 2, and 21 is divisible by 7 and 3.

A natural number expressed as the product of its factors is said to be in **factored form.** Thus 21 can be expressed in factored form as $3 \cdot 7$. To write a number in factored form is called **factoring.** There may be more than one way to factor a number. For example, 12 can be factored as $3 \cdot 4$ or as $2 \cdot 6$, or even as $12 \cdot 1$. However, there is only one way to factor a natural number as a product of primes, excluding rearrangements of the factors. The *prime factorization* of 12 is $2 \cdot 2 \cdot 3$. Notice that if 1 were a prime then another prime factorization of 12 would be $1 \cdot 2 \cdot 2 \cdot 3$. This is the reason that we do not include 1 as a prime number.

Other examples of prime factorizations are

$$18 = 2 \cdot 3 \cdot 3 \quad \text{and} \quad 40 = 2 \cdot 2 \cdot 2 \cdot 5.$$

PROPERTIES OF ADDITION AND MULTIPLICATION

A basic property of addition and multiplication, called the **commutative law,** states the following:

The order in which the terms of a sum (or factors of a product) occur does not change the sum (or product).

Thus, it is always true that

The *order* of the terms has been changed.

$$a + b = b + a$$

and

The *order* of the factors has been changed.

$$a \cdot b = b \cdot a.$$

For example,

$$5 + 3 = 3 + 5$$

and

$$5 \cdot 3 = 3 \cdot 5.$$

Another useful property of addition and multiplication, called the **associative law,** states the following:

> **The way in which three terms in a sum (or three factors in a product) are grouped for addition (or multiplication) does not change the sum (or product).**

Thus, it is always true that

The terms are *grouped* differently.

$$(a + b) + c = a + (b + c)$$

and

The factors are *grouped* differently.

$$(a \cdot b) \cdot c = a \cdot (b \cdot c).$$

For example,

$$(2 + 3) + 4 = 2 + (3 + 4)$$

and

$$2 \cdot (3 \cdot 4) = (2 \cdot 3) \cdot 4.$$

In examples above, we used parentheses to indicate grouping. In $(2 + 3) + 4$, the parentheses indicate that the 2 and 3 are added first. In $2 + (3 + 4)$, the parentheses indicate that the 3 and 4 are added first. Brackets [ ] can be used in the same way that we use parentheses.

ALGEBRAIC EXPRESSIONS

An **algebraic expression,** or simply, an **expression,** is any meaningful collection of numbers, variables, and signs of operation. For example,

$$4x, \qquad 3x + y, \qquad \text{and} \qquad 2(x + 3)$$

are algebraic expressions. An important part of algebra involves translating word phrases into algebraic expressions. Here are some simple examples.

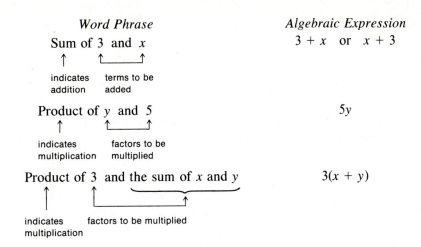

By the commutative law, $5 \cdot y = y \cdot 5$ and $3 \cdot (x + y) = (x + y) \cdot 3$. However, it is customary to write products with the numeral first, as shown in the examples above.

EXERCISES 1.1

■ *Which of the following are prime numbers?*

Sample Problems

a. 5   b. 15   c. 29

Ans.
a. 5 is prime (because it is exactly divisible only by 1 and itself).
b. 15 is not prime (because it is divisible by 3 and 5).
c. 29 is prime (because it is exactly divisible only by 1 and itself).

1. a. 4	b. 7	c. 9	d. 11
2. a. 8	b. 13	c. 14	d. 17
3. a. 21	b. 23	c. 25	d. 29
4. a. 31	b. 33	c. 36	d. 37

■ *List all prime numbers between (not including) the given numbers.*

Sample Problems

a. 4 and 18   b. 82 and 98

Ans. 5, 7, 11, 13, 17   *Ans.* 83, 89

5. 1 and 15	**6.** 16 and 25	**7.** 26 and 35
8. 36 and 45	**9.** 46 and 65	**10.** 66 and 100

■ *Write in prime factored form.*

Sample Problems

a. 24

b. 36

Ans. $24 = 2 \cdot 2 \cdot 2 \cdot 3$

Ans. $36 = 2 \cdot 2 \cdot 3 \cdot 3$

11. 45	**12.** 18	**13.** 28	**14.** 8
15. 27	**16.** 54	**17.** 64	**18.** 32
19. 210	**20.** 308	**21.** 156	**22.** 187

23. Each of the following phrases indicates a mathematical operation. Separate the phrases into two groups so that those in one group indicate addition and those in the other group indicate multiplication.

increased by, times, add, multiply, more than, product, exceeded by, sum, multiplied by

24. Each of the following expressions indicates a mathematical operation. Separate them into two groups as in Exercise 23.

$$(3)(4) \qquad 5 + 3 \qquad a \cdot b \qquad a + b$$

$$xy \qquad 4 + x \qquad (3)(y) \qquad c + 2$$

■ *Fill in the blank according to the indicated law.*

Sample Problems

a. Commutative law
$7 + 10 = 10 + \underline{?}$

b. Associative law
$(6 \cdot 4) \cdot 3 = 6 \cdot (4 \cdot \underline{?})$

Ans. $7 + 10 = 10 + 7$

Ans. $(6 \cdot 4) \cdot 3 = 6 \cdot (4 \cdot 3)$

25. Associative law
$(3 + 6) + 9 = \underline{?} + (6 + 9)$

26. Associative law
$(x + 3) + y = x + (3 + \underline{?})$

27. Commutative law
$6 \cdot 8 = 8 \cdot \underline{?}$

28. Commutative law
$5 + 7 = 7 + \underline{?}$

29. Associative law
$(3 \cdot x) \cdot y = 3 \cdot (x \cdot \underline{?})$

30. Commutative law
$3 + 2x = 2x + \underline{?}$

31. Commutative law
$(8 \cdot 9) \cdot 3 = (9 \cdot \underline{?}) \cdot 3$

32. Commutative law
$(5 + y) + x = x + (5 + \underline{?})$

■ *Write the following word phrases as algebraic expressions.*

Sample Problems

a. Sum of three and four.

b. Product of 2 and x.

Ans. $3 + 4$ or $4 + 3$

Ans. $2 \cdot x$, $2x$, $2(x)$, or $(2)(x)$

33. Sum of five and three.

34. Three times eight.

35. Product of 4 and y.

36. Six multiplied by x.

37. x increased by six.

38. Product of four and y.

39. Twice x.   **40.** Four more than t.

41. Sum of length (l) and width (w).   **42.** Five more than the cost (c).

43. Five times the cost (c).   **44.** Principal (P) times rate (r).

Sample Problems

a. The product of x and the sum of y and three.

b. Two more than the product of nine and z.

Ans.

a. $x(y + 3)$ or $(y + 3)x$ b. $2 + 9z$ or $9z + 2$

45. Two times the sum of five and y.

46. Four times the sum of six and x.

47. Five added to the product of four and x.

48. Nine more than the product of three and y.

49. Two more than the sum of 4 and t.

50. Four more than the product of 5 and t.

51. Product of b and the sum of c and six.

52. Product of a and the sum of 5 and y.

53. Product of five and x, times the sum of three and z.

54. The sum of 2 and x, times the sum of 3 and y.

Recall from your study of arithmetic that "percent" indicates a fraction in hundredths. For example, 30% means thirty hundredths, or 0.30.

■ *Write the following as algebraic expressions using decimals.*

Sample Problems

a. 30% of 250.   b. 65% of x.

Ans. 0.30 (250)   *Ans.* 0.65x

55. 50% of 68.   **56.** 10% of 480.

57. 37% of the cost (c).   **58.** 49% of the total (T).

59. 8% of your salary (s).   **60.** 4% of the original price (p).

1.2 ALGEBRAIC EXPRESSIONS: DIFFERENCES AND QUOTIENTS

DIFFERENCES

In subtracting one number from another, say, $5 - 3$, we are seeking a number 2, which when added to 3 equals 5. In general,

the difference $a - b$ is the number d such that $b + d = a$.

For example,

$$9 - 5 = 4 \quad \text{because} \quad 5 + 4 = 9,$$
$$20 - 13 = 7 \quad \text{because} \quad 13 + 7 = 20.$$

Note that while the sum of two whole numbers $a + b$ is always a whole number, an expression such as $a - b$ does not always represent a whole number. For example,

$$2 - 5 \quad \text{and} \quad 5 - 8$$

do not represent whole numbers.

QUOTIENTS

In dividing one number by another, say, $6 \div 3$, we are seeking the number 2, which when multiplied by 3 equals 6. We call 6 the **dividend** and 3 the **divisor.** We can indicate division by using the division symbol or a fraction bar.

$$
\begin{array}{c}
\text{division symbol} \\
\text{dividend} \longrightarrow \downarrow \text{divisor} \\
6 \div 3 \\
\underbrace{}_{\text{quotient}}
\end{array}
\qquad
\begin{array}{c}
\text{dividend} \\
\text{fraction bar} \longrightarrow \dfrac{6}{3} \left.\vphantom{\dfrac{6}{3}}\right\} \text{quotient} \\
\text{divisor}
\end{array}
$$

In general, the quotient is the number that, when multiplied by the divisor, gives the dividend. That is,

> the quotient $a \div b$ or $\dfrac{a}{b}$ is the number q, such that $b \cdot q = a$.

For example,

$$12 \div 4 = 3 \quad \text{or} \quad \frac{12}{4} = 3 \quad \text{because} \quad 4 \cdot 3 = 12$$

and

$$18 \div 2 = 9 \quad \text{or} \quad \frac{18}{2} = 9 \quad \text{because} \quad 2 \cdot 9 = 18.$$

Note that while the product of two whole numbers $a \cdot b$ is always a whole number, a quotient such as $a \div b$ or $\dfrac{a}{b}$ does not always represent a whole number. For example,

$$8 \div 5 \quad \text{or} \quad \frac{8}{5} \quad \text{and} \quad 10 \div 3 \quad \text{or} \quad \frac{10}{3}$$

do not represent whole numbers.

Also, notice that the commutative and associative properties do not apply to subtraction or division. That is,

$$5 - 3 \neq 3 - 5; \qquad\qquad 6 \div 3 \neq 3 \div 6$$
$$(10 - 3) - 2 \neq 10 - (3 - 2); \qquad (12 \div 6) \div 2 \neq 12 \div (6 \div 2)$$

As with addition and multiplication, subtraction and division can be indicated in a number of different ways using English phrases.

1. "Take away," "less than," "diminished by," "deducted from," "subtract," "subtracted from," "less," and "difference of" all indicate subtraction.
2. "Divided by" and "quotient of" indicate division.

ALGEBRAIC EXPRESSIONS

Here are some examples of algebraic expressions which involve differences and quotients.

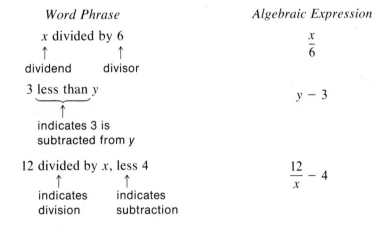

Common Error When translating a phrase such as "*a* subtracted from *b*," the order is important. We must write

$$b - a, \quad \text{not} \quad a - b,$$

since *a* is *subtracted from b*. For example, to translate

"Seven is subtracted from 4 times a number"

we represent the number by a symbol, say x, and then write the phrase as shown on page 10.

7 subtracted from 4 times a number

$$4 \cdot x - 7.$$

If we are asked to write symbolically the phrase "divide the sum of x and y by 7," we can write

$$(x + y) \div 7 \quad \text{or} \quad 7\overline{)x + y}$$

which show that 7 is to be divided into the sum of x and y and not merely into one or the other. In algebra, however, we prefer to show a quotient in fraction form as

$$\frac{x + y}{7} \quad \text{or} \quad (x + y)/7,$$

where the bar beneath the $x + y$ indicates grouping in the same way as parentheses.

QUOTIENTS INVOLVING ZERO

In the case where the dividend is 0 and the divisor is a nonzero number, the quotient is zero. For example,

$$0 \div 5 = 0 \quad \text{or} \quad \frac{0}{5} = 0 \quad \text{because} \quad 5 \cdot 0 = 0.$$

On the other hand a symbol such as $\frac{5}{0}$ is meaningless, since there is no number that, when multiplied by 0, gives 5. Such a symbol is "undefined." Nor do we use the symbol $\frac{0}{0}$ to represent a number, because the product of 0 and any number is 0. Thus we say that $\frac{0}{0}$ is "indeterminate."

If a is any natural number,

1. $\dfrac{0}{a} = 0.$ 2. $\dfrac{a}{0}$ is undefined.   3. $\dfrac{0}{0}$ is indeterminate.

EXERCISES 1.2

■ *Find the difference. If the difference is not a whole number, so state.*

Sample Problems   a. $10 - 6$ b. $5 - 8$

Ans.

a. 4 because $6 + 4 = 10$ b. Not a whole number.

1. $8 - 2$ **2.** $10 - 4$ **3.** $13 - 17$ **4.** $4 - 7$

5. $15 - 6$ **6.** $11 - 2$ **7.** $12 - 12$ **8.** $18 - 18$

9. $2 - 11$ **10.** $4 - 9$ **11.** $23 - 13$ **12.** $21 - 12$

■ *Find the quotient. If the quotient is undefined, so state.*

Sample Problems   a. $\dfrac{18}{6}$ b. $\dfrac{0}{7}$ c. $\dfrac{7}{0}$ d. $\dfrac{6}{6}$

Ans.

a. $\dfrac{18}{6} = 3$ because $6 \cdot 3 = 18$ b. $\dfrac{0}{7} = 0$ because $7 \cdot 0 = 0$

c. $\dfrac{7}{0}$ is undefined   d. $\dfrac{6}{6} = 1$ because $6 \cdot 1 = 6$

13. $\dfrac{15}{5}$ **14.** $\dfrac{20}{4}$ **15.** $\dfrac{18}{0}$ **16.** $\dfrac{32}{0}$

17. $\dfrac{0}{13}$ **18.** $\dfrac{0}{17}$ **19.** $\dfrac{14}{0}$ **20.** $\dfrac{0}{0}$

21. $\dfrac{24}{3}$ **22.** $\dfrac{27}{9}$ **23.** $\dfrac{25}{5}$ **24.** $\dfrac{12}{4}$

25. Each of the following word phrases indicates a mathematical operation. Separate the phrases into two groups so that those in one group indicate subtraction and those in the other group indicate division.

take away, divide, less than, difference, subtract, quotient, decreased by, divided by, diminished by, subtracted from, less

26. Each of the following expressions indicates a mathematical operation. Separate them into two groups as in Exercise 25.

$$6 \div 3 \qquad 4 - 3 \qquad 3\overline{)6} \qquad \frac{6}{3}$$

$$a \div b \qquad a - b \qquad \frac{a}{b} \qquad b\overline{)a}$$

■ *Write the following word phrases as algebraic expressions.*

Sample Problems   a. Six divided by two.   b. Nine decreased by x.

c. Quotient of 5 divided by x.   d. Eight subtracted from y.

(continued)

Ans.

a. $\dfrac{6}{2}$ or $6 \div 2$

b. $9 - x$

c. $\dfrac{5}{x}$ or $5 \div x$

d. $y - 8$

27. Six divided by x. **28.** y divided by 4.
29. Eight decreased by y. **30.** t decreased by 6.
31. b subtracted from a. **32.** g subtracted from h.
33. a divided by b. **34.** t less r.
35. Distance (d) divided by time (t). **36.** Cost (c) less rebate (r).

Sample Problems

a. 25 divided by the sum of 3 and y.
b. Product of 3 and x, less the sum of 3 and y.

Ans.

a. $\dfrac{25}{3 + y}$ or $25 \div (3 + y)$

b. $(3 \cdot x) - (3 + y)$

37. Ten divided by the sum of x and four.
38. Nine less the quotient of b divided by four.
39. y less the sum of four and x.
40. y diminished by the product of five and t.
41. Divide the sum of x and y by five.
42. Sum of p and q, less the product of four and y.
43. Sum of r and s, divided by the product of six and z.
44. Subtract the sum of x and y from the quotient of y divided by four.

■ *Translate the following word phrases into algebraic expressions. Use x as the variable.*

Sample Problems

a. Two times a number plus five
$2 \quad \cdot \quad x \quad + \quad 5$

b. Three less than four times a number.
$4 \cdot x \quad - \quad 3$

Ans. $2x + 5$

Ans. $4x - 3$

45. Six less than twice a number.
46. Ten less than three times a number.
47. The quotient of twice a number divided by 5.

48. The product of three and a number, diminished by 2.

49. Nine subtracted from twice a number.

50. Five subtracted from four times a number.

51. Two subtracted from the product of 4 and a number.

52. One subtracted from the product of 3 and a number.

53. The quotient of 7 divided by a number, added to 5.

54. The quotient of 5 divided by a number, less 4.

55. The product of a number and four more than the number.

56. The sum of a number and three less than twice the number.

57. Three times a number, divided by the sum of the number and 7.

58. The quotient of a number divided by 2, diminished by the number.

1.3   ORDER OF OPERATIONS; NUMERICAL EVALUATION

In Section 1.1 we observed that certain properties of addition and multiplication apply to algebraic expressions involving only one operation. We now consider expressions that involve more than one operation.

ORDER OF OPERATIONS

The expression $4 + 6 \cdot 2$ can be interpreted two ways. We can see it as meaning either

$$(4 + 6) \cdot 2 \quad \text{or} \quad 4 + (6 \cdot 2),$$

in which case the result is either

$$10 \cdot 2 = 20 \quad \text{or} \quad 4 + 12 = 16.$$

To avoid such confusion, we agree to perform multiplications and divisions before additions and subtractions, regardless of the order in which the operations appear. Thus, in the example above,

$$4 + 6 \cdot 2,$$

we first multiply to get

$$4 + 12,$$

and then add to obtain

$$16.$$

Also, if an expression includes any operations inside parentheses, we agree to perform those operations first. Thus, in the expression

$$3(5 + 7),$$

we first perform the addition inside parentheses to get

$$3(12),$$

and then multiply to obtain

$$36.$$

Expressions that appear above or below a fraction bar are also given precedence. For example, to simplify

$$\frac{8 - 4}{2} + 2 \cdot 9$$

we first simplify the $8 - 4$ above the fraction bar to obtain

$$\frac{4}{2} + 2 \cdot 9.$$

Then we divide and multiply to get

$$2 + 18,$$

and finally we add, to obtain

$$20.$$

We can summarize these ideas as follows.

Order of Operations

1. Perform any operations inside parentheses, or above or below a fraction bar.
2. Perform all multiplication and division operations in the order in which they occur from left to right.
3. Perform additions and subtractions in order from left to right.

NUMERICAL EVALUATION

The process of substituting given numbers for variables and simplifying the arithmetic expression according to the order of operations given in this section is called *numerical evaluation*. For example, we can evaluate $6x - 2y$ for $x = 4$ and $y = 3$ by first substituting 4 for x and 3 for y and then performing the operations in the correct order to obtain

$$6(4) - 2(3) = 24 - 6$$
$$= 18.$$

As another example, we evaluate $P + Prt$ for $P = 1000$, $r = .05$, and $t = 2$ as follows:

$$P + Prt = 1000 + 1000(.05)(2)$$
$$= 1000 + 100$$
$$= 1100.$$

EXERCISES 1.3

■ *Simplify.*

Sample Problem

$8 + 3(4 + 1)$

Simplify the quantity in the parentheses.

$8 + 3(5)$

Multiply.

$8 + 15$

Add.

Ans. 23

1. $(2)(4) - 3$ **2.** $(4)(3) - 5$ **3.** $2 + 4 \cdot 3$

4. $9 - 3 \cdot 2$ **5.** $4(0) + 5$ **6.** $5 - 3(0)$

7. $4(3 + 5)$ **8.** $3(7 - 4)$ **9.** $(5 + 4)(3)$

10. $(6 - 1)(4)$ **11.** $(3 + 2)(5 - 1)$ **12.** $(5 - 3)(3 + 1)$

13. $3 + 2(6 - 1)$ **14.** $5 + 3(2 + 3)$ **15.** $12 - 2(1 + 3)$

16. $20 - 3(6 - 4)$ **17.** $4(3) - 2(5)$ **18.** $6(5) + 4(1)$

Sample Problem

$$\frac{2 + 7}{5 - 2} - \frac{2 + 2}{4}$$

Simplify above and below the fraction bar.

$$\frac{9}{3} - \frac{4}{4}$$

Divide as indicated.

$3 - 1$

Subtract.

Ans. 2

19. $\frac{4(3)}{2} - 2$ **20.** $10 - \frac{12(3)}{9}$ **21.** $\frac{5 + 7}{4} - 1$

22. $\frac{16 - 4}{6} + 5$ **23.** $\frac{3(8)}{12} - \frac{6 + 4}{5}$ **24.** $\frac{5 + 9}{7} + \frac{2(8)}{4}$

25. $\frac{3(3) + 5}{2} + \frac{2(5) - 4}{3}$ **26.** $\frac{9(2) - 4}{7} + \frac{3(2) + 2}{4}$

27. $\dfrac{3(5-1)}{6} + \dfrac{2(3)}{6}$

28. $\dfrac{(4)(9)}{3+1} - \dfrac{7(11-6)}{5}$

29. $\dfrac{2(3)+4}{6-1} - \dfrac{8(3)}{3(4)} + \dfrac{6(6)}{6+6}$

30. $\dfrac{9(3)+3}{5(2)} + \dfrac{2+4(2)}{5(2)} - \dfrac{8(8)}{16}$

31. $\left[\dfrac{(8-4)\cdot 2}{4-2}\right]\left[\dfrac{(16-2)\cdot 3}{4+3}\right]$

32. $\left[\dfrac{6+8\cdot 4}{8-6}\right]\left[\dfrac{3\cdot 3\cdot 4}{3+3\cdot 5}\right]$

33. $\dfrac{3[12-(2+6)]}{3(5)-3(10-6)}$

34. $\dfrac{[2(4+3)-4]+8}{16-[4(5)-(10-3)]}$

Sample Problem

$8 + 3 \cdot 4 - 2 \cdot 7$

      Perform all multiplications.

$8 + 12 - 14$

      Add and subtract in order from left to right.

$20 - 14$

Ans. 6

35. $6 \cdot 5 - 2 + 3 \cdot 2$

36. $12 \div 2 + 1 - 3 \cdot 2$

37. $8 \cdot 2 + 4 \cdot 3 - 5 \cdot 2$

38. $16 - 2 \cdot 3 + 4 \cdot 12 \div 8$

39. $9 - 3 \cdot 6 \div 9 + 4 \cdot 2 - 3$

40. $12 \div 3 + 3 \cdot 4 - 2 + 2$

■ *If x = 3, find the value of each expression.*

Sample Problems

a. $4 + 2x$

b. $\dfrac{4x+6}{2}$

      Substitute 3 for x.

$4 + 2(3)$

$\dfrac{4(3)+6}{2}$

$4 + 6$

      Simplify.

$\dfrac{12+6}{2}$

$\dfrac{18}{2}$

Ans. 10

Ans. 9

41. $2x$

42. $5x$

43. $3x - 2$

44. $3x + 1$

45. $4(x-2)$

46. $3(x+3)$

47. $2 + 4x$

48. $15 - 2x$

49. $\dfrac{3+6x}{3}$

50. $\dfrac{3(1+2x)}{3}$

51. $\dfrac{3(2x-2)}{2}$

52. $\dfrac{4(x-1)}{2}$

■ *If x = 2 and y = 3, find the value of each expression.*

Sample Problems

a. $x(y + 1) + 3$

$(2)[(3) + 1] + 3$

$2(4) + 3$

$8 + 3$

Ans. 11

b. $xy + 3y$ Substitute values.

$(2)(3) + 3(3)$

                                   Simplify.

$6 + 9$

Ans. 15

53. xy **54.** $2(xy)$ **55.** $2xy$ **56.** $x(2y)$

57. $3(y - x)$ **58.** $4y - x$ **59.** $5(y + x)$ **60.** $2y + x$

61. $2y - 2x$ **62.** $4y + 3x$ **63.** $\dfrac{3x + y}{y}$ **64.** $\dfrac{x + 2y}{x}$

65. $5y - 2xy + 3x$ **66.** $12x - 3xy - y$

67. $\dfrac{5y + 1}{3x - 2} + \dfrac{3x + 2y}{xy}$ **68.** $\dfrac{24 - 2x}{2 + y} - \dfrac{4x + 1}{3y}$

■ *Evaluate each expression.*

Sample Problems

a. $\dfrac{k}{V}$, where $k = 88$ and $V = 11$.

$\dfrac{88}{11}$

Ans. 8

b. lw, where $l = 8$ and $w = 4$.

$(8)(4)$

Ans. 32

69. $F \cdot d$, where $F = 2000$ and $d = 6$.

70. $2l + 2w$, where $l = 64$ and $w = 41$.

71. $a + 16t$, where $a = 12$ and $t = 4$.

72. $12(b + c)$, where $b = 16$ and $c = 4$.

73. $\dfrac{bh}{2}$, where $b = 500$ and $h = 200$.

74. lwh, where $l = 14$, $w = 6$, and $h = 2$.

75. $\dfrac{h(b + c)}{2}$, where $h = 6$, $b = 4$, and $c = 6$.

76. $a + (n - 1)d$, where $a = 3$, $n = 7$ and $d = 2$.

1.4   EQUATIONS WITH ONE OPERATION

In algebra we develop certain techniques that help solve problems stated in words. These techniques involve first rewriting word sentences in the form of symbols. For example,

Three added to a number equals seven

may be rewritten as

$$? + 3 = 7, \qquad n + 3 = 7, \qquad x + 3 = 7,$$

and so on, where the symbols $?$, n, and x represent the number we want to find. We call such symbolic sentences **equations.** The terms to the left of an equals sign make up the **left-hand member** of the equation; those to the right make up the **right-hand member.** Thus, in the equation $x + 3 = 7$, the left-hand member is $x + 3$, and the right-hand member is 7.

Equations may be true or false, just as word sentences may be true or false. The equation

$$x + 3 = 7$$

will be false if any number except 4 is substituted for the variable. A value of the variable for which the equation is true is called a **solution** of the equation. We can determine whether or not a given number is a solution of a given equation by substituting the number in place of the variable and determining the truth or falsity of the result. The equations we will consider in this chapter have at most one solution.

EQUIVALENT EQUATIONS

In some instances it may be possible to determine the solution of an equation ''by inspection.'' For example, the solution of the equation

$$x + 3 = 7$$

is 4, because $4 + 3 = 7$. If a solution is not evident by inspection, then we need some mathematical ''tools'' in order to solve the equation. These tools depend on the notion of *equivalent equations.*

Equivalent equations are equations that have identical solutions. Thus,

$$3x + 3 = x + 13, \qquad 3x = x + 10, \qquad 2x = 10, \qquad \text{and} \qquad x = 5$$

are equivalent equations, because 5 is the only solution of each of them.

In solving any equation, we transform a given equation whose solution may not be obvious to an equivalent equation whose solution is easily noted.

Thus, the goal of solving an equation is to write an equivalent equation where the variable stands alone on one side of the equals sign, and the solution is on the other side of the equals sign.

SOLVING EQUATIONS

Since an equation is simply a statement that the left-hand and right-hand members are different names for the same number, the same quantities added to or subtracted from each member will produce another equality. Thus, if

$$x + 3 = 7,$$

then

$$x + 3 - 3 = 7 - 3,$$

or

$$x = 4.$$

Note that 4 is also a solution of the original equation $x + 3 = 7$. In general:

> **If the same quantity is added to or subtracted from both sides of an equation, the solution is unchanged.**

In symbols:

$$\text{If } a = b, \quad \text{then} \quad a + c = b + c \quad \text{and} \quad a - c = b - c.$$

We can sometimes use the addition or subtraction property to transform a given equation to an equivalent equation of the form $x = a$, whose solution is evident. We simply "undo" operations that involve the variable. For example, in the equation

$$x + 6 = 11,$$

6 is *added* to x, so we *subtract* 6 from each member of the equation to get

$$x + 6 - 6 = 11 - 6,$$

or

$$x = 5,$$

where the solution is obviously 5. Since each equation obtained in the process is equivalent to the original equation, 5 is also a solution of $x + 6 = 11$.

The solution can be checked by substituting 5 for x in the original equation:

$$5 + 6 \overset{?}{=} 11; \quad \text{Yes.}$$

Now consider the equation

$$3x = 12.$$

Since x is *multiplied* by 3, we *divide* each member of the equation by 3 to obtain the equations

$$\frac{3x}{3} = \frac{12}{3},$$

or

$$x = 4,$$

whose solutions are also 4. (We have used slash bars to emphasize that dividing $3x$ by 3 yields x.) We can easily check that 4 is also the solution to the original equation $3x = 12$. In general:

> **If both sides of an equation are divided by the same (nonzero) quantity, the solution is unchanged.**

In symbols:

$$\text{If} \quad a = b, \quad \text{then} \quad \frac{a}{c} = \frac{b}{c} \quad (c \neq 0),$$

where the symbol $\neq$ is read "is not equal to."

Finally, consider the equation

$$\frac{x}{4} = 3.$$

In this equation x is *divided* by 4, so we multiply each member of the equation by 4 to obtain the equations

$$(4)\frac{x}{4} = (4)3,$$

$$x = 12,$$

whose solutions are also 12. In general:

> **If both sides of an equation are multiplied by the same quantity, the solution is unchanged.**

In symbols:

$$\text{If} \quad a = b, \quad \text{then} \quad a \cdot c = b \cdot c \quad (c \neq 0).$$

In the exercises that follow, you will solve equations by using one of the three properties illustrated above. To help you choose the appropriate property, you should ask yourself what operation has been performed on the variable, and how you can "undo" it.

The *symmetric property of equality* is also helpful in the solution of equations. This property states:

$$\text{If} \quad a = b \quad \text{then} \quad b = a.$$

This enables us to interchange the members of an equation for convenience or clarity. Thus:

$$\text{If} \quad 4 = x + 2 \quad \text{then} \quad x + 2 = 4.$$

$$\text{If} \quad x + 3 = 2x - 5 \quad \text{then} \quad 2x - 5 = x + 3.$$

$$\text{If} \quad d = rt \quad \text{then} \quad rt = d.$$

EXERCISES 1.4

■ *Determine whether each equation is true for the indicated value of the variable; that is, determine whether the given number is, or is not, the solution of the given equation.*

Sample Problems

a. $3x = 12,$ for $x = 4$

$3(4) \overset{?}{=} 12;$ yes

Ans. 4 is the solution

b. $x - 7 = 5,$ for $x = 9$

$9 - 7 \overset{?}{=} 5;$ no

Ans. 9 is not the solution

1. $x + 9 = 12,$ for $x = 3$

3. $3a = 6 + a,$ for $a = 4$

5. $0 = 7 - y,$ for $y = 7$

7. $\dfrac{x}{5} = 3,$ for $x = 15$

9. $8 + a = 6,$ for $a = 2$

2. $x - 4 = 10,$ for $x = 6$

4. $4 - a = 2a,$ for $a = 2$

6. $9 + y = 9,$ for $y = 0$

8. $\dfrac{x}{3} = x - 4,$ for $x = 6$

10. $a - 7 = 3a,$ for $a = 12$

■ *Find the solution of each equation by inspection.*

Sample Problems

a. $x + 6 = 10$

Ans. 4 (because $4 + 6 = 10$)

b. $5 \cdot x = 15$

Ans. 3 (because $5 \cdot 3 = 15$)

11. $x + 4 = 15$

14. $y - 10 = 16$

17. $3a = 12$

20. $\dfrac{y}{12} = 1$

12. $x + 7 = 10$

15. $8 - x = 5$

18. $4a = 28$

21. $\dfrac{x}{3} = 0$

13. $y - 5 = 9$

16. $4 - x = 0$

19. $\dfrac{y}{5} = 35$

22. $\dfrac{x}{6} = 3$

■ *Write an equation equivalent to the given equation.*

Sample
Problem $x - 4 = 7$,   by adding 4 to each member.

$x - 4 + 4 = 7 + 4$

Ans. $x = 11$

23. $x + 4 = 11$,   by subtracting 4 from each member.
24. $x - 3 = 5$,   by adding 3 to each member.
25. $6 - y = 0$,   by adding y to each member.
26. $10 = 7 + y$,   by subtracting 7 from each member.
27. $12 = z + 8$,   by subtracting 8 from each member.
28. $z - 9 = 16$,   by adding 9 to each member.

Sample
Problem $\dfrac{x}{3} = 4$,   by multiplying each member by 3.

$3 \cdot \dfrac{x}{3} = 3(4)$

Ans. $x = 12$

29. $\dfrac{x}{2} = 8$,   by multiplying each member by 2.

30. $\dfrac{y}{3} = 5$,   by multiplying each member by 3.

31. $6 = \dfrac{z}{5}$,   by multiplying each member by 5.

32. $4 = \dfrac{x}{4}$,   by multiplying each member by 4.

33. $\dfrac{y}{3} = 7$,   by multiplying each member by 3.

34. $\dfrac{z}{2} = 6$,   by multiplying each member by 2.

Sample
Problem $4x = 12$,   by dividing each member by 4.

$\dfrac{4x}{4} = \dfrac{12}{4}$

Ans. $x = 3$

35. $6x = 12,$ by dividing each member by 6.
36. $7y = 28,$ by dividing each member by 7.
37. $24 = 4z,$ by dividing each member by 4.
38. $21 = 7x,$ by dividing each member by 7.
39. $5y = 20,$ by dividing each member by 5.
40. $6z = 18,$ by dividing each member by 6.

■ *Solve each equation.*

Sample Problem

$$x + 7 = 12$$
$$x + 7 - 7 = 12 - 7$$ Subtract 7 from each member.

Ans.* $x = 5$

41. $x + 5 = 10$ **42.** $5 + y = 9$ **43.** $z - 6 = 5$
44. $x - 4 = 0$ **45.** $9 = 4 + z$ **46.** $12 = 3 + y$

Sample Problem

$$5y = 15$$
$$\frac{5y}{5} = \frac{15}{5}$$ Divide each member by 5.

Ans. $y = 3$

47. $5x = 15$ **48.** $6y = 24$ **49.** $12 = 4y$
50. $35 = 7x$ **51.** $12 = 3y$ **52.** $25 = 5z$

Sample Problems   a. $\frac{x}{2} = 15$ b. $\frac{x}{5} = 9$

Multiply each member in problem a by 2 and in problem b by 5.

$$2\left(\frac{x}{2}\right) = 2(15) \qquad 5\left(\frac{x}{5}\right) = 5(9)$$

Ans. $x = 30$ *Ans.* $x = 45$

* The solution of the original equation is the number 5; however, the answer is often displayed in the form of the equation $x = 5$.

53. $\dfrac{x}{2} = 10$ **54.** $\dfrac{x}{3} = 6$ **55.** $3 = \dfrac{y}{6}$ **56.** $8 = \dfrac{y}{4}$

57. $\dfrac{z}{3} = 4$ **58.** $\dfrac{z}{7} = 4$ **59.** $6 = \dfrac{x}{3}$ **60.** $5 = \dfrac{x}{9}$

61. $\dfrac{y}{2} = 5$ **62.** $\dfrac{y}{4} = 6$ **63.** $\dfrac{z}{4} = 0$ **64.** $\dfrac{z}{5} = 0$

1.5   EQUATIONS WITH TWO OPERATIONS

In Section 1.4 we learned to solve equations in which only one operation was involved. Now we consider equations such as

$$2 + 3x = 20 \tag{1}$$

that contain two operations. According to our conventions on order of operations, the expression $2 + 3x$ means that the variable x is first multiplied by 3, and the result is then added to 2. In order to find an equivalent equation of the form $x = a$ (i.e., to solve the equation), we must reverse these steps to undo the two operations. Thus we first subtract 2 from each member of Equation (1) to obtain

$$2 + 3x - 2 = 20 - 2,$$

from which

$$3x = 18.$$

Next we divide both sides of this last equation by 3 to obtain

$$\frac{3x}{3} = \frac{18}{3},$$

which gives the solution

$$x = 6.$$

In general, to solve an equation that involves more than one operation, "undo" the operations in reverse order. For example, consider the equation

$$4x - 5 = 7. \tag{2}$$

The expression $4x - 5$ means that x is first multiplied by 4, and then 5 is subtracted from the result. So to solve we must first add 5 to both members of Equation (2) to obtain

$$4x - 5 + 5 = 7 + 5,$$

from which

$$4x = 12,$$

and then divide both sides by 4 to obtain

$$\frac{4x}{4} = \frac{12}{4}$$

to find the solution

$$x = 3.$$

EXERCISES 1.5

■ *Solve each equation.*

Sample Problem

$3x - 4 = 8$

Add 4 to each member.

$3x - 4 + 4 = 8 + 4$

$3x = 12$

Divide each member by 3.

$$\frac{3x}{3} = \frac{12}{3}$$

Ans. $x = 4$

1. $6x - 13 = 5$ **2.** $3x + 2 = 8$ **3.** $7p + 6 = 13$
4. $2p - 9 = 7$ **5.** $5 = 2 + 3y$ **6.** $20 = 7y - 1$
7. $2t - 2 = 6$ **8.** $4t - 3 = 9$ **9.** $2z - 8 = 0$
10. $0 = 3z - 21$ **11.** $3x - 3 = 3$ **12.** $4x + 5 = 21$
13. $15 = 9 + 2a$ **14.** $37 = 2 + 7a$ **15.** $6x + 5 = 5$
16. $8x + 3 = 3$ **17.** $15t + 35 = 50$ **18.** $12t - 39 = 45$

Sample Problem

$$\frac{3x}{4} = 6$$

Multiply each member by 4.

$$4 \cdot \frac{3x}{4} = 4\,(6)$$

$3x = 24$

Divide each member by 3.

$$\frac{3x}{3} = \frac{24}{3}$$

Ans. $x = 8$

19. $\dfrac{2a}{5} = 8$ **20.** $\dfrac{4a}{5} = 12$ **21.** $8 = \dfrac{2b}{3}$ **22.** $4 = \dfrac{2b}{5}$

23. $\dfrac{2c}{4} = 10$ **24.** $\dfrac{4c}{5} = 8$ **25.** $8 = \dfrac{4x}{7}$ **26.** $10 = \dfrac{5x}{3}$

27. $15 = \dfrac{3y}{5}$ **28.** $12 = \dfrac{4y}{3}$ **29.** $0 = \dfrac{5z}{7}$ **30.** $\dfrac{6z}{5} = 0$

31. $\dfrac{5p}{6} = 15$ **32.** $\dfrac{9p}{2} = 18$ **33.** $\dfrac{13t}{7} = 26$ **34.** $\dfrac{25t}{8} = 50$

Sample Problem

$$\dfrac{x}{3} + 5 = 7$$

Subtract 5 from each member.

$$\dfrac{x}{3} + 5 - 5 = 7 - 5$$

$$\dfrac{x}{3} = 2$$

Multiply each member by 3.

$$3 \cdot \dfrac{x}{3} = 3(2)$$

Ans. $x = 6$

35. $\dfrac{x}{4} + 2 = 3$ **36.** $\dfrac{x}{5} + 4 = 9$ **37.** $\dfrac{a}{12} - 2 = 3$ **38.** $\dfrac{a}{7} - 3 = 5$

39. $16 = 1 + \dfrac{y}{3}$ **40.** $6 = 3 + \dfrac{y}{8}$ **41.** $6 + \dfrac{b}{2} = 10$ **42.** $9 + \dfrac{b}{4} = 12$

43. $7 = \dfrac{z}{6} - 4$ **44.** $13 = \dfrac{z}{3} - 5$ **45.** $12 = \dfrac{t}{8} + 12$ **46.** $3 = \dfrac{t}{5} + 3$

47. $\dfrac{x}{6} - 12 = 12$ **48.** $\dfrac{x}{2} - 14 = 14$

Sample Problem

$$\dfrac{3x}{4} - 2 = 7$$

Add 2 to each member.

$$\dfrac{3x}{4} - 2 + 2 = 7 + 2$$

$$\dfrac{3x}{4} = 9$$

Multiply each member by 4.

$$4 \cdot \dfrac{3x}{4} = 4(9)$$

$$3x = 36$$

Divide each member by 3.

$$\frac{3x}{3} = \frac{36}{3}$$

Ans. $x = 12$

49. $\dfrac{2x}{3} - 5 = 7$ **50.** $\dfrac{5x}{2} - 4 = 6$ **51.** $\dfrac{3c}{5} + 2 = 8$ **52.** $\dfrac{7c}{4} + 2 = 16$

53. $1 = \dfrac{4a}{5} - 7$ **54.** $1 = \dfrac{2a}{9} - 9$ **55.** $2 = \dfrac{5x}{3} + 2$ **56.** $13 = \dfrac{2x}{7} + 13$

57. $\dfrac{12t}{7} - 30 = 18$ **58.** $\dfrac{9t}{11} - 23 = 4$ **59.** $6 + \dfrac{3y}{4} = 36$ **60.** $4 + \dfrac{5y}{8} = 19$

1.6   INTRODUCTION TO WORD PROBLEMS

Most problems that deal with practical applications of mathematics are expressed in words. In order to solve a word problem whose solution is not immediately evident, we must write an equation that models the information given in the problem. We can then use the techniques of the previous sections to solve the equation.

TRANSLATING WORD SENTENCES

In Section 1.1 we translated English phrases such as "the sum of a number and 5" into algebraic expressions using variables. In a similar manner we can translate word sentences into algebraic equations. In each of the following examples, the unknown quantity is "a number" and is represented by the variable x.

The sum of a number and five is 32.

$$x \quad + \quad 5 \quad = 32$$

Three increased by twice a number is 15.

$$3 \quad + \quad 2 \cdot x \quad = 15$$

Two times a number added to nine is 25.

$$2 \cdot x \quad + \quad 9 \quad = 25$$

CONSTRUCTING MODELS

We were able to model the foregoing problems by translating each sentence directly into an equation. It is not always so easy to construct a model from the given information. Consider the following problem:

Jerry needs an additional $35 in order to fly to New York. The air fare to New York is $293. How much money does Jerry have?

We begin the process of constructing a model by first writing in words the quantity we wish to find and then assigning a variable to represent that quantity.

Amount of money Jerry has: x

Next we express the given information as an equation involving the variable.

$$x \quad + \quad 35 \quad = \quad 293$$

$$\underbrace{}_{\substack{\text{amount} \\ \text{Jerry has}}} \quad \underbrace{}_{\substack{\text{amount} \\ \text{he needs}}} \quad \underbrace{}_{\text{air fare}}$$

Finally we solve the equation by subtracting 35 from each member.

$$x + 35 - 35 = 293 - 35$$

$$x = 258$$

So Jerry has $258.

We used the following steps in solving the example above. You will find it helpful to follow these steps carefully as you first try to solve word problems.

Steps for Solving Word Problems

1. Express the unknown quantity in words; then represent it by a variable.
2. Express the information in the problem as an algebraic equation involving the variable.
3. Solve the equation.

USE OF FORMULAS

In the previous examples, we wrote equations about the unknown quantities by using information given in the problems themselves. For some applications, known relationships among the pertinent variables may provide us with an appropriate model. Equations which express such relationships are called **formulas.** For example, problems concerning uniform motion often involve the formula $d = rt$, where d stands for the distance traveled in time t at rate of speed r. Thus, if a car travels for 3 hours at 50 miles per hour, then the distance it travels is

$$d = (50)(3) = 150 \text{ miles.}$$

EXERCISES 1.6

■ *Translate the following word sentences into equations. Use x as the variable.*

Sample Problem

The sum of 8 and twice a certain number is equal to 26.

$$8 + 2 \cdot x = 26$$

Ans. $8 + 2x = 26$

1. The sum of a number and 21 is 59.

2. Six more than a certain number is 22.

3. Eight is 6 less than twice a certain number.

4. Eight less than twice a certain number is 6.

5. If a certain number is added to 15, the sum is 53.

6. If four times a number is increased by one, the sum is 29.

7. Three times a number less 12 is equal to 9.

8. 89 is five more than 6 times a number.

9. If twice a number is diminished by 7, the result is 33.

10. Twice a number is added to 8, yielding 26.

■ *For the following word problems:*
a. *Write in words the unknown quantity, and assign a variable to represent it,*

b. *write an equation that models the problem, and*

c. *solve the equation.*

Sample Problem

Martha paid $26 less for a suit at a discount store than her mother paid at a boutique for the same suit. If Martha paid $89 for the suit, how much did her mother pay?

a. Price Martha's mother paid: x

b. $\quad \underbrace{89}_{\substack{\text{price} \\ \text{Martha paid}}} = \underbrace{x}_{\substack{\text{price Martha's} \\ \text{mother paid}}} - \underbrace{26}_{\text{less \$26}}$

c. $89 + 26 = x - 26 + 26$

$\qquad 115 = x$

Ans. Martha's mother paid $115.

11. Ruth needs $21 more to pay next semester's tuition. If tuition is $150, how much money does Ruth have?

12. Steve needs $8 more in order to buy his textbooks. If his books cost $63, how much does Steve have?

13. A used car costs $3400 less than the new version of the same model. If the new car costs $9200, how much does the used car cost?

14. A two-bedroom house costs $20,000 more than a one-bedroom house in the same neighborhood. The two-bedroom house costs $105,000. How much does the one-bedroom house cost?

15. John earned three times as much as Sam over the summer. If John earned $861, how much did Sam earn?

16. A restaurant bill is divided equally by seven people. If each person paid $8.50, how much was the bill?

Sample Problem

Emily spends 40% of her monthly income on rent. If her rent is $360 a month, how much does Emily make?

a. Monthly income: x

b. $\underbrace{0.40}_{40\%} \quad \underbrace{(x)}_{\substack{\text{monthly} \\ \text{income}}} \quad = \quad \underbrace{360}_{\substack{\text{monthly} \\ \text{rent}}}$

c. $\dfrac{0.40x}{0.40} = \dfrac{360}{0.40}$

$x = 900$

Ans. Emily's monthly income is $900.

17. The Dodgers won 60% of their games last season. If they won 96 games, how many games did they play?

18. Fifty-three percent of the electorate voted for candidate Phil I. Buster. If candidate Buster received 106,000 votes, how many citizens voted?

19. Sixty-five percent of the students at a small college receive financial aid. If 377 students receive financial aid, what is the enrollment of the college?

20. Iris got a 6% raise. Her new salary is $21 a week more than her old salary. What was her old salary?

Sample Problem

Mitch bought a compact disc player for $269 and a number of compact discs at $14 apiece. If the total bill before tax was $367, how many discs did he buy?

a. Number of discs: x

b.   269   +   14   · x = 367

price of   price of   number
player   one disc   of discs

#25. 40 + d = 82

40 + 14x = 82

c. 269 + 14x − 269 = 367 − 269

$$14x = 98$$

$$\frac{14x}{14} = \frac{98}{14}$$

$$x = 7$$

Ans. Mitch bought 7 compact discs.

21. Alice works 36 hours a week, which is 4 more than twice as many hours as Neil works. How many hours a week does Neil work?

22. Herman is 6 years less than 5 times his daughter Laura's age. If Herman is 34, how old is Laura?

23. Grace's new apartment rents for $60 less than twice the cost of her old apartment. If the new apartment costs $540 a month, what was the rent on her old apartment?

24. Lois is $12 short of the price of 4 apple trees. If Lois has $64, how much does one apple tree cost?

25. Bernice rents a sofa for 2 weeks at a cost of $82. If the rental agency charges $40 plus a daily fee, how much is the daily fee?

26. A car rental company charges $38 a day plus a mileage fee of 15 cents a mile. If the total bill on a rental is $65, how many miles was the car driven?

27. A woman buys a car for a total cost of $10,200. She paid $1200 down and will pay the rest in 36 equal monthly installments. How much is each installment?

28. Gregory purchased stereo equipment on time, putting down $100 and agreeing to monthly payments of $30. If the total cost of the equipment is $640, how many monthly payments will he make?

29. A plumber charges $40 for a house call plus $18 an hour. If Esther's plumbing bill is $130, how long did the plumber work?

30. Marvin's telephone company charges $13 a month plus 15 cents a minute. If Marvin's bill last month was $25, how many minutes did he talk on the phone?

Sample Problem

The distance from Los Angeles to San Francisco is approximately 420 miles. How long will it take a car traveling at 60 mph to go from Los Angeles to San Francisco?   (*Hint:* Distance equals rate times time.)

a. Time to travel from Los Angeles to San Francisco: t

(continued)

b. $420 = 60 \cdot t$

$\underbrace{420}_{\text{distance}} \quad \underbrace{60}_{\text{rate}} \; \underbrace{t}_{\text{time}}$

c. $\dfrac{420}{60} = \dfrac{60t}{60}$

$7 = t$

Ans. It will take 7 hours.

31. Agnes drove 364 miles in 7 hours. How fast did she drive?

32. A sprinter runs 220 yards in 20 seconds. What is the sprinter's speed?

33. A jet flies 2800 miles at a speed of 560 miles per hour. How long does the trip take?

34. How long will it take a cyclist traveling 13 miles an hour to cover 234 miles?

35. The area of a rectangle is given by the formula $A = lw$, where l stands for the length of the rectangle, and w for its width. If a rectangular garden plot is 12 feet wide, how long must it be in order to provide 180 square feet of gardening space?

36. A roll of carpet contains 400 square feet of carpet. If the roll is 16 feet wide, how long is the piece? (*Hint:* See problem 35.)

37. The perimeter of a rectangle is given by the formula $P = 2l + 2w$, where l stands for the length of the rectangle, and w for its width. A farmer has 500 yards of fencing material with which to enclose a rectangular pasture. He would like the pasture to be 75 yards wide. How long will it be?

38. Inez has 124 centimeters of wood strip with which to make a picture frame. If the frame must be 28 centimeters wide, how long will it be? (*Hint:* See problem 37.)

39. The area of a triangle is given by the formula $A = bh/2$, where b stands for the base of the triangle, and h for its height. A triangular sail requires 12 square meters of fabric. If the base of the sail measures 4 meters, how tall is the sail?

40. A city park has the shape of a triangle and covers 30,000 square feet. If the base of the triangle is 150 feet wide, what is the height of the triangle? (*Hint:* See problem 39.)

CHAPTER SUMMARY

[1.1]   The numbers 1, 2, 3, 4, . . . are called **natural numbers.** These numbers (together with zero) are also called **whole numbers.** Letters such as $a, b, c, \ldots x,$ y, z that are used to represent numbers are called **variables.**

Two numbers added or subtracted together are called **terms.** Two numbers multiplied together are called **factors.**

A natural number greater than 1 that is exactly divisible only by itself and 1 is called a **prime number.** Every natural number can be written as a product of **prime factors** in exactly one way.

The following properties apply to the operations of addition and multiplication:

$$\left.\begin{array}{r} a + b = b + a \\ a \cdot b = b \cdot a \end{array}\right\} \quad \text{Commutative laws.}$$

$$\left.\begin{array}{r} (a + b) + c = a + (b + c) \\ (a \cdot b) \cdot c = a \cdot (b \cdot c) \end{array}\right\} \quad \text{Associative laws.}$$

Any meaningful collection of numbers, variables, and signs of operation is called an **algebraic expression.**

[1.2] The difference $a - b$ is the number d such that $b + d = a$.

The quotient $a \div b$ or a/b is the number q such that $(b)(q) = a$; the divisor b cannot equal zero. Division by zero is undefined.

[1.3] When a mathematical expression contains more than one operation, the operations *must be performed in a specified order* (see page 14).

We can *evaluate* algebraic expressions by replacing the variables with numbers and simplifying the resulting expression.

[1.4] Symbolic versions of word sentences are called **equations.** A value of the variable for which an equation is true is called a **solution** of the equation.

Equations that have identical solutions are called **equivalent equations.** An equation whose solution is not evident by inspection can be transformed to an equivalent equation whose solution is evident. We do this by applying one or more of the following properties:

If $a = b$, then

$$a + c = b + c, \qquad a - c = b - c,$$

$$\frac{a}{c} = \frac{b}{c} \quad (c \neq 0), \qquad \text{and} \qquad a \cdot c = b \cdot c \quad (c \neq 0).$$

By the symmetric property of equality, the members of an equation can be interchanged without any changes of sign. Thus:

$$\text{If} \quad a = b, \qquad \text{then} \qquad b = a.$$

[1.5] Equations which involve more than one operation are solved by "undoing" the operations in the reverse order.

[1.6] The steps on page 28 can be helpful in solving word problems.

Formulas sometimes provide appropriate models for word problems.

■ *The symbols introduced in this chapter appear on the inside front cover.*

CHAPTER REVIEW

1. Write each natural number in prime factored form.

    a. 108   b. 580   c. 1183

2. Express each word phrase in symbols.

    a. The product of 6 and x.

    b. Divide the sum of 4 and y by 6.

    c. The product of y and the sum of 3 and x.

3. Translate each word phrase into an algebraic expression. Use x as the variable.

    a. Three times a number, divided by 6.

    b. The product of 4 and a number, increased by 5.

    c. Five less than 8 times a number.

■ *Simplify.*

4. a. $8 + 4(0)$ b. $8 + \dfrac{4}{0}$ c. $8 + \dfrac{0}{4}$

5. a. $3 + 2(5)$ b. $18 - 2(1 + 3)$ c. $3 + 2 \cdot 3 - 2$

6. a. $\dfrac{12 - 2(5)}{2}$ b. $6 + \dfrac{(7 - 2)6}{3}$ c. $16 + 32 \div 8 - 2 \cdot 3$

7. a. $\dfrac{2 \cdot 12 - 2 \cdot 4}{2(12 - 2 \cdot 4)}$ b. $18 \div (2 \cdot 3 - 3) \div 3$ c. $6 - 3(12 - 2 \cdot 4) \div 4$

8. If $a = 1$ and $b = 2$, find the value of each expression.

    a. $2(a + b)$ b. $2a + b$ c. $2ab$

9. What are the parts of an algebraic expression separated by plus and minus signs called?

10. Write an equation expressing the following:

    a. A number added to 3 equals 27.

    b. Twice a number less 5 equals 17.

    c. Three times a number divided by 4 equals 6.

11. Which of the following are equivalent equations?

    a. $2x - 5 = 15$ b. $2x - 4 = 14$ c. $2x - 4 = 16$

■ *Solve.*

12. a. $3x = 15$ b. $\dfrac{a}{6} = 12$ c. $y - 8 = 8$

13. a. $2x + 3 = 19$ b. $7a - 2 = 26$ c. $37 = 4y - 3$

14. a. $\dfrac{3a}{7} = 21$ b. $\dfrac{b}{9} - 1 = 4$ c. $2 + \dfrac{6c}{5} = 20$

15. a. $3a - 9 = 0$ b. $\dfrac{3b}{9} = 0$ c. $3c + 9 = 9$

■ *For the following word problems:*

a. *Write in words the unknown quantity and assign a variable to represent it,*

b. *write an equation that models the problem, and*

c. *solve the equation.*

16. A particular turntable costs $38 more at Sam's Stereo than at Al's Audio. If Sam sells the turntable for $123, what is Al's price?

17. Gus and Judy made a down payment of $16,800 on their new house. This was 15% of the selling price of the house. What was the selling price?

18. Stella's cable TV bill for February was $16.05. If the basic service is $7.95 per month, and each movie channel costs $1.35 per month, how many movie channels does Stella take?

19. Robin drove 423 miles at an average speed of 47 miles per hour. How long did it take?

20. Leslie bought 332 feet of chain link fence, which is just enough to enclose her property. If the property is 60 feet wide, how long is it?

2 ELEMENTARY NOTIONS OF ALGEBRA USING INTEGERS

2.1 INTEGERS AND THEIR GRAPHS

INTEGERS

On occasion, we use natural numbers to represent physical quantities such as money (5 dollars), temperature (20 degrees), and distance (10 miles). Since this representation does not differentiate between gains and losses, degrees above or below zero, or distances in opposite directions from a starting point, mathematicians have found it convenient to represent these ideas symbolically by the use of plus (+) and minus (−) signs. For example, we may represent:

A loss of five dollars as −$5.
A gain of five dollars as +$5.
Ten degrees below zero as −10°.
Ten degrees above zero as +10°.
Ten miles to the west of a starting point as −10 miles.
Ten miles to the east of a starting point as +10 miles.

Nonzero numbers that are preceded by a minus sign are called **negative numbers.** Nonzero numbers that are preceded by a plus sign are called **positive numbers.** Note that zero is not considered positive or negative.

The numbers

$$\ldots, -3, -2, -1, 0, +1, +2, +3, \ldots$$

are called **integers.** When a numeral is written without a sign, for example, 3, 5, 9, we assume that a positive sign is intended.

We refer to the numbers

$$. . . , -6, -4, -2, 0, 2, 4, 6, . . .$$

as **even integers.** And we refer to the numbers

$$. . . , -5, -3, -1, 1, 3, 5, . . .$$

as **odd integers.**

NUMBER LINE

We can use a **number line** as shown in Figure 2.1 to represent the integers.

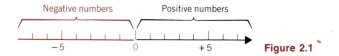

Figure 2.1

To indicate a particular integer on the number line, we **graph** the integer by placing a dot in the appropriate position on the line.

For example, the graphs of −6, −3, 0, 4, and 7 would appear as shown in Figure 2.2.

Figure 2.2

To construct a number line:

1. Draw a straight line.
2. Decide on a convenient unit of scale and mark off units of this length on the line, beginning on the left.
3. Below the line, label enough of these units to establish the scale, usually two or three points. The point representing 0 is called the *origin.*
4. Add a small arrow pointing to the right to indicate that numbers are larger to the right.
5. Above the line, label the numbers to be graphed. Graph the numbers by placing dots at the appropriate places on the line.

The number line is particularly useful in visualizing the order of two integers.

> **An integer whose graph lies to the left of the graph of a second integer is less than the second integer.**

For example, in Figure 2.3, the graph of 3 is *to the left* of the graph of 7. Therefore, 3 *is less than* 7. We could also state this relationship as "7 *is greater than* 3."

Figure 2.3

In Figure 2.4, the graph of −6 is *to the left of* the graph of −2. Therefore, −6 *is less than* −2, and −2 *is greater than* −6.

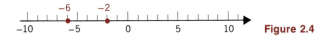

Figure 2.4

ORDER SYMBOLS

We use special symbols to indicate the order relationship between numbers:

$<$ means "is less than";

$>$ means "is greater than."

For example,

$2 < 5$ is read "2 is less than 5,"

and

$5 > 2$ is read "5 is greater than 2."

Notice that the point of the symbols $<$ and $>$ always points to the smaller number.

$2 < 5$	$5 > 2$
Points to ⟋↑	↑⟍ Points to
smaller number.	smaller number.

Thus, for

$-4 < -2$, read "negative four *is less than* negative two";

$-3 < 3$, read "negative three *is less than* three"; and

$-1 < 0$, read "negative 1 *is less than* zero."

This notion of the order of integers is consistent with our physical experiences. A temperature of $-4°$ is lower or less than one of $-2°$, $-3°$ is less than $3°$, and $-1°$ is less than $0°$.

THE SYMBOL "–"

We have now used the symbol "–" to indicate a sign of operation (subtraction) and to indicate negative numbers. We also use the symbol "–" to mean the "opposite" or "negative" of a number. Thus:

$-x$ means the "opposite" of x.

If x is positive, then $-x$ is negative. For example,

$$-(5) = -5 \quad \text{and} \quad -(10) = -10.$$

If x is negative, then $-x$ is positive. For example,

$$-(-5) = +5 \quad \text{and} \quad -(-10) = +10.$$

For any a, positive or negative,

$$-(-a) = a.$$

ABSOLUTE VALUE

In Figure 2.5, observe that while $+8$ and -8 are in opposite directions from the origin, each of the two numbers is the same *distance* (eight units) from the origin.

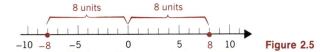

Figure 2.5

The positive number 8, which specifies the distance between $+8$ and the origin, or between -8 and the origin, is called the **absolute value** of $+8$ and of -8.
    Thus:

The absolute value of a number a is the distance between a and 0 on the number line.

Notice that since distance is never negative, *the absolute value of a number is never negative.*

We designate absolute value by vertical bars. For example, for

$|-5| = 5$,   read "the absolute value of negative five is equal to five";

$|7| = 7$,   read "the absolute value of seven is equal to 7";

$|0| = 0$,   read "the absolute value of zero is equal to zero."

We can define the absolute value of a more formally by:

$$|a| = \begin{cases} a \text{ if } a \text{ is greater than or equal to } 0. \\ -a \text{ if } a \text{ is less than } 0. \end{cases}$$

This definition is consistent with our earlier observation that the absolute value is never negative. For example, if $a = -3$, then a is less than 0 so $|a| = -a$, or

$$|-3| = -(-3) = 3.$$

EXERCISES 2.1

■ *Graph the numbers on a number line.*

Sample Problem

$-8, -6, -2, 1, 5,$ and 9

Ans.

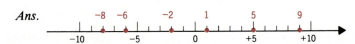

1. $0, -3, 5, -5, 2,$ and -1 **2.** $-4, 3, -2, -1, 6,$ and 9
3. $-2, 4, 7, -5, -3,$ and 3 **4.** $-4, -7, 0, -2, 4,$ and 8

■ *Replace the comma in each pair with the proper symbol:* $<$ *or* $>$.

Sample Problems

a. $1, -3$ b. $-9, -6$

Ans.
a. $>$ (the graph of 1 is to the right of the graph of -3)
b. $<$ (the graph of -9 is to the left of the graph of -6)

5. $4, 8$ **6.** $7, 2$ **7.** $0, 4$ **8.** $0, -4$
9. $3, -7$ **10.** $-4, 5$ **11.** $-5, -9$ **12.** $-2, -8$
13. $-5, -2$ **14.** $-8, -1$ **15.** $-5, 7$ **16.** $4, -5$
17. $0, -4$ **18.** $8, 0$

■ *Locate the graphs of the numbers on a number line.*

19. The natural numbers less than 10.

20. The integers between -3 and 7.

21. The even integers between -3 and 11.

22. The odd integers between -5 and 5.

23. The odd integers between -5 and -1.

24. The odd integers between -15 and 0.

■ *Simplify.*

Sample Problems

a. $|-7|$　　　　　b. $-|-5|$　　　　　c. $-|2|$

$\qquad\qquad\qquad\qquad = -(5)\qquad\qquad\qquad = -(2)$

Ans. 7　　　　　*Ans.* -5　　　　　*Ans.* -2

25. $|6|$　　**26.** $|-2|$　　**27.** $|-10|$　　**28.** $|5|$

29. $-|9|$　　**30.** $-|-4|$　　**31.** $-|-6|$　　**32.** $-|7|$

■ *Replace the comma in each pair with the proper symbol: $<$, $>$, or $=$.*

33. $|-7|, 8$　　　　**34.** $|-2|, -2$　　　　**35.** $|-5|, |5|$

36. $|-3|, |3|$　　　　**37.** $-|-4|, -5$　　　　**38.** $-|-8|, -6$

39. $|6|, 0$　　　　**40.** $|-12|, 0$　　　　**41.** $|0|, 6$

42. $|0|, -3$　　　　**43.** $|-7|, |-6|$　　　　**44.** $|-2|, |4|$

■ *Simplify.*

Sample Problems

a. $|6| - |-3|$　　　　　　b. $3|-4| - \dfrac{|8|}{2}$

　　　　　　　　　　　　　　　　Evaluate absolute values.

$\qquad 6 - 3 \qquad\qquad\qquad\qquad 3(4) - \dfrac{8}{2}$

　　　　　　　　　　　　　　　　Simplify.

$\qquad\qquad\qquad\qquad\qquad\qquad 12 - 4$

Ans. 3　　　　　　*Ans.* 8

45. $|-8| - |4|$　　　　　**46.** $|-2| + |-7|$

47. $\dfrac{|-6|}{3} + 5$　　　　　**48.** $|16| - \dfrac{8}{|4|}$

49. $3|-6|$　　　　　**50.** $5|9|$

51. $2 + 5|-3|$　　　　　**52.** $|-12| - |-3||-2|$

Omit 45-56

53. $\dfrac{|7 - 3| - 2}{7 - |12 - 5|}$ **55.**

54. $5 + |10 - 2|$

56. $|-3||6 - 2|$

■ *Choose the correct word to make a true statement.*

57. If x is a positive number, then $|x|$ is (*positive/negative*).
58. If x is a negative number, then $|x|$ is (*positive/negative*).
59. If x is a positive number, then $-x$ is (*positive/negative*).
60. If x is a negative number, then $-x$ is (*positive/negative*).

2.2   SUMS OF INTEGERS

To see what is meant by the sum of two integers, we can think of such numbers as representing gains and losses, (+) numbers denoting gains and (−) numbers denoting losses.

	+5 gain	−5 loss	−5 loss	+5 gain
	+3 gain	−3 loss	+3 gain	−3 loss
Sum:	+8 gain	−8 loss	−2 loss	+2 gain

In algebra, we usually write sums horizontally. Thus, we would write the sum of +5 and +3 as

$$(+5) + (+3) = +8,$$

and we would write the sum of −5 and −3 as

$$(-5) + (-3) = -8.$$

For the sum of −5 and +3, we would have

$$(-5) + (+3) = -2,$$

and for the sum of +5 and −3,

$$(+5) + (-3) = 2.$$

The examples above suggest a rule for the addition of integers.

To add integers with like signs:

1. Add the absolute values of the numbers.
2. The sum has the same sign as the numbers.

To add integers with unlike signs:

1. Find the absolute value of each and subtract the lesser absolute value from the greater absolute value.
2. The sum has the same sign as the number with the greater absolute value.

For example,

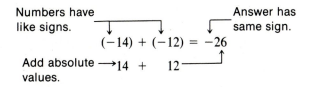

Numbers have like signs.   Answer has same sign.

$$(-14) + (-12) = -26$$

Add absolute values. $\longrightarrow 14 + 12$

Numbers have unlike signs.   Since $|-15|$ is greater than $|+9|$, the sum is negative.

$$(-15) + (+9) = -6$$

Subtract the lesser absolute value $\longrightarrow 15 - 9$ from the greater absolute value.

Numbers have unlike signs.   Since $|+15|$ is greater than $|-9|$, the sum is positive.

$$(+15) + (-9) = +6$$

Subtract the lesser absolute value $\longrightarrow 15 - 9$ from the greater absolute value.

Notice that we use the symbols $+$ and $-$ in two very different ways: they may indicate the operations of addition and subtraction, or they may indicate whether a number is positive or negative. In the foregoing example, the $+$ or $-$ symbols inside parentheses indicate the sign of the accompanying integer. We use parentheses for clarification whenever two $+$ or $-$ symbols occur in succession. However, when no confusion will arise we may delete the parentheses to simplify the expression. For example,

$$(+5) + (+3) = 5 + 3 = 8,$$

$$(-5) + (-3) = -5 + (-3) = -8,$$

and

$$(-5) + (+3) = -5 + 3 = -2.$$

The commutative and associative laws of addition hold with respect to integers. For example,

$$2 + (-3) = -3 + 2,$$

and

$$[2 + (-3)] + 4 = 2 + [(-3) + 4].$$

EXERCISES 2.2

■ *Add.*

Sample Problems

a. -4
$\underline{-7}$

b. -5
$\underline{+2}$

(continued)

a. Since the signs are alike, add the absolute values.

$$|-4| + |-7| = 4 + 7 = 11$$

The sum has the same sign as the numbers -4 and -7.

Ans. -11

b. Since the signs are unlike, subtract the lesser absolute value from the greater absolute value.

$$|-5| = 5; \quad |+2| = 2; \qquad 5 - 2 = 3$$

Since $|-5|$ is greater than $|+2|$, the sum of $(-5) + (+2)$ is negative.

Ans. -3

1. $+3$ $+4$	**2.** $+2$ $+8$	**3.** -6 -3	**4.** -4 -9
5. $+7$ -3	**6.** -4 $+8$	**7.** -8 $+2$	**8.** $+4$ -6
9. $+8$ 0	**10.** -6 0	**11.** -7 $+7$	**12.** $+8$ -8

Sample Problems

a. $(-5) + (-7)$ 　　　　　　　 b. $(+6) + (-2)$

a. Since the signs are alike, add the absolute values.

$$|-5| + |-7| = 5 + 7 = 12$$

The sum has the same sign as the numbers, -5 and -7.

Ans. -12

b. Since the signs are unlike, subtract the lesser absolute value from the greater absolute value.

$$|+6| = 6; \quad |-2| = 2; \qquad 6 - 2 = 4$$

Since $|+6|$ is greater than $|-2|$, the sum is positive.

Ans. $+4$ or 4

13. $(+7) + (+3)$	**14.** $(+7) + (+2)$	**15.** $(-3) + (-7)$
16. $(-6) + (+3)$	**17.** $(+8) + (-3)$	**18.** $(-8) + (+3)$
19. $(-6) + (+2)$	**20.** $(-15) + (+1)$	**21.** $(+5) + 0$
22. $0 + (-5)$	**23.** $(-6) + (+6)$	**24.** $(-8) + (+8)$

■ *Simplify wherever possible, and add.*

Sample Problems

a. $(-7) + (-4)$
$-7 + (-4)$

Ans. -11

b. $(-9) + (+12) + (-2)$
$-9 + 12 + (-2)$
$3 + (-2)$

Ans. 1

25. $(+5) + (+3)$ **26.** $(-6) + (-2)$ **27.** $(-8) + (0)$

28. $(-5) + (+3)$ **29.** $(+7) + (-2)$ **30.** $(-5) + (-6)$

31. $(-5) + (+8) + (+3)$ **32.** $(+7) + (-7) + (+2)$

33. $(-4) + (-5) + (+6)$ **34.** $(-5) + (-3) + (-4)$

35. $(+4) + (0) + (-7)$ **36.** $(-9) + (+9) + (0)$

37. $6 + 0 + (-8) + (+5)$ **38.** $4 + (0) + (-7) + (-6)$

39. $(-7) + (0) + 7 + (+3)$ **40.** $(-5) + 6 + (+6) + (-3)$

■ *Simplify and add.*

a. $5 + [-3 + (-4)]$
$5 + [-7]$

Ans. -2

b. $[(-3) + (-6)] + [6 + (-4) + 1]$
$[-9] + [3]$

Ans. -6

41. $-6 + (-3 + 5)$ **42.** $-12 + [4 + (-8)]$

43. $[5 + (-8)] + (-4)$ **44.** $(-11 + 2) + 7$

45. $(-5 + 3) + [1 + (-3) + 4]$ **46.** $7 + (-6 + 2) + (-3)$

47. $(6 + 2) + (-4 + 3) + (-9 + 2)$ **48.** $-2 + [-6 + (-3)] + (8 - 1)$

49. $[-3 + (-2)] + [2 + (-6)] + [(-4) + (-2)]$

50. $[6 + (-5) + 8] + [(-4) + (+9) + (-3)]$

2.3 DIFFERENCES OF INTEGERS

When we discussed the difference of two natural numbers in Section 1.2, we saw that $5 - 3$ is the number 2 that, when added to 3, gives 5. In general, $a - b$ is the number that, when added to b, gives a. The same idea holds for the difference of two integers. For example,

$$(+9) - (+6) = +3 \quad \text{because} \quad (+6) + (+3) = +9.$$

Notice that the sum $(+9) + (-6)$ is also equal to +3, so that we have

$$(+9) - (+6) = (+9) + (-6).$$

Thus, subtracting a *positive* number gives the same result as *adding its opposite, a negative number*. It is also true that subtracting a negative number gives the same result as *adding its opposite, a positive number*. For example,

$$(+6) - (-4) = (+6) + (+4) = +10,$$

and

$$(-5) - (-3) = (-5) + (+3) = -2.$$

This suggests the following rule for subtraction.

To subtract an integer *b* from an integer *a*:

1. Change the subtraction to addition.
2. Change the sign of *b*.
3. Proceed as in addition.

In symbols:

$$a - b = a + (-b)$$

where $-b$ is the opposite of b.

The following examples illustrate the application of the subtraction rule.

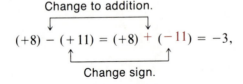

$$(+8) - (+11) = (+8) + (-11) = -3,$$

$$(+7) - (-3) = (+7) + (+3) = 10,$$

$$(-7) - (-3) = (-7) + (+3) = -4.$$

We can simplify differences by deleting the parentheses as we did for sums. We will first rewrite the problem as an equivalent addition problem and then delete the + symbol that indicates addition. In the remaining expression each + or − symbol indicates the sign of the number that follows. For example,

2.3 DIFFERENCES OF INTEGERS

$$(+9) - (+6) = (+9) + (-6)$$

———Delete.

$$= 9 - 6 = 3,$$

where we view the expression $9 - 6$ as the sum of 9 and -6.

In the next example, we begin by changing each subtraction to an equivalent addition.

$$5 - (+7) + (-3) - (-2) = 5 + (-7) + (-3) + (+2).$$

Once this is done, we can delete all $+$ or $-$ symbols that indicate operations, since all operations are now additions, and remove all parentheses:

———Delete addition symbols.

$$5 + (-7) + (-3) + (+2) = 5 - 7 - 3 + 2.$$

Now each $+$ or $-$ symbol indicates the sign of the number that follows, and the operations are understood to be additions. Thus

$$5 - 7 - 3 + 2 = 5 - 7 - 3 + 2$$
$$= -2 - 3 + 2$$
$$= -5 + 2 = -3.$$

Here is another example.

———Change signs.

$$-6 + (-3) - (-8) - 4 = -6 + (-3) + (+8) - 4$$

———Delete.

$$= -6 - 3 + 8 - 4$$
$$= -9 + 8 - 4$$
$$= -1 - 4 = -5.$$

In problems that contain brackets, we perform the operations inside brackets first. For example,

———Change signs.

$$4 - 3 - [-6 + (-5) - (-2)] = 4 - 3 - [-6 + (-5) + (+2)]$$

———Delete.

(continued)

$$= 4 - 3 - [-6 - 5 + 2]$$
$$= 4 - 3 - [-9]$$
$$= 4 - 3 + 9 = 10.$$

We can also find the difference between two integers written in vertical form by changing the sign on the number being subtracted and then adding the two numbers. For example, to subtract $+11$ from $+8$ in a vertical form we have

$$
\begin{array}{r} +8 \\ \text{subtract } +11 \\ \hline \end{array}
\quad \text{is changed to the sum} \quad
\begin{array}{r} +8 \\ \text{add } -11. \\ \hline -3 \end{array}
$$

EXERCISES 2.3

■ *Subtract by first changing each problem to an equivalent addition problem.*

Sample Problems

a. $(+5) - (+2)$ b. $(-6) - (-4)$

                                    Express as a sum; change signs.

$(+5) + (-2)$ $(-6) + (+4)$

                                         Add.

Ans. 3   *Ans.* -2

1. $(+8) - (+2)$	**2.** $(+7) - (+1)$	**3.** $(+2) - (-6)$
4. $(+3) - (-9)$	**5.** $(-4) - (-7)$	**6.** $(-8) - (-6)$
7. $(-5) - (-2)$	**8.** $(-8) - (-1)$	**9.** $(7) - (2)$
10. $(8) - (5)$	**11.** $(3) - (8)$	**12.** $(1) - (7)$

■ *Subtract the bottom number from the top number.*

Sample Problems

a. $+6$ $+6$
   $+8$ Change to -8
             and add. -2

b. -5 -5
   -8 Change to $+8$
             and add. $+3$

Ans. -2 *Ans.* 3

13. $+4$ $+4$	**14.** $+8$ $+8$	**15.** $+7$ $+2$	**16.** $+5$ $+3$

17. $+6$	**18.** $+3$	**19.** -6	**20.** 0
$\underline{+9}$	$\underline{+8}$	$\underline{0}$	$\underline{-6}$
21. -8	**22.** -3	**23.** -4	**24.** -2
$\underline{-2}$	$\underline{-1}$	$\underline{-8}$	$\underline{-4}$

■ *Simplify.*

Sample Problems

a. $(-5) + (-3) - (-4)$ b. $(6) - (5) + (-2)$

Express as a sum; change signs.

$(-5) + (-3) + (+4)$ $(6) + (-5) + (-2)$

Delete operation symbols.

$-5 - 3 + 4$ $6 - 5 - 2$

Add.

Ans. -4 *Ans.* -1

25. $(-5) + (-4) - (+3)$ **26.** $(-6) - (-2) + (-4)$
27. $(+9) - (+3) + (-4)$ **28.** $(-2) + (+6) - (-2)$
29. $(+6) - (+2) - (-4)$ **30.** $(+6) - (-2) - (-3)$
31. $(4) - (2) + (-2)$ **32.** $(-2) + (6) + (2)$
33. $(-2) + (-2) - (-2) + (2)$ **34.** $(4) - (-4) + (4) - (4)$
35. $(-4) - (-2) + (5) + (1)$ **36.** $(7) - (4) + (-3) + (-1)$
37. $6 - 8 + (-3) + (+2)$ **38.** $4 - (-9) + 6 + (-3)$
39. $11 + 1 - (+7) - (-7)$ **40.** $(+5) + 5 - (-5) - 5$

Sample Problem

$(-3 + 8) - [(-3) + 4 - (-6)]$

Write expression inside brackets as a sum; change signs.

$(-3 + 8) - [-3 + 4 + 6]$

Add within parentheses and brackets.

$(5) - [7]$

Add 5 and -7.

Ans. -2

41. $-6 - (10 - 2)$ **42.** $3 - (6 - 11)$
43. $5 - [8 - (-1)]$ **44.** $-2 - [(-4) - (-3)]$
45. $(12 - 10) - (8 - 15)$ **46.** $[6 - (-3)] - (-4 - 7)$
47. $9 - [3 - (-2) + 6]$ **48.** $6 - 5 - [-2 + (-4) + 6]$
49. $8 - [(6 - 3) - (-2 + 6)]$ **50.** $[(-7 + 1) + (-6 - 2)] - (-4)$

2.4 PRODUCTS AND QUOTIENTS OF INTEGERS

PRODUCTS

You may recall that multiplication is a form of addition. That is, we can think of 3(2) as the sum of three 2's (2 + 2 + 2), and we can think of 3(−2) as the sum of three −2's [(−2) + (−2) + (−2)], which is −6. Thus, it would appear that the product of a positive number and a negative number should be a negative number, and that the product of two positive numbers is positive.

Now let's investigate the product of two negative numbers. First consider the following sequence of products:

$$4(-2) = -8$$
$$3(-2) = -6$$
$$2(-2) = -4$$
$$1(-2) = -2$$
$$0(-2) = 0.$$

If we continue the sequence with the product −1(−2), it seems plausible (from the sequence above) that the number 2 is the next number in the sequence following the equals sign. That is, the sequence continues with

$$-1(-2) = 2,$$
$$-2(-2) = 4,$$
$$-3(-2) = 6,$$

and so on. It appears (at least intuitively) that the product of two negative numbers is a positive number. In fact, this is the case and we adopt the following rule.

To find the product of two integers:

1. Multiply the absolute values of the numbers.

2. Determine the sign of the product:
  if the factors have like signs, the product is positive;
  if the factors have unlike signs, the product is negative.

In symbols:

$$(+a)(+b) = +(ab), \qquad (-a)(-b) = +(ab),$$
$$(+a)(-b) = -(ab), \quad \textbf{and} \quad (-a)(+b) = -(ab).$$

For example,

$$(+5)(+9) = +45, \qquad (-5)(-9) = +45,$$
$$(+5)(-9) = -45, \quad \textbf{and} \quad (-5)(+9) = -45.$$

Also, the product of zero and any integer is zero. For example,

$$(-5) \cdot 0 = 0 \qquad \text{and} \qquad (+5) \cdot 0 = 0.$$

The commutative and associative laws of multiplication hold for integers. For example, by the commutative law,

$$(-2)(3) = 3(-2).$$

And by the associative law,

$$[(-2)(3)](4) = (-2)[(3)(4)].$$

We can determine the sign of the product of any number of factors. If we have an *odd* number of negative factors, the product is negative; if we have an *even* number of negative factors, the product is positive. For example,

one negative factor⌐   ⌐ negative product
$$(-2)(3)(4) = (-6)(4) = -24$$

three negative factors⌐   ⌐ negative product
$$(-2)(-3)(-4) = 6(-4) = -24$$

two negative factors⌐   ⌐ positive product
$$(-2)(3)(-4) = (-6)(-4) = +24$$

four negative factors⌐   ⌐ positive product
$$(-2)(-3)(-2)(-4) = (-12)(-4) = +48.$$

QUOTIENTS

We define the quotient of two integers the same way we define the quotient of two natural numbers (Section 1.2). However, the sign of the quotient has to be consistent with the rule of signs for the multiplication of integers. Recall that for natural numbers, the quotient a/b is the number q such that $b \cdot q = a$. Let us examine the quotient of two integers by considering some numerical cases.

$$\frac{+6}{+3} = +2, \qquad \text{because} \qquad (+3)(+2) = +6,$$

$$\frac{+6}{-3} = -2, \qquad \text{because} \qquad (-3)(-2) = +6,$$

$$\frac{-6}{+3} = -2, \qquad \text{because} \qquad (+3)(-2) = -6,$$

$$\frac{-6}{-3} = +2, \qquad \text{because} \qquad (-3)(+2) = -6.$$

These examples suggest the following rule.

> **To find the quotient of two integers:**
> 1. Find the quotient of the absolute values of the numbers.
> 2. Determine the sign of the quotient:
>   if the dividend and divisor have like signs, the quotient is positive;
>   if the dividend and divisor have unlike signs, the quotient is negative.

In symbols:

$$\frac{+a}{+b} = +\left(\frac{a}{b}\right), \qquad \frac{-a}{-b} = +\left(\frac{a}{b}\right),$$

$$\frac{+a}{-b} = -\left(\frac{a}{b}\right), \quad \text{and} \quad \frac{-a}{+b} = -\left(\frac{a}{b}\right).$$

For example,

$$\frac{+12}{+3} = +4, \qquad \frac{-12}{-3} = +4,$$

$$\frac{+12}{-3} = -4, \quad \text{and} \quad \frac{-12}{+3} = -4.$$

QUOTIENTS INVOLVING ZERO

As for whole numbers, a quotient equals zero if the dividend is zero and the divisor is not zero. And as always, *division by 0 is undefined.* For example,

$$\frac{0}{3} = 0 \quad \text{and} \quad \frac{0}{-3} = 0,$$

whereas

$$\frac{3}{0} \quad \text{and} \quad \frac{-3}{0} \quad \text{are undefined,}$$

and

$$\frac{0}{0} \quad \text{is indeterminate.}$$

ORDER OF OPERATIONS

The order of operations for natural numbers discussed in Section 1.3 also applies to simplifying expressions containing integers. Because additions and subtractions are the last operations to be performed, it is useful to separate an expression into its terms. For example, the expression $4 - 3(-4)$ is composed of two terms, 4 and $3(-4)$, each of which must be simplified before they can be combined. Thus

$$4 - 3(-4) = 4 - (-12)$$
$$= 4 + 12$$
$$= 16.$$

This idea can also be used in evaluating algebraic expressions. For example, we evaluate the expression $2x - 3xy$ for $x = -5$ and $y = -2$ by first substituting -5 for x and -2 for y, and then simplifying the resulting expression to obtain

$$2(-5) - 3(-5)(-2) = -10 - 3(10)$$
$$= -10 - 30$$
$$= -40.$$

Common Error

Negative integers should be enclosed in parentheses when substituted into algebraic expressions. In the previous example, $2x$ should be written $2(-5)$ and not $2 - 5$.

EXERCISES 2.4

■ *Multiply.*

Sample Problems

a. $(-3)(-7)$ b. $(-4)(7)$

Determine sign of product.

$+$ $-$

Multiply absolute values of factors.

21   28

Ans. $+21$ or 21   *Ans.* -28

1. $(5)(-3)$	**2.** $(-2)(4)$	**3.** $(-5)(-6)$	**4.** $(4)(8)$
5. $(5)(0)$	**6.** $0(-4)$	**7.** $(4)(-2)$	**8.** $(-3)(-3)$
9. $(-5)(3)$	**10.** $(-8)(-4)$	**11.** $(2)(-7)$	**12.** $(6)(7)$
13. $(-6)(0)$		**14.** $(0)(8)$	
15. $(2)(-3)(4)$		**16.** $(-2)(-3)(-4)$	
17. $(-6)(-1)(3)$		**18.** $(4)(-1)(-7)$	
19. $(-4)(0)(6)$		**20.** $(-5)(4)(0)$	
21. $(-2)(5)(4)(-3)$		**22.** $(3)(-5)(-2)(-2)$	
23. $(-4)(-2)(-1)(-1)$		**24.** $(-3)(-2)(1)(-2)$	

■ *Divide.*

a. $\dfrac{-12}{-2}$ b. $\dfrac{15}{-3}$

                                  Determine the sign of the quotient.

  +   −   Divide the absolute values of the numerator and denominator.

  6   5

Ans. +6   *Ans.* −5

25. $\dfrac{-25}{-5}$ 26. $\dfrac{-32}{-8}$ 27. $\dfrac{-27}{9}$ 28. $\dfrac{12}{-4}$

29. $\dfrac{0}{7}$ 30. $\dfrac{0}{-4}$ 31. $\dfrac{8}{0}$ 32. $\dfrac{-5}{0}$

33. $-12 \div 2$ 34. $-20 \div (-5)$ 35. $15 \div (-3)$ 36. $-35 \div 7$

■ *Simplify.*

a. $-2(3) - (-3)(-5)$ b. $\dfrac{-6 + 2}{-2} - \dfrac{15}{-3}$

                              Simplify each term.

   $-6 - 15$ $\dfrac{-4}{-2} - \dfrac{15}{-3}$

Ans. −21 $2 - (-5)$

                      $2 + 5$ Rewrite as a sum and add.

                 Ans. 7

37. $5(-4) - 3(-6)$ 38. $(-6)(-7) + 6(-3)$

39. $6 - 2(-5) + 7$ 40. $5 - (-4)3 - 7(-2)$

41. $-3(8) + (-6)(-2) - 5(2)$ 42. $5 - (-4)(3 - 7) - 2$

43. $-5[4 - 2(3)] + 6$ 44. $2(-8) - [4(-3) - 2]$

45. $1 + 3[-2(-3 + 8)] - 4 - 3$ 46. $3(-2) - 2[6 + 5(-3) - 1] - (3 + 2)$

47. $\dfrac{-12}{4} + \dfrac{3 - 8}{5}$ 48. $\dfrac{15}{-3} - \dfrac{4 - 8}{8 - 12}$

49. $\dfrac{8(-7 + 4)}{6} - \dfrac{6 - 4(-5)}{9 - (-4)}$ 50. $\dfrac{-11(3) - 7(-5)}{-13(-2) + 8(-3)} + \dfrac{5}{3 - 4}$

■ *Evaluate the following expressions.*

Sample Problems

a. $12 - 3xy$
    for $x = -3$, $y = 2$

$12 - 3(-3)(2)$

$12 - (-18)$

$12 + 18$

Ans. 30

b. $\dfrac{4a + 6b}{7}$ for $a = -1$, $b = -4$

Substitute values.

$\dfrac{4(-1) + 6(-4)}{7}$

Perform multiplications.

$\dfrac{-4 + (-24)}{7}$

Add.

$\dfrac{-28}{7}$

Divide.

Ans. -4

51. $-3x + 2y$ for $x = -5$, $y = -4$

52. $-2y - 7x$ for $x = -8$, $y = -3$

53. $ab - 2a - 5b$ for $a = 9$, $b = -2$

54. $3a - 5ab + 2$ for $a = 1$, $b = -7$

55. $\dfrac{8 - 6xy}{x}$ for $x = -2$, $y = 4$

56. $\dfrac{9y}{x} - \dfrac{xy}{2}$ for $x = -3$, $y = 12$

57. $\dfrac{x - y}{x + y}$ for $x = 2$, $y = -6$

58. $\dfrac{7 - xy}{xy + 9}$ for $x = -5$, $y = 5$

59. $\dfrac{4(ab - 3a)}{3} - (a - 2b)$ for $a = 3$, $b = -5$

60. $\dfrac{4 - (b - a)}{3b} - \dfrac{3a - 4b}{a + 4b}$ for $a = 12$, $b = -2$

2.5   EQUATIONS AND WORD PROBLEMS USING INTEGERS

In Chapter 1 we used the notion of equivalent equations to solve equations involving whole numbers: by adding or subtracting the same expression from each member of an equation, or by multiplying or dividing each member by the same expression, we produced simpler equations with the same solution as the original. Now that we can perform the four arithmetic operations on integers, we can use these ideas to solve equations involving integers.

To solve the equation

$$7 - 3x = -14$$

we first subtract 7 from each member to obtain

$$7 - 3x - 7 = -14 - 7$$

from which

$$-3x = -21.$$

Next we divide both sides by -3 to get

$$\frac{-3x}{-3} = \frac{-21}{-3}$$

which gives the solution

$$x = 7.$$

Now consider the equation

$$-6 - \frac{2x}{3} = 8.$$

We begin by adding 6 to both sides of the equation to obtain

$$-6 - \frac{2x}{3} + 6 = 8 + 6,$$

from which

$$-\frac{2x}{3} = 14.$$

Next we multiply both sides by 3 to get

$$3\left(-\frac{2x}{3}\right) = 3(14)$$

or

$$-2x = 42.$$

Finally we divide both sides by -2 to get

$$\frac{-2x}{-2} = \frac{42}{-2}$$

Which gives the solution

$$x = -21.$$

APPLICATIONS

Many applications of equations involve the use of negative numbers. Consider the following problem:

> A small manufacturing firm lost $20,000 in its first year of operation. If their costs for the first year were $85,000, how much revenue did the company bring in?

The techniques from Chapter 1 for constructing and solving equations still apply: we first write in words what we want to find and represent this unknown quantity by a variable. Then we write an equation that expresses the information given in the problem.

Company's first-year revenue: x

$$\underbrace{x}_{\text{revenue}} - \underbrace{85{,}000}_{\text{costs}} = \underbrace{-20{,}000}_{\text{profit}}$$

Notice that the $20,000 loss is represented as a negative profit. We solve the equation by adding 85,000 to both members.

$$x - 85{,}000 + 85{,}000 = -20{,}000 + 85{,}000$$

$$x = 65{,}000$$

The company's revenue was $65,000.

EXERCISES 2.5

■ *Find the solution of each equation by inspection.*

Sample Problems

a. $x + 6 = 4$

b. $5 \cdot x = -15$

Ans. -2 (because $-2 + 6 = 4$)

Ans. -3 (because $5 \cdot (-3) = -15$)

1. $x + 4 = -1$ **2.** $x + 7 = -2$ **3.** $x - 5 = -4$

4. $x - 10 = -5$ **5.** $8 - x = -5$ **6.** $4 - x = -7$

7. $3x = 12$ **8.** $-4x = 28$ **9.** $6x = -36$

10. $-7x = 42$ **11.** $-5x = -25$ **12.** $-4x = -32$

13. $\dfrac{x}{5} = -35$ **14.** $\dfrac{x}{-8} = 4$ **15.** $\dfrac{x}{-3} = 0$

16. $\dfrac{x}{12} = 0$ **17.** $\dfrac{x}{-4} = 1$ **18.** $\dfrac{x}{-1} = 4$

■ *Solve.*

Sample Problems

$$5 - 2x = -9$$

Subtract 5 from both sides.

$$5 - 2x - 5 = -9 - 5$$
$$-2x = -14$$

Divide both sides by -2.

$$\frac{-2x}{-2} = \frac{-14}{-2}$$

Ans. $x = 7$

19. $x + 6 = 4$

20. $x - 5 = -4$

21. $z + 8 = 0$

22. $z - 7 = 0$

23. $6x = -24$

24. $-3x = -15$

25. $\dfrac{a}{-4} = 7$

26. $\dfrac{a}{-9} = -2$

27. $\dfrac{2y}{3} = -10$

28. $\dfrac{-4a}{5} = -12$

29. $36 = \dfrac{-3b}{5}$

30. $-20 = \dfrac{5b}{7}$

31. $3c - 7 = -13$

32. $5c + 12 = 2$

33. $30 - 7x = 2$

34. $-3 + 4x = -15$

35. $-5 = -2 - 3x$

36. $-20 = 1 - 7t$

Sample Problem

$$4 - \frac{3x}{5} = 10$$

Subtract 4 from both sides.

$$4 - \frac{3x}{5} - 4 = 10 - 4$$

$$-\frac{3x}{5} = 6$$

Multiply both sides by 5.

$$5\left(-\frac{3x}{5}\right) = 5(6)$$
$$-3x = 30$$

Divide both sides by -3.

$$\frac{-3x}{-3} = \frac{30}{-3}$$

Ans. $x = -10$

37. $4 = 1 - \dfrac{3x}{7}$ **38.** $-4 = 2 + \dfrac{3y}{5}$

39. $0 = \dfrac{5x}{2} + 10$ **40.** $0 = 6 - \dfrac{2y}{3}$

41. $7 = -5 + \dfrac{2y}{3}$ **42.** $\dfrac{3x}{2} - 4 = 2$

■ *For the following word problems:*
a. *Write in words the unknown quantity and represent it by a variable,*
b. *write an equation that models the problem, and*
c. *solve the equation.*

Sample Problem

The temperature rose 27° since yesterday. Today it is 11°. What was the temperature yesterday?

a. Yesterday's temperature: x
b. $x + 27 = 11$
c. $x + 27 - 27 = 11 - 27$
       $x = -16$

Ans. The temperature yesterday was $-16°$.

43. Last night the temperature dropped 13° to a new low of $-7°$. What was the temperature yesterday?

44. The temperature is now 14°, up 23° since this morning. What was the temperature this morning?

45. A 4-day warming trend, during which the mercury rose 8° per day, brought a new temperature of 26°. How cold was it 4 days ago?

46. Today the thermometer says it is 56°. If the weather cools 4° per day, how long will it be before the first freeze (32°)?

47. Eric is on a diet to reduce his current weight of 186 lbs. to 162 lbs. If he loses 4 pounds a week, how long will it take him to reach his desired weight?

48. Linda wants to weigh 128 pounds for her graduation in 7 weeks. If she weighs 149 pounds now, how much must she lose each week?

49. A sporting goods store took in $85,000 last year. After subtracting its expenses, the store experienced a loss of $11,500. How much were its expenses?

50. An appliance manufacturer lost $56,000 on a new line of blenders, reducing its overall profit to $112,000. How much did the company make on its other products?

51. As a promotional incentive, a clothing manufacturer sold 2600 new suits at a loss, reducing her revenue from $367,000 to $315,000. What was the loss on each suit?

52. The city government purchased 500 typewriters at a discount, thus reducing its equipment expenditures from $192,300 to $178,800. How much did they save on each typewriter?

53. A transatlantic airline sustained a loss of $1,630,000 after paying its expenses of $26,480,000. If the airline made 2000 flights, what was its revenue per flight?

54. A bookstore owes its creditors $350,000. If it clears $8000 a week to put towards the loan, how long will it be before the debt is reduced to $30,000?

CHAPTER SUMMARY

[2.1]   The numbers

$$\ldots, -3, -2, -1, 0, +1, +2, +3, \ldots$$

are called **integers.**

A **number line** can be used to represent the relative order of a collection of integers. On a number line, *an integer whose graph lies to the left of the graph of a second integer is less than the second integer.* The point on the number line associated with zero is called the **origin.**

The **absolute value** of an integer a is the nonnegative number $|a|$ which specifies the distance between a and the origin.

[2.2]   To *add* two integers with like signs:

1. Add the absolute values of the numbers.
2. The sum has the same sign as the numbers.

To *add* two integers with unlike signs:

1. Find the absolute value of each and subtract the lesser absolute value from the greater absolute value.
2. The sum has the same sign as the number with the greater absolute value.

[2.3]   To *subtract* an integer b from an integer a, change the subtraction to an addition, change the sign of b, and proceed as in addition. Thus

$$a - b = a + (-b).$$

[2.4]   To find the **product** of two integers, multiply the absolute values of the numbers. If the factors have like signs, the product is positive; if the factors have unlike signs, the product is negative. If one of the factors is zero, the product is zero.

To find the **quotient** of two integers, find the quotient of the absolute values of the numbers. If the dividend and divisor have like signs, the quotient is positive; if the dividend and divisor have unlike signs, the quotient is negative. If zero is divided by any nonzero number, the quotient is zero. *Division by 0 is undefined.*

To evaluate an expression for given numbers, substitute the given number(s) for the variable(s) and follow the correct order of operations.

[2.5]   Equations and word problems involving integers are solved with the techniques of equivalent equations and modeling introduced in Chapter 1.

■ *The symbols introduced in this chapter appear on the inside of the front cover.*

CHAPTER REVIEW

1. Graph the integers $-6, -2, -1, 2, 5, 7$ on a number line.
2. Graph the odd integers between -6 and 4 on a number line.
3. Arrange the integers $2, -3, 5, -4, 0$ in order from the smallest to the largest.
4. Replace the comma in each pair with the proper symbol: $<, >$, or $=$.
   a. $-5, 2$ b. $|-3|, 3$ c. $|-2|, 0$

■ *Add.*

5. a. $+3$ b. -2 c. -6
      $\underline{-5}$ $\underline{-4}$ $\underline{7}$
6. a. $(-2) + (-1)$ b. $(-3) + (0)$ c. $(-8) + (5)$
7. Subtract the bottom number from the top number.
   a. 6 b. -7 c. 0
      $\underline{8}$ $\underline{-3}$ $\underline{-2}$
8. Subtract.
   a. $(-8) - (-3)$ b. $6 - (-10)$ c. $5 - (+7)$
9. Multiply.
   a. $(-5)(-3)$ b. $(6)(-4)$ c. $(-7)(0)(-3)$
10. Divide.
   a. $\dfrac{-12}{3}$ b. $\dfrac{-16}{-2}$ c. $\dfrac{8}{-1}$

■ *Simplify.*

11. a. $4 - 7 + 9 - 3 - 8$

  b. $-3 - (-7) + 5 - 6 + (-9) - 4$

  c. $2(-8) - 4(-3)(-2) + (-5)(3)$

12. a. $\dfrac{-(-3)(8)}{4(-6)}$ **b.** $\dfrac{32 - 4(-3)}{(-6)5 + 2(13)}$ **c.** $\dfrac{5(-3) + 1}{7(-2)} - \dfrac{3(4 - 16)}{(-9)(-2)}$

13. a. $3|-6| - |-4|$ **b.** $\dfrac{|-24|}{3(-4)}$ **c.** $\dfrac{|-2| - (-2)}{-2(-2)}$

14. If $x = -2$, $y = 3$, and $z = -4$, find the value of each expression.

  a. $\dfrac{2xy}{-z}$ **b.** $\dfrac{3z - x}{x}$ **c.** $\dfrac{yz - y}{5}$

■ *Solve.*

15. a. $x - 7 = -13$ **b.** $-4y = 20$ **c.** $\dfrac{z}{-6} = -2$

16. a. $3 - 5x = -17$ **b.** $\dfrac{3y}{2} + 10 = -2$ **c.** $4 - \dfrac{2z}{5} = 8$

■ *For the following word problems:*
a. *Write in words the unknown quantity and assign a variable to represent it,*
b. *write an equation that models the problem, and*
c. *solve the equation.*

17. Today it is $-23°$, a drop of $17°$ from yesterday's low. How cold was it yesterday?

18. Robert reduced from 231 pounds to 187 pounds by losing 4 pounds per week. For how long did Robert diet?

19. A greeting card company initially made $13,000, but after the holidays they sold their remaining cards at a loss of $2 a box and ended up with a net profit of $7800. How many boxes of cards did they sell at a loss?

20. The Ohio State Alumni Association bought 127 bus tickets to Pasadena, reducing its travel budget from $25,963 to $17,835. What is the bus fare from Ohio State University to Pasadena?

CUMULATIVE REVIEW

1. Quantities that are added are called ? ; quantities that are multiplied are called ? .

2. The ? of an integer is the distance between the integer and the origin.

3. In the simplification of algebraic expressions, ? operations are performed before ? operations.

4. Two equations that have the same solution are called ? equations.

5. The signed numbers, together with 0, make up the ? .

6. Graph the odd integers between -8 and 8 on a number line.

7. Graph on a number line the first six natural numbers exactly divisible by 4.

8. Simplify:

   a. $6 - 0$ b. $0 - 6$ c. $0(-6)$

   d. $\dfrac{6}{0}$ e. $\dfrac{0}{6}$ f. $\dfrac{0}{0}$

9. Simplify:

   a. $-(-6)$ b. $|-6|$

   c. $-|-6|$ d. $-|6|$

10. Which is greater, $|-5|$ or 3?

11. Express each word phrase in symbols:

    a. The product of -5 and x.

    b. Twelve divided by the sum of a and -3.

    c. Six less than the sum of 8 and b.

12. Translate each English sentence into an algebraic equation.

    a. Three times an integer plus 12 is equal to -18.

    b. The sum of an integer and 7, divided by -4, equals 5.

    c. Six less than four times an integer is equal to 22.

■ *Simplify.*

13. a. $-2 + 3 - 7 - 9$ b. $(-6) + 7 - (-3) - 2$ c. $16 - 20 + (-5) + 8$

14. a. $1 - 3(-2 + 5)$ b. $2(-3) - 2(5 - 1)$ c. $2(-3)(-2)(+5) - 1$

15. a. $18 + 6 \div 3 - 4(-2)$ b. $(18 + 6) \div (3 - 4) - 2$

    c. $(18 + 6) \div 3 - (4 - 2)$

16. Evaluate for $x = 3$, $y = -4$.

   a. $\dfrac{3xy}{6}$ b. $-3x - 4y + 2$ c. $(5 - x)(3 + y)$

17. Evaluate for $a = -1$, $b = 6$, and $c = -2$.

   a. $ab - bc - ac$ b. $(b - a)(c - b)$ c. $a + cb(c - a)$

■ *Solve the following equations.*

18. a. $\dfrac{2a}{3} = -16$ b. $\dfrac{-4b}{7} = -8$ c. $\dfrac{-c}{5} = 12$

19. a. $4x + 20 = 0$ b. $-3x + 8 = 8$ c. $-6x - 6 = 6$

20. a. $2y - 9 = -15$ b. $2 - \dfrac{3y}{4} = -7$ c. $-1 = \dfrac{6y}{7} + 5$

■ *For the following applied problems:*

a. *Write in words the unknown quantity and assign a variable to represent it,*

b. *write an equation that models the problem, and*

c. *solve the equation.*

21. The Chula Vista chapter of Pounds Away lost a total of 666 pounds last year. If the chapter has 37 members, what was the average weight loss per member?

22. The average temperature in Duluth in February is $-16°$. If the temperature increases an average of $3°$ per week, how long will it be before the average temperature in Duluth is $62°$?

23. Alicia hired a caterer for her birthday party for an initial fee of $125 plus an hourly rate. If the party began at 8 P.M. and lasted until 2 A.M., and Alicia paid the caterer $209, what was the hourly rate?

24. Last January Jorge owed his father $568, but he made monthly payments on the debt, and this January he owes $184. What were his monthly payments?

25. An entrepreneur borrowed $75,000 to market solar-powered eggbeaters. She applied all the revenue from her first shipment of eggbeaters, priced at $12.95 each, to the debt, but still owes her creditors $10,250. How many eggbeaters were in the first shipment?

3 MORE ABOUT EQUATIONS; INEQUALITIES

In this chapter we learn algebraic techniques that will enable us to solve more applied problems. For example, consider the equation

$$2x + 7 = 4x - 3,$$

in which there are *two* terms that contain the unknown. In order to solve equations like this one, we must be able to add and subtract expressions involving variables.

3.1 COMBINING LIKE TERMS

We can add integers by using a counting procedure. If we wish to add 3 to 5, we can first count out five units, then, starting with the next unit, count out three more, giving us the number 8 as the sum. Now suppose we wish to add three 2's to five 2's, that is, $5(2) + 3(2)$. We can add these *like quantities* by counting five 2's, arriving at the number 10, and then counting three more 2's, to make a total of eight 2's or 16. This addition is shown on the number line in Figure 3.1.

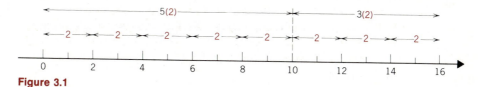

Figure 3.1

LIKE TERMS

In algebra, where terms are usually made up of both numerals and variables, we have to decide what constitutes *like quantities* so that we can apply the idea of addition just developed. We could add 2's as we did because they represented a common unit in each number. For variables, we know that $x = x$ and $ab = ab$, regardless of the numbers that these letters represent. It is also clear that, in general, $x \neq y$, $a \neq b$, $x \neq xy$, and so forth. We therefore define **like terms** to be any terms that are *exactly alike* in their variable factors. Like terms may differ only in numerical factors. Thus,

$$2x \quad \text{and} \quad 3x, \quad\quad \text{and} \quad\quad -4a \quad \text{and} \quad 7a$$

are like terms, whereas

$$2x \quad \text{and} \quad 3y, \quad\quad \text{and} \quad\quad 2x \quad \text{and} \quad 3$$

are not like terms because the variable factors are different.

Any product of factors in a term is called the **coefficient** of the remaining factors in the term. For example, in $3xy$, 3 is the coefficient of xy, x is the coefficient of $3y$, y is the coefficient of $3x$, and $3x$ is the coefficient of y. In $3xy$, the number 3 is called the **numerical coefficient.** In a term such as xy or x, the numerical coefficient is understood to be 1.

ADDING LIKE TERMS

We can illustrate the addition of like terms on a number line, by considering the unit of distance to be equal to the variable part of each term. Thus, we can represent $2x + 3x = 5x$ as shown in Figure 3.2.

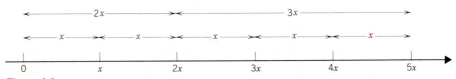

Figure 3.2

In view of our definition for like terms and the discussion above, we state the following rule:

> **To add like terms, add their numerical coefficients.**

For example,

$$5(2) + 3(2) = (5 + 3) \cdot 2 = 8(2),$$
$$2x + 3x = (2 + 3)x = 5x,$$

and

$$-4y - 3y = (-4 - 3)y = -7y.$$

We sometimes refer to this addition process as **combining like terms.**

If a variable is preceded by a negative sign, the coefficient -1 is understood. Thus,

$$-x = -1 \cdot x,$$

$$-a = -1 \cdot a,$$

and so on. For example,

$$-a + 4a = (-1 + 4)a = 3a.$$

Many expressions contain both like terms and unlike terms. In such expressions we can combine only the like terms. For example,

$$-6a + 2 - 5 + 3a + 5a = 2a - 3.$$

EQUIVALENT EXPRESSIONS

We say that two expressions are **equivalent** if they name the same number for *all* replacements of the variable. Although expressions may name the same number for some replacements of the variable, they are not necessarily equivalent—they must name the same number for *all* replacements. For example,

$$6 + 2x \qquad \text{and} \qquad 8x$$

name the same number for $x = 1$, but when any other value is used for x, say $x = 2$, we get

$$6 + 2x = 6 + 2(2) \qquad \text{and} \qquad 8x = 8(2) = 16$$

$$= 6 + 4 = 10;$$

Since $10 \neq 16$, the expressions $6 + 2x$ and $8x$ are not equivalent.

Common Error

Notice that,

$$2 + 3y \neq 5y$$

since 2 and $3y$ are *not* like terms.

As in addition of integers it is convenient to rewrite sums that involve parentheses without using parentheses. We can therefore write $(+5x) + (-3x)$ simply as $5x - 3x$. We can extend this idea to expressions in which two or more terms are grouped by parentheses.

> **In expressions involving only addition, parentheses that are preceded by a plus sign may be dropped; each term within the parentheses keeps its original sign.**

For example,

$$(2x + 3) + (5x - 7) = 2x + 3 + 5x - 7$$
$$= 7x - 4.$$

SUBTRACTING LIKE TERMS

We may subtract like terms in the same way that we added like terms:

To subtract like terms, subtract their numerical coefficients.

For example,

$$7x - 3x = (7 - 3)x = 4x,$$

and

$$6a - (-8a) = [6 - (-8)]a = 14a.$$

As in addition, expressions that contain parentheses should be simplified before like terms are combined. However, we must be careful when removing parentheses preceded by a minus sign. In the expression

$$(5x - 2) - (7x - 4),$$

the minus sign preceding $(7x - 4)$ applies to *each term* inside the parentheses. Therefore, when we remove the parentheses we must change the signs of $7x$, which is understood to be $+$, and of -4. Thus,

$$(5x - 2) - (7x - 4) = 5x - 2 - 7x + 4 \qquad \text{Each sign is changed.}$$
$$= 5x - 2 - 7x + 4$$
$$= -2x + 2.$$

As a general rule:

If an expression is inside parentheses preceded by a minus sign, the sign of each term is changed when the expression is written without parentheses.

Common Error

Many errors occur when differences are involved. You must be careful to change signs properly. Thus,

$$(3x - 4) - (x + 2) \neq 3x - 4 - x + 2$$

Sign has *not* been changed. ⌐

THE DISTRIBUTIVE LAW

The process that we have been using to combine like terms is an application of an important property called the **distributive law.** Expressed in symbols,

$$a \cdot c + b \cdot c = (a + b) \cdot c \tag{1}$$

In this form, the sum $a \cdot c + b \cdot c$ is expressed as the product $(a + b) \cdot c$. Another form of the distributive law,

$$a \cdot (b + c) = a \cdot b + a \cdot c \tag{2}$$

enables us to write a product of the form $a \cdot (b + c)$ as a sum. According to the order of operations, we perform additions inside parentheses before multiplications. However, if this is inconvenient or impossible (for instance if the terms within parentheses are not like terms), then we can obtain an equivalent expression without parentheses by using form (2) of the distributive law. For example,

$$2(3x + 1) = 2(3x) + 2(1)$$
$$= 6x + 2.$$

If an expression in parentheses is preceded by a minus sign, we may interpret this to mean that the expression is multiplied by -1:

$$-(3x - 4) = -1(3x - 4)$$
$$= -1(3x) - 1(-4)$$
$$= -3x + 4.$$

Common Error

The distributive law does not apply to expressions in which only one term appears in parentheses. Thus,

$$4(7 + a) = 4 \cdot 7 + 4 \cdot a = 28 + 4a, \quad \text{Distributive Law applies}$$

but

$$4(7a) = (4)(7)(a) = 28a. \quad \text{Distributive Law does not apply.}$$

EXERCISES 3.1

■ *Add or subtract like terms.*

Sample Problems

a. $4x - (-6x)$

$4x + 6x$

b. $2a - (-3a) + (-6a)$

Write without parentheses.

$2a + 3a - 6a$

Add numerical coefficients.

Ans. $10x$

Ans. $-a$

1. $4y + 2y$ **2.** $3y + 4y$ **3.** $-6x + 2x$

4. $7x - 9x$ **5.** $7a + (-5a)$ **6.** $-6a + (-4a)$

7. $12x - (-3x)$ **8.** $-4x - (-5x)$ **9.** $(-6z) + (+4z)$

10. $(-7z) + (-3z)$ **11.** $-8b + (+8b)$ **12.** $4b - (+4b)$

13. $3x + 4x + x$ **14.** $-2x + 6x - x$

15. $a + 5a - 3a$ **16.** $-a - 4a + 6a$

17. $6y - 8y + 2y - 3y$ **18.** $-3y + 7y - 6y - 4y$

19. $2x + (3x) - (-x)$ **20.** $(2x) - (-5x) - (3x)$

21. $(-8g) + (-3g) - (-7g)$ **22.** $(6g) - g - (-2g)$

23. $-4s + (-3s) - 2s + 7s$ **24.** $s + (-5s) - (-6s) + 2s$

Sample Problems

a. $3x - (-2) - 5x$

$3x + 2 - 5x$

b. $-5 + 2a - (-3a) + 7$

Write without parentheses.

$-5 + 2a + 3a + 7$

Combine like terms.

Ans. $2 - 2x$

Ans. $5a + 2$

25. $6t + 3 - 4t$ **26.** $2 + 5t - 8t$

27. $3 + 4y - (-8y) - 3$ **28.** $7 - 4y + (-6y) - 8$

29. $-2 + 4x - (+6x) + 3x$ **30.** $5x - (-4x) + 2 - 6x$

31. $4 - 3a + (-6a) + 2$ **32.** $9 - 8a - 7 - (-8a)$

■ *Simplify.*

Sample Problems

a. $4y - (6y - y)$

$4y - 5y$

b. $(2 - 5b) - (3 + 2b)$

Write without parentheses.

$2 - 5b - 3 - 2b$

Combine like terms.

Ans. $-y$

Ans. $-1 - 7b$

33. $3a + (4a - a)$

34. $9a + (2a - 7a)$

35. $6b - (2b + b)$

36. $5b - (3b - b)$

37. $(5x - 3x) + x$

38. $(5x + x) - 3x$

39. $4x + (5x + 2)$

40. $6x + (4x - 1)$

41. $3y - (1 + 3y)$

42. $7 - (5 - 4y)$

43. $2 - (3 - 2r)$

44. $5r - (-2 - 8r)$

45. $2x - 1 + (x + 1)$

46. $3x - 2 + (2x + 1)$

47. $2y + 3 - (y - 4)$

48. $(2y + 3) - y - 4$

49. $6a + 5 - (2a - 5) - 2a$

50. $(3a - 2) - 2a - (5 - 2a)$

51. $(4x - 7) - (3x - 2)$

52. $(8x + 9) - (-2x + 6)$

53. $(2x + 7) - (6x - 1) + (4x - 2) - (-2x + 3)$

54. $(7 - 2x) - (2x + 8) - (-3 - x) + (-5x + 1)$

■ *Express the following word phrases as algebraic expressions.*

Sample Problem

$5a$ less the quantity $3a - 6$.

Ans. $5a - (3a - 6)$

55. Sum of $-2y$ and the quantity $y + 1$.

56. Sum of $-x$ and the quantity $3x + 2$.

57. y less the quantity $4 + y$.

58. x less the quantity $6 - 2x$.

59. The sum of $3a$ and -4 subtracted from 12.

60. The sum of $-2a$ and 8 subtracted from 6.

■ *Use the distributive law to write the expression without parentheses.*

Sample Problems

a. $-5(3a - 4)$

$-5(3a) - 5(-4)$

Ans. $-15a + 20$

b. $6(2x + 1) - 3(x - 2)$

Apply the distributive law.

$6(2x) + 6(1) - 3(x) - 3(-2)$

Simplify.

$12x + 6 - 3x + 6$

Combine like terms.

Ans. $9x + 12$

61. $3(x - 4)$

62. $4(x + 1)$

63. $5(2y - 2)$

64. $2(3y + 6)$

65. $-2(x + 8)$

66. $-3(x - 7)$

67. $-5(4 - 5a)$

68. $9(6 - 2x)$

69. $-(5b - 3)$

70. $-(-8b - 5)$

71. $-6(-6 + 2t)$

72. $-7(3t + 3)$

73. $-6(x + 1) + 2x$

74. $4x - 9(1 - x)$

75. $7(y - 5) + 3(5 - y)$

76. $5(y + 3) - 2(y - 3)$

77. $-4(3 + 2z) - 3(2z + 1)$

78. $-2(2 - z) - 2(2z - 1)$

79. $3(x - 2) - 2(x + 3) + 2x$

80. $3 - 3(2x + 1) - 3(x - 1)$

81. $-3(2x - 5) + 2x - 5(x + 2)$

82. $8(1 - 2x) + 3x - 4(2x - 2)$

83. $6 - 2(a - 3) - 4(1 + 3a) + 12a - 2$

84. $7 - 5a + 4(2 - 3a) - 6a - 2(4a - 3)$

■ *Use a numerical example to show that the following pairs of expressions are not equivalent.*

Sample Problem

$7 - 4x$; $3x$

Let x be 3.

$$7 - 4x = 7 - 4(3)$$
$$= 7 - 12 = -5; \qquad 3x = 3(3) = 9$$

Ans. Thus, $7 - 4x$ and $3x$ are not equivalent.

85. $2 + 7x$; $9x$

86. $5 - 3x$; $2x$

87. $-(a - 3)$; $-a - 3$

88. $4 - (5 - a)$; $4 - 5 - a$

89. $2(3x)$; $2 \cdot 3 \cdot 2 \cdot x$

90. $2(3 + x)$; $2 \cdot 3 + x$

3.2 SOLUTION OF EQUATIONS

Equations that arise in applications often involve more than one term that contains the variable. To solve such an equation, we obtain an equivalent equation in which all terms containing the variable are in one member, and all constant terms are in the other. For example, given the equation

$$2x + 1 = x - 2,$$

we first add -1 to (or subtract 1 from) each member to get

$$2x + 1 - 1 = x - 2 - 1,$$
$$2x = x - 3.$$

If we now add $-x$ to (or subtract x from) each member, we get

$$2x - x = x - 3 - x,$$
$$x = -3,$$

where the solution -3 is obvious.

Since each equation obtained in the process is equivalent to the original equation, -3 is also a solution of $2x + 1 = x - 2$. The solution can be checked by substitution:

$$2(-3) + 1 \overset{?}{=} (-3) - 2,$$

$$-5 = -5.$$

There may be several different ways to obtain a suitable equivalent equation. Sometimes one method is better than another. For example, consider the equation

$$2x = 3x - 9. \tag{1}$$

If we first add $-3x$ to each member, we get

$$2x - 3x = 3x - 9 - 3x,$$

$$-x = -9,$$

where the variable has a negative coefficient. Although we can see by inspection that the solution is 9, because $-(9) = -9$, we can avoid the negative coefficient by adding $-2x$ and $+9$ to each member of Equation (1). In this case, we get

$$2x - 2x + 9 = 3x - 9 - 2x + 9,$$

$$9 = x,$$

from which the solution 9 is obvious. If we wish, we can write the last equation as

$$x = 9$$

by the symmetric property of equality.

REMOVING PARENTHESES

Often we must solve equations in which the variable occurs within parentheses. We can solve these equations in the usual manner after we have simplified them by applying the distributive law to remove the parentheses. For instance, to solve the equation

$$3x - 2(4 - x) = 25 - 6x$$

we first apply the distributive law to the term $-2(4 - x)$ to get

$$3x - 8 + 2x = 25 - 6x.$$

Then we combine like terms in the left member, obtaining

$$5x - 8 = 25 - 6x.$$

Finally we add $6x$ and 8 to each member, which gives

$$5x - 8 + 6x + 8 = 25 - 6x + 6x + 8$$

or

$$11x = 33,$$

and divide both sides by 11 to arrive at the solution

$$\frac{\cancel{11}x}{\cancel{11}} = \frac{33}{11}$$

$$x = 3.$$

The techniques we have learned thus far are sufficient to solve the equations in this chapter. Although there is no specific order in which the properties should *always* be applied, the following sequence of steps is appropriate for many problems.

Steps to solve equations:

1. When applicable, use the distributive law to remove parentheses.
2. Combine like terms in each member of the equation.
3. Using the addition or subtraction property, write the equation with all terms containing the unknown in one member and all terms not containing the unknown in the other.
4. Combine like terms in each member.
5. Use the multiplication property to remove fractions.
6. Use the division property to obtain an equation of the form $x = a$.

As another example consider an equation that contains decimal fractions:

$$0.09x + 0.06(13,000 - x) = 930.$$

Such equations arise in applied problems in the following sections. In order to simplify the equation, we first multiply each member by 100 to eliminate the decimals.

$$0.09x + 0.06(13,000 - x) = 930.00$$

$$9x + 6(13,000 - x) = 93,000$$

Notice that *each term* of the equation is multiplied by 100. We now proceed according to the steps just listed.

$$9x + 78,000 - 6x = 93,000$$

$$3x + 78,000 = 93,000$$

$$3x = 15,000$$

$$x = 5000$$

EXERCISES 3.2

■ *Solve each equation.*

Sample Problem

$5x - 7 = 2x - 4x + 14$

Combine like terms, $2x - 4x$.

$5x - 7 = -2x + 14$

Add $+2x$ and $+7$ to each member; combine like terms.

$5x - 7 + 2x + 7 = -2x + 14 + 2x + 7$

$7x = 21$

Divide each member by 7.

$$\frac{7x}{7} = \frac{21}{7}$$

Ans. $x = 3$

1. $7 - 2c = c + 1$
2. $c + 2 = 6 - 3c$
3. $6 - x = 6 + 2x$
4. $8 + x = 8 - 5x$
5. $4x - 3 = 2x + 5$
6. $6x - 5 = 2x + 7$
7. $0 = 4x + 5 - 3x$
8. $0 = 5x - 4x - 6$
9. $-6t = 3t$
10. $7r = 5r$
11. $3x - 14 = 5x - 4x + 2$
12. $2x - 3 + 2x = 4 - x + 8$
13. $2y - 3 + 3y = 4y + 2$
14. $6z + 2 - 7z = 10 - 2z$
15. $6a - 4 + 2 = 3a + 1$
16. $5y + 1 - y = 10 + y$
17. $3x + 4 - 5x + 2 = 0$
18. $5x + 7 - 2x - 16 = 0$
19. $0 = 7 - 2x + 3 - 3x$
20. $0 = 3x + 5 - 7x + 3$
21. $-x(2 + 5) = 2x + 16 - x$
22. $3y(7 - 2) + 17 = 16y + y - 1$

Sample Problem

$4(3x - x) = 3(5x - 3x) + x + 6$

Combine like terms in parentheses.

$4(2x) = 3(2x) + x + 6$

Perform the multiplications.

$8x = 6x + x + 6$

Combine like terms $6x$ and x.

$8x = 7x + 6$

Add $-7x$ to each member.

$8x - 7x = 7x + 6 - 7x$

Combine like terms.

Ans. $x = 6$

23. $2(4t - t) + 6 = 2(2t + t) + 8 - t$
24. $-3(x - 3x) + 5 = -4(3x - x) + 7 + 13x$

25. $\dfrac{4x - 2x}{2} + 3(x + 2x) = 2(3x + x) + x$

26. $5(2x + x) - \dfrac{3(2x + x)}{9} = 2(3x + 4x) + x$

Sample Problem

$\dfrac{5x - 2x}{4} = -6$

Simplify above the fraction bar. Multiply each member by 4.

$\cancel{4} \cdot \dfrac{3x}{\cancel{4}} = 4(-6)$

$3x = -24$

Divide each member by 3.

$\dfrac{\cancel{3}x}{\cancel{3}} = \dfrac{-24}{3}$

Ans. $x = -8$

27. $\dfrac{4t - t}{6} = 5$ **28.** $\dfrac{7t - t}{18} = -2$ **29.** $\dfrac{5x + 2x}{4} = 7$

30. $\dfrac{8x - 3x}{4} = -10$ **31.** $\dfrac{4y + 6y}{20} = \dfrac{7 + 2}{3}$ **32.** $\dfrac{y - 3y}{3} = \dfrac{2 - 18}{4}$

33. $\dfrac{9 + 16}{5} = \dfrac{3x + 5x}{24}$ **34.** $\dfrac{49 - 25}{-6} = \dfrac{2y + 4y}{9}$

Sample Problem

$\dfrac{4x - 2x}{3} + 3 = 5$

Combine like terms, $4x - 2x$.

$\dfrac{2x}{3} + 3 = 5$

Add -3 to each member; simplify.

$\dfrac{2x}{3} + 3 - 3 = 5 - 3$

$\dfrac{2x}{3} = 2$

Multiply each member by 3.

$\cancel{3}\left(\dfrac{2x}{\cancel{3}}\right) = 3(2)$

Divide each member by 2.

$\dfrac{2x}{\cancel{2}} = \dfrac{6}{\cancel{2}}$

Ans. $x = 3$

35. $\dfrac{4x - 2x}{2} + 4 = 12 - 5$

36. $\dfrac{x - 3x}{4} + 2 = 11 - 2$

37. $\dfrac{3x + 2x}{2} + 2 = \dfrac{10 + 4}{2}$

38. $\dfrac{5x - 3x}{4} + 2 = \dfrac{8 + 2}{5}$

39. $\dfrac{4x + x}{3} - 3 = 4 + 3$

40. $\dfrac{3z - z}{6} + 3 = 2 + 9$

41. $18 + \dfrac{3(5x - 4x)}{2} = 15$

42. $-3 + \dfrac{7(4x - 3x)}{2} = 25$

43. $14 + \dfrac{3(5x - 3x)}{9} = -4$

44. $\dfrac{2(7x - 5x)}{5} - 6 = 22$

45. $\dfrac{3(8x - 6x)}{5} + 3 = 15$

46. $\dfrac{2(5x - 4x)}{3} - 5 = 13$

47. $\dfrac{5(4x - x)}{3} + 8x = x - 36$

48. $\dfrac{4(5x + x)}{2} + 4x = 4x - 24$

49. $\dfrac{8(2t + 5t)}{7} + \dfrac{2(3t + t)}{4} = 8(2t + t) + 28$

50. $\dfrac{6(5u - u)}{8} - \dfrac{5(u + 5u)}{3} = 3(2u - u) + 30$

Sample Problem

$4(5 - y) + 3(2y - 1) = 3$

Apply the distributive law.

$20 - 4y + 6y - 3 = 3$

Solve for y.

$2y + 17 = 3$

$2y = -14$

Ans. $y = -7$

51. $3(x - 5) = 6$

52. $5(3x - 2) = 35$

53. $2(4y + 5) = 2$

54. $6(3y - 4) = -60$

55. $28 = 4(1 + 2x)$

56. $0 = 7(8 - 2x)$

57. $-3(2x + 1) - 4 = -1$

58. $-5(2x - 3) + 2 = 47$

59. $7(y - 3) = 2y - 31$

60. $4(5 - y) = 10 - 6y$

61. $3y + 35 = 4(2y + 5)$

62. $5x - 64 = -2(3x - 1)$

63. $-x - (8 + x) = 2$

64. $3(7 + 2x) = 30 + 7(x - 1)$

65. $5x - (x + 2) = 7 + (x + 3)$

66. $4(y - 1) = 5(y - 2)$

67. $-2y + 5(y + 1) = 25 + 7y$

68. $25 + 5y = -2(y - 4) - 18$

69. $(a - 1) - (a + 2) = 2a - 3$

70. $3(2a - 1) + 2(a + 5) = 15$

71. $(b + 5) - (b - 1) = 3b$

72. $5a - 4(1 - a) = 41 - 6a$

73. $b = 4(b + 6) + 3b$ **74.** $b = (b + 1) - (b - 5)$

75. $3(2x - 4) - 4(1 - 3x) = 2$ **76.** $-6(3y + 2) + 5(y - 3) = -1$

77. $-2(3a - 4) - 8(a - 1) = 3(2a - 8)$

78. $5(6 - 3b) = -3(3b + 4) - (4b - 8)$

79. $7(1 - 2x) + 3(-2x - 1) = 6 - 2(7x - 11)$

80. $2x - 3(5x + 2) - 6 = 3(x - 4) -4(x - 3)$

Sample Problem

$$0.08(x - 200) + 65 = 0.15x$$

Multiply each member by 100.

$$0.08(x - 200) + 65.00 = 0.15x$$

Apply the distributive law.

$$8x - 1600 + 6500 = 15x$$

Solve for x.

$$4900 = 7x$$

Ans. $x = 700$

81. $0.5(x - 30) + 0.2x = 55$ **82.** $0.6x + 0.3(x - 6) = 9$

83. $0.4(x - 30) - 0.2(x - 40) = 6$ **84.** $0.7(x + 20) - 0.5(x + 30) = 13$

85. $0.8(2x + 5) - 0.5(11) = 0.3(x + 8)$ **86.** $0.1(3x - 1) - 0.4(13) = 0.9(3x - 5)$

87. $0.25(x + 3) - 0.45(x - 3) = 0.30$ **88.** $0.15(x - 6) + 0.30x = 0.90$

89. $0.12x + 0.08(8000 - x) = 840$ **90.** $0.09(x + 250) + 0.06x = 90$

3.3 PREPARING FOR WORD PROBLEMS

In many applied problems there will be more than one unknown quantity. In this chapter we represent each of the unknowns in terms of a *single variable*. For example, in the statement

One number is six more than another

there are two unknown quantities. If we let x represent the *smaller* number, then $x + 6$ represents the other (larger) number. If we let x represent the *larger* number, than $x - 6$ represents the other (smaller) number.

In the next section we construct and solve equations for problems that involve two or more unknowns; in the examples and exercises in this section we give our attention *only* to the first step in that important process: representing algebraically the unknown quantities.

In the following examples we *first write a short word phrase to describe each unknown quantity*. Next we represent one of the unknowns by a variable, and then express any other unknown quantities in terms of this variable.

A board is cut into two pieces so that one piece is twice as long as the other. If the board is twelve feet long, how long is each piece?

	Step 1	*Step 2*
The length of the shorter piece:		x
The length of the longer piece:		$2x$

A skirt and blouse together cost $60. If the skirt cost $8 more than the blouse, how much did each cost?

	Step 1	*Step 2*
The price of the skirt:		x
The price of the blouse:		$60 - x$

The length of a rectangle is three feet less than twice its width. If it takes 66 feet of fencing to enclose the rectangle, what are its dimensions?

	Step 1	*Step 2*
The width of the rectangle:		x
The length of the rectangle:		$2x - 3$

[The sum of three consecutive whole numbers is 57. Find the numbers.]

	Step 1	*Step 2*
The first whole number:		x
The second whole number:		$x + 1$
The third whole number:		$x + 2$

EXERCISES 3.3

■ *For each problem:*

a. *Write a short word phrase describing each unknown quantity,*

b. *assign a variable to represent one of the unknowns, and express any other unknowns in terms of this variable.*

Do not solve.

Sample Problem

We have twice as many cats as dogs. All together we keep 21 animals. How many cats do we have?

Ans.

a. The number of dogs: x b.
   The number of cats: $2x$

1. There are 10 times as many students as professors at Tilden College. The student body and the faculty together number 539. How many students are there?

2. For every smoker in the restaurant there are 3 nonsmokers. If there are 68 people in the restaurant, how many smoke?

3. There are 50 more sheep than goats in the pasture. If there are 98 animals grazing, how many are goats?

4. There are 8 fewer men than women in the scuba-diving class. If the total enrollment is 34, how many are women?

5. Delbert and Francine together make $25,000. If Delbert can double his salary, their combined income will be $34,000. How much does each earn now?

6. An apple and a glass of milk together contain 260 calories. A glass of milk and two apples contain 410 calories. How many calories are there in an apple?

7. The difference in the prices of two cameras is $300. If I can buy the cheaper one at a 20% discount, it will cost only half as much as the more expensive one. What is the price of the expensive camera?

8. There is 12 years' difference in Paul's and Betty's ages. Twenty years ago, Paul was twice as old as Betty. How old is each now?

9. There are twice as many dimes as quarters in the bowl. If the bowl contains $7.65, how many quarters are there?

10. I have 4 more nickels than dimes. If I have $1.85, how many nickels do I have?

11. Jill bought 50 postage stamps; some are 22¢ stamps and some are 14¢ stamps for postcards. If she paid $9.40, how many of each kind of stamp did she buy?

12. The teller counted 18 bills in fives and tens for a total of $115. How many fives were counted?

13. Sheila invested $1000 more at 7% than she invested at 9%. After one year she earned $710 in interest. How much did she invest at each rate?

14. Fernando invested $20,000, part at 8% and part at 5%. If his interest after one year was $1180, how much did he invest at each rate?

15. The discount on a TV is 20% of the list price. If the discount is $128, what is the list price of the TV?

16. A company's profit on their product is 15% of the item's selling price. If the profit is $1.05 on each item, what is the selling price?

17. A rope is cut into two pieces so that one piece is 2 meters less than five times the length of the other piece. If the rope was originally 46 meters long, how long is each piece?

18. The length of a vegetable garden is 6 feet more than twice its width. If 42 feet of fencing will enclose the garden, what are its dimensions?

19. Ron drives 20 miles per hour faster than Steve. If Steve takes 9 hours to drive the same distance that Ron drove in 6 hours, how fast does each drive?

20. Carla drives 15 miles per hour slower than Jane. If they set off in opposite directions, after four hours they will be 340 miles apart. How fast does each drive?

21. Harry traveled three times as far as Wayne. Together they covered 384 miles. How far did Harry travel?

22. Wanda walked 2 miles more than twice as far as Renée in the walkathon. That was 8 miles further than Renée walked. How far did Wanda walk?

23. Jason and Melissa drive at constant speeds in opposite directions. After an hour they are 90 miles apart. If Melissa stops but Jason drives for another hour in the same direction, they will be 148 miles apart. How far did Melissa drive?

24. Frankie and Johnny travel at constant speeds in the same direction. After an hour they are 30 miles apart. Frankie turns around, but Johnny continues in the same direction, and after another hour they are 140 miles apart. How far did Frankie drive?

25. A 10-gallon jug is filled with a solution of oil and vinegar. The oil costs $12 a gallon, and the vinegar costs $6 a gallon. If the jug is worth $108, how much vinegar is in the solution?

26. A vat holds 50 liters and is filled with a mixture of club soda and prune juice. Club soda costs $3.50 a gallon, and prune juice costs $8 a gallon. If the vat is worth $310, how much prune juice is in the mixture?

27. Randolph wants to buy five pounds of chocolates for Carlotta. Caramel centers cost $5 a pound and cream-filled cost $7 a pound. If Randolph has $29, how can he spend exactly that amount on chocolates?

28. A nursery wants to sell a 50-bulb mixture of tulips and daffodils. Tulip bulbs cost 13¢ apiece, and daffodils cost 18¢. If the mixture is to sell for $7.50, how many of each bulb should be included?

29. The sum of three consecutive odd integers is 45. Find the integers.

30. The sum of three consecutive even integers is −66. Find the integers.

Sample Problem

One integer is three more than another. If x represents the smaller integer, represent in terms of x:

a. The larger integer.

b. Five times the smaller integer.

c. Five times the larger integer.

Ans. $x + 3$

Ans. $5x$

Ans. $5(x + 3)$

31. One integer is four more than a second integer. If x represents the smaller integer, represent in terms of x:
   a. The larger integer.
   b. Five times the smaller integer.
   c. Five times the larger integer.

32. One integer is six less than a second integer. If n represents the larger integer, represent in terms of n:
   a. The smaller integer.
   b. Two times the smaller integer.
   c. Two times the larger integer.

33. One integer is six more than another. If n represents the smaller integer, represent in terms of n, three times the larger integer.

34. One integer is five less than another. If n represents the larger integer, represent in terms of n, six times the smaller integer.

Sample Problem

The sum of two integers is 13. If x represents the smaller integer, represent in terms of x:

a. The larger integer.

b. Five times the smaller integer.

c. Five times the larger integer.

Ans. $13 - x$

Ans. $5x$

Ans. $5(13 - x)$

35. The sum of two integers is 27. If n represents the smaller integer, represent in terms of n:
   a. The larger integer.
   b. Three times the smaller integer.
   c. Three times the larger integer.

36. The sum of two integers is 39. If n represents the larger integer, represent in terms of n:
   a. The smaller integer.
   b. Six times the smaller integer.
   c. Four times the larger integer.

37. The difference of two integers is 16. If n represents the smaller integer, represent in terms of n:
   a. The larger integer.
   b. Five times the smaller integer.
   c. Two times the larger integer.

38. The difference of two integers is 21. If n represents the larger integer, represent in terms of n:
   a. The smaller integer.
   b. Two times the larger integer.
   c. Three times the smaller integer.

| | | Number | |
| | Value of | of | Value of |
Denomination	One Coin	Coins	Coins
Quarters	0.25	x	$0.25x$
Dimes	0.10	$x + 16$	$0.10(x + 16)$

Step 4   Write an equation, using the principle stated above.

$$\begin{bmatrix} \text{value of} \\ \text{quarters} \end{bmatrix} + \begin{bmatrix} \text{value of} \\ \text{dimes} \end{bmatrix} = \begin{bmatrix} \text{value of} \\ \text{collection} \end{bmatrix}$$

$$0.25x \quad + 0.10(x + 16) \quad = \quad 5.80$$

Step 5   Solve the equation. First multiply each member by 100.

$$0.25x + 0.10(x + 16) = 5.80$$
$$25x + 10x + 160 = 580$$
$$35x + 160 = 580$$
$$35x = 420$$
$$x = 12$$

Step 6   **Ans.** There are 12 quarters and 12 + 16 or 28 dimes.

EXERCISES 3.4

■ *Solve each of the following problems completely. Follow the six steps outlined on page 84.*

Sample Problem   One integer is five more than a second integer. Three times the smaller integer plus twice the larger equals 45. Find the integers.

Steps 1–2   The smaller integer: x

The larger integer: $x + 5$

Step 3   A sketch is not appropriate.

Step 4

Three times the smaller integer plus twice the larger equals 45.

$$3 \cdot x + 2 \cdot (x + 5) = 45$$

(continued)

Step 5
$$3x + 2(x + 5) = 45$$
$$3x + 2x + 10 = 45$$
$$5x + 10 = 45$$
$$5x = 35$$
$$x = 7$$

Step 6   **Ans.** The smaller integer is 7, and the larger is 7 + 5 or 12.

1. One integer is two less than a second integer. The larger integer plus four times the smaller equals 17. Find the integers.

2. One integer is three more than a second integer. Four times the second integer plus twice the first equals 42. Find the integers.

3. The gasoline tank of one truck holds 4 gallons more than the tank of another truck. If it takes 144 gallons of gasoline to fill each tank three times, how many gallons does each tank hold?

4. In a mathematics textbook containing 368 pages, there were 28 pages more of exercises than there were of explanatory material. How many pages of each were there in the book?

Sample Problem   The sum of two consecutive even integers is 46. What are the integers?

Steps 1–2   The smaller even integer: x

The next consecutive even integer: $x + 2$

Step 3   A sketch is not appropriate.

Step 4 $x + (x + 2) = 46$

Step 5 $2x + 2 = 46$
$$2x = 44$$
$$x = 22$$

Step 6   **Ans.** The smaller integer is 22; the next consecutive even integer is 22 + 2 or 24.

5. The sum of two consecutive even integers is 26. Find the integers.

6. The sum of two consecutive even integers is 86. Find the integers.

7. The sum of two consecutive odd integers is 32. Find the integers.

8. The sum of two consecutive odd integers is 28. Find the integers.

9. Find three consecutive integers whose sum is −33.

10. The sum of three consecutive integers is 57. Find the integers.

11. Find three consecutive odd integers whose sum is −21.

12. Find three consecutive even integers whose sum is 84.

13. The sum of three consecutive even integers equals four times the smallest integer. What are the integers?

14. The sum of three consecutive odd integers equals 1 less than four times the smallest integer. What are the integers?

Sample Problem

There were 2480 votes cast in an election. The winning candidate received 142 votes more than the losing candidate. How many votes did each candidate receive?

Steps 1–2   The number of votes for the winner: x

The number of votes for the loser: $x - 142$

Step 3   A sketch is not appropriate.

Step 4 $x + (x - 142) = 2480$

Step 5 $2x - 142 = 2480$

$2x = 2622$

$x = 1311$

Step 6   **Ans.** The winner received 1311 votes, and the loser received 1311 − 142 or 1169 votes.

15. At a recent election, the winning candidate received 50 votes more than his opponent. If there were 4376 votes cast in all, how many votes did each candidate receive?

16. There were 12,822 votes cast in a recent election. The winning candidate received 132 votes more than his opponent. How many votes did each candidate receive?

17. A board 144 centimeters long is cut into two pieces so that one piece is 24 centimeters longer than the other piece. How long is each of the two pieces?

18. A board 112 centimeters long is cut into two pieces so that one piece is three times as long as the other. How long is each of the two pieces?

19. A cable is 20 feet long. Where should it be cut so that one piece will be 4 feet shorter than the other?

20. Where should a 30-foot cable be cut so that twice the longer piece is equal to three times the shorter?

21. A board 24 feet long is cut into three pieces of which the second is three times as long as the first, and the third is 4 feet longer than the first. How long is each of the three pieces?

22. A board 12 feet long is cut into three pieces so that the second is three times as long as the first, and the third is 2 feet longer than the first. How long is each of the three pieces?

23. A rope 51 feet long is cut into three pieces so that the second piece is 3 feet longer than the first, and the third is twice as long as the first. How long is each of the three pieces?

24. A rope 63 meters long is cut into three pieces so that the longest piece is three times the length of the shortest piece, and the other piece is 3 meters longer than the shortest piece. How long is each of the three pieces?

25. Four pieces of equal length are cut from a 340-centimeter board, and a 24-centimeter piece remains. How long is each of the four equal pieces?

26. If a 36-ounce solution fills four glass containers of the same size with 4 ounces of the solution left over, how many ounces will each container hold?

27. A man drove from town A to town B and then returned to town A. Leaving town A again to return to town B, the man found that after 5 kilometers on the road, he had traveled a total of 19 kilometers since he first left town A. How far is it from town A to town B?   (*Hint:* Make a sketch.)

28. Two ships leave port at the same time, traveling in the same direction. In one hour, one ship sails three times as far as the other. If the ships are then 14 kilometers apart, how far has the slower ship sailed?   (*Hint:* Make a sketch.)

29. Two cars leave town A at the same time, traveling in the same direction. In one hour, one car travels four times as far as the other. If the cars are then 54 miles apart, how far has the slower car traveled?

30. A woman drove from town A to town B. Leaving town B to return to town A, the woman found that after 13 kilometers on the road, she had traveled a total of 37 kilometers since she first left town A. How far is town A from town B?

Sample Problem

Student tickets to the rodeo cost $2.00, and general admission is $3.50. If 418 tickets were sold, bringing in receipts of $1196, how many of each type of ticket were sold?

Steps 1–2   The number of general admission tickets: x

The number of student tickets: $418 - x$

Step 3

Type of Ticket	Cost of One Ticket	Number of Tickets	Receipts from Tickets
General	3.50	x	$3.50x$
Student	2.00	$418 - x$	$2.00(418 - x)$

Step 4

$$\begin{bmatrix} \text{receipts from} \\ \text{general} \\ \text{tickets} \end{bmatrix} + \begin{bmatrix} \text{receipts from} \\ \text{student} \\ \text{tickets} \end{bmatrix} = \begin{bmatrix} \text{receipts} \\ \text{from all} \\ \text{tickets} \end{bmatrix}$$

$$3.50x \quad + \quad 2.00(418 - x) \quad = \quad 1196.00$$

Step 5

$$3.50x + 2.00(418 - x) = 1196.00$$
$$350x + 83600 - 200x = 119{,}600$$
$$150x + 83600 = 119{,}600$$
$$150x = 36{,}000$$
$$x = 240$$

Step 6   **Ans.** 240 general tickets were sold and $418 - 240$ or 178 student tickets.

31. A woman had $1.80 in change. The change was entirely in the form of dimes and nickels. If she had three more dimes than nickels, how many of each coin did she have?

32. A man had $1.45 in change. The money consisted of quarters and dimes only. If he had four fewer quarters than he had dimes, how many of each coin did he have?

33. A woman had $1.14 in change consisting of pennies, nickels, and dimes. She had six more nickels than pennies and six fewer dimes than pennies. How many of each coin did she have?

34. A man had $1.47 in change consisting of pennies, nickels, and quarters. He had three more pennies than quarters and one more nickel than pennies. How many of each coin did he have?

35. One thousand tickets were sold at a football game. Adults paid $4.25 each for their tickets, and children paid $1.50 each. If the total recepts for the game were $3150, how many tickets of each kind were sold?

36. Three hundred tickets were sold at a baseball game. Adults paid $3.50 each for their tickets, and children paid $1.50 each. If the total receipts for the game were $750, how many tickets of each kind were sold?

37. Two trucks are carrying material to a road construction job. One truck can carry 4 tons more per trip than the other. If the smaller truck makes five trips and the larger truck makes seven trips, they can deliver a total of 112 tons of material. What is the capacity of each truck?

38. A 10-ton truck and a 12-ton truck are carrying material to a road construction job. If the smaller truck makes three trips more than the larger truck, how many trips does each make if together they deliver 140 tons?

39. Delbert and Francine together make $25,000. If Delbert can double his salary, their combined income will be $34,000. How much does each earn now?

40. An apple and a glass of milk together contain 260 calories. A glass of milk and two apples contain 410 calories. How many calories are there in an apple?

3.5   MORE WORD PROBLEMS

In this section we continue our study of modeling and problem solving. Certain investment problems use the fact that the amount of interest (i) earned in one year (at simple interest) equals the product of the interest rate (r) and the amount of money (p) invested, or $i = r \cdot p$. For example, $1000 invested for one year at 9% yields

$$i = (0.09)(1000) = \$90.$$

The following examples further illustrate this principle.

> Howard invested $40,000, part at 6% and the remainder at 8%. Find the amount invested at each rate if the total yearly income is $2720.

Steps 1–2   We first describe each unknown quantity in words. Then we express each in terms of a single variable.

Amount invested at 6%: x

Amount invested at 8%: $40,000 - x$

Step 3   Next, we make a table showing the amounts of money invested, the rates of interest, and the amounts of interest earned on each investment.

Amount Invested	Interest Rate	Amount of Interest
x	0.06	$0.06x$
$40,000 - x$	0.08	$0.08(40,000 - x)$

Step 4   Now we can write an equation relating the interest from each investment and the total interest received.

$$\underbrace{0.06x}_{\substack{\text{interest from} \\ \text{first investment}}} + \underbrace{0.08(40,000 - x)}_{\substack{\text{interest from} \\ \text{second investment}}} = \underbrace{2720}_{\text{total income}}$$

Step 5   To solve for x, first multiply each member by 100.

$$0.06x + 0.08(40,000 - x) = 2720.00$$
$$6x + 320,000 - 8x = 272,000$$
$$-2x = -48,000$$
$$x = 24,000$$

Step 6   **Ans.** $24,000 was invested at 6%, and $40,000 − $24,000 or $16,000 was invested at 8%.

Tables are also useful in solving mixture problems.

$\begin{bmatrix}\text{How much candy worth 80¢ a kilogram (kg) must a grocer blend with 60 kg of} \\ \text{candy worth \$1 a kilogram to make a mixture worth 90¢ a kilogram?}\end{bmatrix}$

Steps 1–2   First we describe the unknown quantity in words. Then we assign a variable to represent the unknown.

Kilograms of 80¢ candy: x

Step 3   Next, we make a table showing the types of candy, the amount of each, and the total value of each.

	Value of 1 kg	Number of Kilograms	Value of Candy
0.80 candy	0.80	x	$0.80x$
1.00 candy	1.00	60	$1.00(60)$
0.90 candy	0.90	$x + 60$	$0.90(x + 60)$

Step 4   Using the entries in the table, we can write an equation. The total value must be the sum of the values of the ingredients.

$$\begin{bmatrix}\text{value of} \\ x\text{ kg} \\ \text{of 80¢ candy}\end{bmatrix} + \begin{bmatrix}\text{value of} \\ 60\text{ kg} \\ \text{of \$1 candy}\end{bmatrix} = \begin{bmatrix}\text{value of} \\ (x+60)\text{ kg} \\ \text{of 90¢ candy}\end{bmatrix}$$

$$0.08x \quad + \quad 1.00(60) \quad = \quad 0.90(60 + x)$$

Step 5   To solve the equation, first multiply each member by 100.

$$0.80x + 1.00(60) = 0.90(60 + x)$$
$$80x + 6000 = 5400 + 90x$$
$$600 = 10x$$
$$60 = x$$

Step 6   **Ans.** The grocer should use 60 kg of the 80¢ candy.

In Section 1.6 we used the formula $d = rt$ to describe uniform motion, where d is the distance traveled by an object in time t at rate of speed r.

> Tom and Jerry live in towns that are 525 miles apart. They arrange to meet at an intermediate point. If Tom drives 15 mph faster than Jerry, and they meet after driving for 5 hours, how fast does each drive?

Steps 1–2   First describe each unknown quantity in words. Then express each in terms of a single variable.

$$\text{Jerry's speed: } x$$

$$\text{Tom's speed: } x + 15$$

Step 3   Make a table showing the rate, the time, and the distance that each person traveled.

	Rate	*Time*	*Distance*
Jerry	x	5	$5x$
Tom	$x + 15$	5	$5(x + 15)$

A sketch of the distances involved is also helpful.

Note that the distance Tom drives plus the distance Jerry drives must equal the total distance between their hometowns.

Step 4   Write an equation expressing the relationship noted above.

$$\underbrace{5x}_{\substack{\text{distance}\\\text{Jerry drives}}} + \underbrace{5(x + 15)}_{\substack{\text{distance}\\\text{Tom drives}}} = \underbrace{525}_{\substack{\text{total}\\\text{distance}}}$$

Step 5   Solve the equation.

$$5x + 5x + 75 = 525$$
$$10x + 75 = 525$$
$$10x = 450$$
$$x = 45$$

Step 6   **Ans.** Jerry's speed is 45 miles per hour, and Tom's speed is $45 + 15$ or 60 miles per hour.

EXERCISES 3.5

■ *Solve each of the following problems completely. Follow the six steps outlined on page 84.*

Sample Problem

Two investments produce an annual interest of $320. $1000 more is invested at 11% than at 10%. How much is invested at each rate?

Step 1–2   Amount invested at 10%: x
Amount invested at 11%: $x + 1000$

Step 3

Amount Invested	Interest Rate	Amount of Interest
x	0.10	$0.10x$
$x + 1000$	0.11	$0.11(x + 1000)$

Step 4

$$\underbrace{0.10x}_{\substack{\text{interest from} \\ \text{first investment}}} + \underbrace{0.11(x + 1000)}_{\substack{\text{interest from} \\ \text{second investment}}} = \underbrace{320}_{\text{total income}}$$

Step 5

$$0.10x + 0.11(x + 1000) = 32{,}0.00$$
$$10x + 11x + 11{,}000 = 32{,}000$$
$$21x + 11{,}000 = 32{,}000$$
$$21x = 21{,}000$$
$$x = 1000$$

Step 6   **Ans.** $1000 was invested at 10%, and $1000 + $1000 or $2000 was invested at 11%.

1. Two investments produce an annual income of $1060. One investment earns 10% and the other earns 11%. How much is invested at each rate if the amount invested at 11% is $6200 more than the amount invested at 10%?

2. An amount of money is invested at 9% and $1200 more than that amount is invested at 10%. How much is invested at each rate if the total income is $1013?

3. An amount of $34,000 is invested, part at 8% and the remainder at 9%. Find the amount invested at each rate if the yearly income on each investment is the same.

4. An amount of money is invested at 8% and twice that amount is invested at 10%. How much is invested at each rate if the total income is $1288?

5. Rhoda invested $100 more in bonds earning 6% than in stocks that earn 8%. The annual income from the stocks was $8 more than the income from the bonds. How much did Rhoda invest in each?

6. Simone has $4800 to put into two accounts: a savings account that earns 5% interest, and a long-term account that earns 7%. If she wants to earn $270 interest in one year, how much must she put in the long-term account?

7. Mario borrowed $30,000 from two banks to open a print shop. One bank charges 12% interest and the other charges 15%. If the annual interest on both loans is $3750, how much did Mario borrow at each rate?

8. Stefan borrowed twice as much on his 14% car loan as on his 4% student loan. The annual interest on the car loan is $1200 more than the interest on the student loan. How much did Stefan borrow on each loan?

Sample Problem

How many quarts of a 20% solution of acid should be added to 10 quarts of a 30% solution of acid to obtain a 25% solution?

Steps 1–2   Number of quarts of 20% solution: x

Step 3

Percent of Acid	Number of Quarts	Amount of Acid in Solution
20%	x	$0.20x$
30%	10	$0.30(10)$
25%	$(x + 10)$	$0.25(x + 10)$

Step 4
$$\begin{bmatrix} \text{Pure acid in} \\ \text{20\% solution} \end{bmatrix} + \begin{bmatrix} \text{Pure acid in} \\ \text{30\% solution} \end{bmatrix} = \begin{bmatrix} \text{Pure acid in} \\ \text{25\% solution} \end{bmatrix}$$

$$0.20x \quad + \quad 0.30(10) \quad = \quad 0.25(x + 10)$$

Step 5
$$0.20x + 0.30(10) = 0.25(x + 10)$$
$$20x + 300 = 25x + 250$$
$$50 = 5x$$
$$10 = x$$

Step 6   **Ans.** Add 10 quarts of 20% solution to produce the desired solution.

9. How many quarts of a 20% solution of acid should be added to 30 quarts of a 50% solution of acid to obtain a 40% solution of acid?

10. How many quarts of a 40% salt solution must be added to 40 quarts of a 10% salt solution to obtain a 20% salt solution?

11. How many ounces of an alloy containing 50% aluminum must be melted with an alloy containing 70% aluminum to obtain 40 ounces of an alloy containing 55% aluminum?

12. How many pounds of an alloy containing 60% copper must be melted with an alloy containing 20% copper to obtain 8 pounds of an alloy containing 30% copper?

13. How many grams of metal worth 50¢ a gram should be mixed with 20 grams of metal worth 32¢ a gram to produce an alloy worth 40¢ a gram?

14. How many pounds of dog food worth 5¢ a pound should a pet store owner mix with 15 pounds of dog food worth 8¢ a pound to produce a mixture worth 6¢ a pound?

15. Fine powder is worth 30¢ a kilogram, and coarse powder is worth 12¢ a kilogram. How many kilograms of the fine powder should be mixed with 50 kg of the coarse powder for the mixture to sell for 20¢ a kilogram?

16. A man uses 60 kg of fine powder worth 30¢ a kilogram and a coarse powder worth 25¢ a kilogram to make a mixture that he wishes to sell for 28¢ a kilogram. How many kilograms of the coarse powder does he use?

Sample Problem

The current in the Columbia River is 6 miles per hour. A boat travels downstream to Portland in 2 hours and returns in 3 hours. What is the speed of the boat in still water?

Steps 1–2   Speed of the boat in still water: x

x = Number of grams

 # 13.

$.50x + .32(20) = .40(x + 20)$

Step 3

	Rate	Time	Distance
Trip downstream	$x + 6$	2	$2(x + 6)$
Trip upstream	$x - 6$	3	$3(x - 6)$

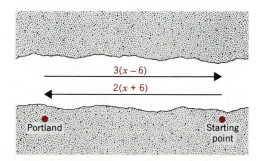

$3(x - 6)$

$2(x + 6)$

Portland   Starting point

(continued)

Step 4 $\underbrace{2(x + 6)}_{\substack{\text{distance traveled} \\ \text{downstream}}} = \underbrace{3(x - 6)}_{\substack{\text{distance traveled} \\ \text{upstream}}}$

Step 5 $2x + 12 = 3x - 18$

$30 = x$

Step 6   **Ans.** The speed of the boat in still water is 30 miles per hour.

17. Two cyclists leave Kansas City and travel in opposite directions. If the first cyclist travels at 14 mph and the second at 18 mph, after how many hours will they be 224 miles apart?

18. A ship sails from London for New York at 44 mph. At the same time a second ship leaves New York bound for London at 26 mph. The distance between London and New York is 3500 miles. How long will it be before the two ships meet?

19. Greg leaves Boston and drives west at 40 mph. Three hours later Karla heads west on the same road at 70 mph. How long does it take Karla to catch up with Greg?

20. Ethel and Gideon leave Atlanta at the same time and drive in opposite directions. Ethel drives 5 mph slower than Gideon. After 6 hours they are 570 miles apart. How fast does each drive?

21. A paddleboat journey down the Mississippi takes 5 hours, and the return trip upstream takes 9 hours. The current in the river flows at 8 mph. What is the speed of the paddleboat in still water?

22. Chester can row 9 miles an hour in still water. He rows for 5 hours upstream from his house to an island and returns the next day in 4 hours. How far is it from Chester's house to the island?

23. A balloonist flies from Fresno to Fullerton with a tailwind of 8 mph in 6 hours, and returns with a headwind of 2 mph in 8 hours. How far is it from Fresno to Fullerton?

24. Jetair flies from Chicago to Los Angeles in 6 hours against prevailing winds of 30 mph, and returns with the wind in 5 hours. What is the speed of the plane in still air?

3.6 INEQUALITIES

ORDER RELATIONSHIPS

In Section 2.1 we saw that for two different numbers, the graph of the smaller number lies to the left of the graph of the larger number on a number line. These order relationships can be expressed by using the following symbols:

$<$	means	"is less than,"
$>$	means	"is greater than."
$\leq$	means	"is less than or equal to,"
$\geq$	means	"is greater than or equal to."

For example,

"1 is less than 3" can be written as $1 < 3$.

"7 is less than 9" can be written as $7 < 9$.

"-3 is greater than -5" can be written as $-3 > -5$.

"2 is less than or equal to x" can be written as $2 \leq x$.

"4 is greater than or equal to y" can be written as $4 \geq y$.

Statements that involve any of the symbols above are called **inequalities.** Inequalities such as

$$1 < 3 \quad \text{and} \quad 7 < 9$$

are said to have the *same sense* because the left-hand member is less than the right-hand member in each case. Inequalities such as

$$7 < 9 \quad \text{and} \quad -3 > -5$$

are said to be of *opposite sense* because in one case the left-hand member is less than the right-hand member and in the other case the left-hand member is greater than the right-hand member.

PROPERTIES OF INEQUALITIES

The equations in one variable that we have solved have had only one solution. However, inequalities generally have an infinite number of solutions. For example, the graphs of the infinite number of integer solutions of the inequality $x > 3$ are shown in Figure 3.3.

Figure 3.3

Sometimes it is not possible to determine the solutions of a given inequality simply by inspection. But using the following properties, we can form an equivalent inequality in which the solution is evident by inspection.

1. **If the same expression is added to each member of an inequality, the result is an equivalent inequality of the *same sense*.**

In symbols,

$$\text{If} \quad a < b, \quad \text{then} \quad a + c < b + c.$$

For example:

a. Because $3 < 5$, then

$$3 + 4 < 5 + 4, \quad \text{or} \quad 7 < 9.$$

b. If $x < 7$, then

$$x + 2 < 7 + 2.$$

c. If $4 < y$, then

$$4 + (-2) < y + (-2).$$

> **2. If each member of an inequality is multiplied (or divided) by the same positive number, the result is an equivalent inequality of the *same* sense.**

In symbols,

$$\text{If} \quad a < b \quad \text{and} \quad c > 0, \quad \text{then} \quad ac < bc \quad \text{and} \quad \frac{a}{c} < \frac{b}{c}.$$

For example:

a. Because $2 < 3$ and $5 > 0$,

$$2(5) < (3)(5), \quad \text{or} \quad 10 < 15.$$

b. If $3x < 12$, then

$$\frac{3x}{3} < \frac{12}{3}, \quad \text{or} \quad x < 4.$$

c. If $z > 0$, then since $5 < 8$,

$$5(z) < 8(z), \quad \text{or} \quad 5z < 8z.$$

> **3. If each member of an inequality is multiplied (or divided) by the same negative number, the result is an equivalent inequality of the *opposite* sense.**

In symbols,

$$\text{If} \quad a < b \quad \text{and} \quad c < 0, \quad \text{then} \quad ac > bc \quad \text{and} \quad \frac{a}{c} > \frac{b}{c}.$$

For example:

a. Because $3 < 5$ and $-2 < 0$.

$$3(-2) > 5(-2), \qquad \text{or} \qquad -6 > -10.$$

b. If $-3x < 12$, then

$$\frac{-3x}{-3} > \frac{12}{-3}, \qquad \text{or} \qquad x > -4.$$

c. If $y < 0$, then since $2 < 5$,

$$2(y) > 5(y), \qquad \text{or} \qquad 2y > 5y.$$

The three previous properties also apply to inequalities of the form $a > b$.

SOLVING INEQUALITIES

Now let us see how the three properties can help us solve inequalities. Consider the inequality

$$\frac{3x}{2} < x + 1, \qquad \text{where } x \text{ is an integer.}$$

Multiplying each member by 2 (a positive number), we have

$$2\left(\frac{3x}{2}\right) < 2(x + 1),$$

$$3x < 2x + 2.$$

Then adding $-2x$ to each member, we get

$$3x + (-2x) < 2x + 2 + (-2x),$$

$$x < 2.$$

The graph of the solutions to this inequality is shown in Figure 3.4.

Figure 3.4

In the previous example, all the inequalities were of the same sense because we applied only Properties 1 and 2. Now consider the inequality

$$-3x + 1 > 7, \qquad \text{where } x \text{ is an integer.}$$

Adding -1 to each member, we get

$$-3x + 1 + (-1) > 7 + (-1),$$

$$-3x > 6.$$

Now we apply Property 3 and divide each member by -3. In this case we have to *reverse* the sense of the inequality.

$$\frac{-3x}{-3} < \frac{6}{-3},$$

$$x < -2.$$

The graph of the solutions to the inequality is shown in Figure 3.5.

Figure 3.5

WORD PROBLEMS

When solving word problems involving inequalities, we follow the six steps outlined on page 84, except that the word *equation* is replaced by the word *inequality*.

EXERCISES 3.6

■ *Complete each statement.*

1. If $x < 5$, then $x + 3 \underline{\ ?\ } 5 + 3$.

2. If $y > 4$, then $y + (-2) \underline{\ ?\ } 4 + (-2)$.

3. If $x > y$, then $7x \underline{\ ?\ } 7y$.

4. If $x < y$, then $-3x \underline{\ ?\ } -3y$.

5. If $\dfrac{x+1}{2} < 5$, then $2\left(\dfrac{x+1}{2}\right) \underline{\ ?\ } 2(5)$.

6. If $-5x < 30$, then $-\dfrac{5x}{5} \underline{\ ?\ } \dfrac{30}{-5}$.

■ *Solve each inequality.*

Sample Problems

a. $x - 8 > 3$

   Add 8 to each member.

$x - 8 + 8 > 3 + 8$

Ans. $x > 11$

b. $x + 6 > 2$

   Add -6 to each member.

$x + 6 - 6 > 2 - 6$

Ans. $x > -4$

7. $x + 3 > 7$ **8.** $x - 5 < 8$ **9.** $y + 10 < -5$

10. $y - 4 < -3$ **11.** $x + 2 \geq 9 - 4$ **12.** $x + 5 \geq 16 - 9$

13. $x - 16 > 3(2 + 1)$ **14.** $x + 9 < 3(4 - 2)$

Sample Problems

a. $5y < -20$

Divide each member by 5.

$$\frac{5y}{5} < \frac{-20}{5}$$

Ans. $y < -4$

b. $\frac{y}{-3} < 4$

Multiply each member by -3 and *reverse* the inequality.

$$-3\left(\frac{y}{-3}\right) > (-3)4$$

Ans. $y > -12$

15. $-3y < 15$ **16.** $-5y > 20$ **17.** $\frac{x}{3} \le 4$ **18.** $\frac{x}{5} \ge 2$

19. $4y > 12$ **20.** $6y < 24$ **21.** $\frac{x}{-3} > 7$ **22.** $\frac{x}{-4} > 8$

23. $-6x \le -72$ **24.** $-7x \ge -56$ **25.** $\frac{t}{-8} < -5$ **26.** $\frac{t}{-2} < -10$

■ *Solve each inequality and graph the solutions. All variables are integers.*

Sample Problem

$$\frac{2x - 6x}{8} > -2$$

Combine like terms, $2x - 6x$.

$$\frac{-4x}{8} > -2$$

Multiply each member by 8.

$$8\left(\frac{-4x}{8}\right) > 8(-2)$$

$$-4x > -16$$

Divide each member by -4 and reverse the inequality.

$$\frac{-4x}{-4} < \frac{-16}{-4}$$

Ans. $x < 4$

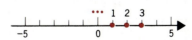

27. $2x + 3 > 7$ **28.** $3x - 4 < 11$ **29.** $-3x + 2 < 11$

30. $-4x - 3 > -11$ **31.** $5x \le 4x - 6$ **32.** $4x \ge 3x + 5$

33. $2 + 3x < 4x - 1$ **34.** $3x + 5 > 2x + 3$ **35.** $0 < 4y + 5 - 3y$

36. $0 > 8y + 6 - 7y$ **37.** $-6 > \dfrac{3y}{5}$ **38.** $12 < \dfrac{2y}{3}$

39. $\dfrac{2x}{3} + 1 < 3$ **40.** $\dfrac{-3x}{4} - 2 > 4$ **41.** $\dfrac{6x - 4x}{3} < 2$

42. $\dfrac{-3z + z}{6} > 4$ **43.** $\dfrac{3x + 2x}{3} - 3 \geq 7$ **44.** $\dfrac{7x - 9x}{4} + 2 \leq 2$

45. $2(3t - 6) < 5 - (t - 4)$ **46.** $4 - 3(2t - 4) < -2(4 - 3t)$

47. $5(-3y + 1) - 2(y + 2) \geq -3(4y - 7)$

48. $-2y + 6(3y - 5) \geq (2y - 6) - (5y - 14)$

49. $-10 - 4(2x + 5) + 9x \leq 3x - 8 - 2(3 - 3x)$

50. $3x - 4(1 + 3x) \leq -2 - 5(2x - 10)$

■ *Follow the six steps outlined on page 84 to solve the following problems.*

Sample Problems

The sum of three consecutive integers is greater than 108. What value could the smallest integer be?

Steps 1–2   The smallest integer: x
                    The next integer: $x + 1$
                    The largest integer: $x + 2$

Step 3   A sketch is not appropriate.

Step 4 $x + (x + 1) + (x + 2) > 108$

Step 5 $3x + 3 > 108$
                     $3x > 105$
                      $x > 35$

Step 6   **Ans.** The smallest integer must be greater than 35.

51. The sum of three consecutive integers is greater than 93. What values are possible for the smallest integer?

52. The sum of three consecutive integers is less than 126. What values are possible for the smallest integer?

53. Four times an integer is at least 10 more than three times the integer. What can the integer be?

54. Six times an integer is at least 3 more than three times the integer. What can the integer be?

55. A 40-foot board is cut into three pieces so that one piece is three times the length of the second piece. The carpenter wants at least 4 feet of the board for the third piece. How long can the shortest piece be?

56. A 31-foot board is cut into three pieces so that one piece is twice the length of the second piece. The carpenter wants at least 4 feet of the board for the third piece. How long can the shortest piece be?

57. A businessman is allowed $50 per day to rent a car. If the daily rate is $25 plus $0.05 per mile, how many miles can he drive so that he does not exceed his allowance?

58. A businesswoman is allowed $91 for food expenses. She has three dinners out and each one costs twice as much as the previous meal. How much could the first meal cost if she is to stay within her budget?

CHAPTER SUMMARY

[3.1]   Terms that are exactly alike in their variable factors are called **like terms.** Any product of factors in a term is called the **coefficient** of the remaining factors. We add like terms by adding their numerical coefficients. We subtract like terms by subtracting their numerical coefficients.

Algebraic expressions that have identical values for all substitutions for any variables they contain are called **equivalent expressions.**

A single example can be used to show that two expressions are not equivalent.

In expressions involving only addition, parentheses that are preceded by a plus sign may be dropped; each term within the parentheses retains its original sign.

In expressions involving addition and subtraction, parentheses preceded by a minus sign may be dropped, provided the sign of *each term* inside the parentheses is changed when the expression is written without parentheses.

Like terms can be combined by applying the **distributive law** in the form

$$ac + bc = (a + b)c$$

Algebraic expressions containing parentheses can be written without parentheses by applying the distributive law in the form

$$a(b + c) = ab + ac.$$

[3.2]   To solve equations in which more than one term contains the variable, we obtain an equivalent equation in which all terms containing the variable are in one member, and all constant terms are in the other.

Equations involving parentheses can be solved in the usual way, except that we should first simplify the equation by applying the distributive law to remove parentheses.

The steps on page 74 can be helpful in solving equations.

[3.3]   In applied problems that involve more than one unknown quantity, we represent each of the unknowns in terms of a single variable.

[3.4–3.5]   The steps on page 84 can be helpful in solving word problems.

[3.6]   We can use the following properties to write equivalent inequalities:

1. **If $a < b$, then $a + c < b + c$.**

2. **If $a < b$ and $c > 0$**   then $ac < bc$ and $\dfrac{a}{c} < \dfrac{b}{c}$.

3. **If $a < b$ and $c < 0$**   then $ac > bc$ and $\dfrac{a}{c} > \dfrac{b}{c}$.

We must reverse the inequality whenever we multiply or divide both members of an inequality by a negative number.

■ *The symbols introduced in this chapter appear on the inside of the front cover.*

CHAPTER REVIEW

■ *Combine like terms.*

1. a. $(3x - 2) + 5x + 6$ b. $(1 - 4y) - 3y - (2 + 6y)$

2. a. $8 - (-3a) + (5 - 3a)$ b. $5z - (2 - 3z) - (-4z)$

■ *Simplify.*

3. a. $4(3x - 2) + 7$ b. $6 - 2(3 - 3x)$

4. a. $-3(5y + 4) + 2(-3y + 2)$ b. $8(6 - 3y) - 2(-4 - 5y)$

■ *Solve.*

5. a. $4x + 3x = 35$ b. $4x - 4 = 2x - 4$ c. $8z + 6z = 2z - 12$

6. a. $\dfrac{2a}{3} = -12$ b. $\dfrac{b + 4b}{3} = 15$ c. $\dfrac{6x - 2x}{3} = -4$

7. a. $\dfrac{-9a - a}{2} = 10$ b. $\dfrac{3x + 5x}{2} = 6 + 3x$ c. $\dfrac{8x - 4x}{2} = \dfrac{8 + 10}{3}$

8. The sum of two numbers is 24. If one of the numbers is represented by x, how can the second number be represented in terms of x?

9. How can the value (in cents) of x dimes be represented in terms of x?

10. How can the value (in cents) of $(x + 3)$ quarters be represented in terms of x?

11. If oranges cost $1.85 per dozen, how can the cost (in cents) of $(x + 4)$ dozen oranges be represented in terms of x?

■ *Solve the following problems completely. Follow the six steps on page 84.*

12. The sum of three consecutive odd integers is five times the smallest integer. Find the integers.

13. A 32-foot board is cut into three pieces, so that one of the pieces is 3 feet longer than a second piece, and the third is 5 feet longer than the second. How long is each piece?

14. The total number of points scored in a college basketball game was 105. If the home team won by 17 points, what was the final score?

15. A small company sells two kinds of pencil sharpeners, electric and mechanical. The electric sharpeners cost $8, and the mechanical variety cost $3 each. Last month they sold 86 sharpeners and brought in $443. How many of each kind of pencil sharpener did they sell?

16. Marvin invested $300 more at 6% than he invested at 8%. His total annual income was $242. How much did he invest at each rate?

17. How many quarts of pure antifreeze must be added to five quarts of a 30% antifreeze solution to produce a 50% antifreeze solution?

18. Cashews cost $1.20 a pound and peanuts cost $0.80 a pound. How many of each should be used to produce 20 pounds of a mixture worth $0.90 a pound?

19. Jake rides relay for the Pony Express, covering his leg of the route in 6 hours and returning home, 8 miles per hour slower, in 9 hours. How far does Jake ride?

■ *Solve the inequalities and graph each solution set.*

20. a. $-5y + 1 > 26$ **b.** $3x + 2 > x - 10$ **c.** $\dfrac{4x - 5x}{4} > 2$

═══════════════════════════════

CUMULATIVE REVIEW

1. Like terms may be combined by adding the _____ of the variable factors of the terms.

2. What are equivalent expressions?

3. What are equivalent equations?

4. Products involving parentheses may be simplified by using the _____ law.

5. We must reverse the sense of an inequality whenever we _____ both sides by a _____ number.

■ *Simplify.*

6. a. $4 + 8 - 9 - 5$ b. $4 + 8 - (9 - 5)$ c. $(4 + 8)(9 - 5)$

7. a. $1 - 2 - 2 + (-2)$ b. $(1 - 2)(-2) - 2$ c. $1 - 2(-2 - 2)$

8. a. $|8| - |10|$ b. $|8 - 10|$ c. $|-8| + |-10|$

9. a. $3x - 2 + (4x - 1)$ b. $-6x - (12 - 4x) + 2$ c. $8x - (-3x) - 7 - 4$

10. a. $4(3x - 5) + 2(3x + 2)$
   b. $5 - 3(6x + 1) + 4x$
   c. $-2 - (7 - 2x) - 2x$

11. Evaluate for $x = -2$, $y = 5$.

   a. $\dfrac{2xy - 3x}{4 + 3x}$ b. $\dfrac{-3(2x - y)}{x + y}$ c. $\dfrac{6x - 3}{y} - \dfrac{8x - xy}{x - 4}$

12. a. If an odd integer is represented by x, how may the next consecutive odd integer be represented in terms of x?

   b. If an even integer is represented by x, how may the next consecutive even integer be represented in terms of x?

13. If the width of a rectangular garden is represented by x, how may the length, which is three times this width, be represented in terms of x?

14. Show by direct substitution that -2 is a solution of $-2a - 3 = 7 + 3a$.

15. Solve the inequalities, and graph their solutions.

   a. $5y - 2 > -22$ b. $\dfrac{-3y + y}{4} \leq 7$

■ *Solve.*

16. a. $7b - 3 = 6 - 2b$ b. $32 - 6b = 4(3 + b)$

17. a. $3(x - 5) = 45$ b. $-(x - 2) = 26 + 3x$

18. a. $7 = -5 + \dfrac{2y}{3}$ b. $\dfrac{3y}{2} - 4 = 2$

19. a. $\dfrac{3(10t - 4t)}{2} - 8 = 5t + 20$ b. $27 - \dfrac{2t - 12t}{5} = 4(t - 5t) - 9$

20. a. $-4(2a + 3) + 2(3a - 6) = 6$ b. $17 - 8(a - 3) = 2a + 3(2 - a)$

■ *Solve the following problems completely. Follow the six steps outlined on page 84.*

21. Miriam pays her babysitter $1.60 an hour plus a $0.75 tip. If she owes the babysitter $7.15 how many hours did he work?

22. Two investments produce an annual interest of $324. An amount of $500 more is invested at 12% than at 10%. How much is invested at each rate?

23. One number is six more than a second number. Ten times the smaller number minus four times the larger number equals six. Find the numbers.

24. One number is 10 more than a second number. Eight times the smaller number added to three times the larger number equals 129. Find the numbers.

25. A woman had $2.65 in change, consisting of eight more nickels than dimes. How many of each coin did she have?

4 POLYNOMIALS AND FORMULAS

4.1 EXPONENTS

It is not unusual for the same factor to occur more than once in a product. A shorthand way of writing such products is by using *exponents*. An **exponent** is a number written to the right and a little above a factor to indicate the number of times this factor occurs in a product. The number to which an exponent is attached is called the **base.** The product is then referred to as a **power** of the factor.

$$2^3 = \underbrace{2 \cdot 2 \cdot 2}_{\text{3 factors of 2}}$$

(exponent and base labels)

In general,

$$a^n = a \cdot a \cdot a \cdots a \quad (n \text{ factors}),$$

n a positive integer. For example:

5^2 means $(5)(5)$, read "five squared" or "5 to the second power";

2^4 means $(2)(2)(2)(2)$, read "two to the fourth power";

x^3 means xxx, read "x cubed" or "x to the third power";

$(x-2)^3$ means $(x-2)(x-2)(x-2)$, read "x minus two, quantity cubed."

We can use exponential notation when we write numbers in prime factored form. For example,

$$25 = 5 \cdot 5 = 5^2$$

and

$$16 = 2 \cdot 2 \cdot 2 \cdot 2 = 2^4.$$

When we write a variable such as x with no exponent indicated, it is understood that the exponent is 1. That is,

$$x = x^1.$$

To indicate that a negative number is raised to a power we must enclose the negative number in parentheses. For example, to express "the square of -5" we write

$$(-5)^2 = (-5)(-5) = 25,$$

and "the cube of -4" is written

$$(-4)^3 = (-4)(-4)(-4) = -64.$$

Common Error   Note that

$$(-3)^2 = (-3)(-3) = 9,$$

but

$$-3^2 = -1 \cdot 3^2 = -9,$$

since the exponent applies only to the 3 and not to the negative sign. Thus, $-3^2 \neq (-3)^2$. In general,

$$-x^2 \neq (-x)^2.$$

It is understood that an exponent is attached only to the *base* and not to other factors in the product. Thus,

$3x^2$, read "three x squared," means $3xx$,
$5x^2y^3$, read "five x squared y cubed," means $5xxyyy$,
$2x^3$, read "two x cubed," means $2xxx$,
$(2x)^3$, read "the quantity $2x$ cubed," means $(2x)(2x)(2x)$.

Common Error   Note that in the expression $2x^3$, the exponent applies only to the factor x and not to the product $2x$. Thus $2x^3$ and $(2x)^3$ are not equivalent expressions. In general,

$$ab^n \neq (ab)^n.$$

EXERCISES 4.1

■ *Write in exponential form.*

Sample Problems

a. $3 \cdot 3 \cdot 3 \cdot xxyyy$

Ans. $3^3x^2y^3$

b. $x(x - 2)(x - 2)$

Ans. $x(x - 2)^2$

1. $6 \cdot 6$

2. $5 \cdot 5 \cdot 5$

3. xxx

4. $yyyy$

5. $(-3)(-3)yy$

6. $3 \cdot 3 \cdot 3xx$

7. $2xxyyy$

8. $5 \cdot 5xxyyz$

9. $2aabccc$

10. $5aaabbc$

11. $(x - 3)(x - 3)$

12. $y(y + 2)(y + 2)$

Sample Problems

a. $xx + xyyy$

b. $3xyy - 2 \cdot 2xxxy$

c. $(5b)(5b) - 5bb$

Ans. $x^2 + xy^3$

Ans. $3xy^2 - 2^2x^3y$

Ans. $(5b)^2 - 5b^2$

13. $3 \cdot 3 + 2 \cdot 2 \cdot 2$

14. $3 \cdot 3 \cdot 3 + 5 \cdot 5$

15. $3xx + 5yyy$

16. $2xxx - 4yzz$

17. $(-3)(-3) - bb$

18. $a(3a)(3a) - (2b)(2b)$

19. $xxx + xxyy$

20. $xxy - xyyy$

21. $3aa - (-3c)(-3c)$

22. $(-2c)(-2c) + 2aa$

23. $(x - y)(x - y) + yy$

24. $xxy - (x + y)(x + y)$

■ *Write equivalent expressions without exponents.*

Sample Problems

a. $2(3x)^2$

b. $3a^2(7a)^2$

Ans. $2(3x)(3x)$

Ans. $3aa(7a)(7a)$

25. $5x^2y^3$

26. $3x^4y^3$

27. $-2a^2b^2c$

28. $7ab^2c^2$

29. $y^2(3x)^2$

30. $6x(2y)^3$

31. $(5a)^2(-2b)^3$

32. $(-3a)^3(-3b)^2$

33. $3x^2$

34. $5x^2$

35. $(3x)^2$

36. $(5x)^2$

Sample Problems

a. $(2x + 1)^2$

b. $y^2(y - 2)^3$

Ans. $(2x + 1)(2x + 1)$

Ans. $yy(y - 2)(y - 2)(y - 2)$

37. $(a - 4)^3$

38. $(2a + 1)^2$

39. $-y^3(3y + 4)^2$

40. $-x^2(2x - 1)^3$

41. $x^2y^3(x - y)^2$

42. $x^3y(y + 2x)^3$

■ *Use a numerical example to show that the following pairs of expressions are not equivalent.*

Sample Problems

a. $4x^2$; $(4x)^2$

b. x^2; $2x$

Ans. Let x be 2.
$$4x^2 = 4 \cdot 2^2$$
$$= 4 \cdot 4 = 16;$$
$$(4x)^2 = (4 \cdot 2)^2$$
$$= 8^2 = 64$$

Ans. let x be 3.
$$x^2 = 3^2 = 9;$$
$$2x = 2 \cdot 3 = 6$$

43. $3y^2$; $(3y)^2$ **44.** $2x^2$; $(2x)^2$

45. a^3; $3a$ **46.** x^4; $4x$

47. $-x^2$; $(-x)^2$ **48.** $-a^4$; $(-a)^4$

■ *Simplify.*

Sample Problems

a. -4^2
   $-1 \cdot 4 \cdot 4$

b. $(-4)^2$
   $(-4)(-4)$

c. $-3 \cdot 2^2$
   $-3 \cdot 2 \cdot 2$

Ans. -16 *Ans.* $+16$ or 16 *Ans.* -12

49. -3^2 **50.** $(-3)^2$ **51.** -3^3 **52.** $(-3)^3$

53. $-3 \cdot 2^2$ **54.** $(-3 \cdot 2)^2$ **55.** $(-1)^2 \cdot 2$ **56.** $(-1)^3 \cdot 2$

57. $-(-2)^2$ **58.** $-(-2)^3$ **59.** $(-1)(-4)^2$ **60.** $(-1)^2(-4)^3$

■ *Write in prime factored form using exponents.*

Sample Problems

a. 360
   $2 \cdot 2 \cdot 2 \cdot 3 \cdot 3 \cdot 5$

b. 847
   $7 \cdot 11 \cdot 11$

Ans. $2^3 \cdot 3^2 \cdot 5$ *Ans.* $7 \cdot 11^2$

61. 54   **62.** 220   **63.** 180   **64.** 441

65. 210   **66.** 800   **67.** 625   **68.** 286

4.2 ORDER OF OPERATIONS

Since an exponent applies only to its base, to simplify the expression $5 \cdot 2^3$ we first evaluate the power 2^3 to get $5 \cdot 8$ and then multiply to get 40. In general, powers should be computed before multiplications. We will therefore amend the order of operations given in Section 1.3 to include expressions involving exponents.

Order of Operations

1. Perform any operations inside parentheses, or above or below a fraction bar.

2. Compute all indicated powers.

3. Perform all other multiplication operations and any division operations in the order in which they occur from left to right.

4. Perform additions and subtractions in order from left to right.

For example, to simplify

$$\frac{8 - 4}{2} + 2 \cdot 3^2,$$

we first simplify the $8 - 4$ above the fraction bar, and the power 3^2 to obtain

$$\frac{4}{2} + 2 \cdot 9.$$

Then we divide and multiply to get

$$2 + 18.$$

And finally we add, to obtain

$$20.$$

NUMERICAL EVALUATION

Recall that to evaluate an algebraic expression for given values of the variables, we first substitute the values into the expression and then simplify according to the order of operations. For example, to evaluate

$$x^3 - 2y^2$$

for $x = -3$ and $y = 4$ we first substitute -3 for x and 4 for y and then perform the operations in the correct order to obtain

$$(-3)^3 - 2(4)^2 = -27 - 2(16)$$
$$= -27 - 32$$
$$= -59.$$

EXERCISES 4.2

■ *Simplify.*

Sample Problem

a. $-7 + 2 \cdot 3^2$

$-7 + 2 \cdot 9$

$-7 + 18$

Ans. 11

b. $1 - 3(-2)^3(-4)^2$ Evaluate powers.

$1 - 3(-8)(16)$ Multiply.

$1 + 384$ Add.

Ans. 385

1. $(-2)(3) - 2^2$

2. $(3)(5) - 2^3$

3. $14 - (-3)^2$

4. $18 - (-4)^2$

5. $4(3 + 5)^2$

6. $3(7 - 4)^2$

7. $5 - 4 \cdot 3^2$

8. $-6 + 2 \cdot 4^2$

9. $(3 + 2)^2(5 - 7)^3$

10. $(5 - 3)^4(3 - 6)^3$ **11.** $3(5 - 3^2)$ **12.** $4(4 + 4^2)$

13. $-2 \cdot 3^2 + 4$ **14.** $-3 \cdot 2^2 - 5$ **15.** $-(2 \cdot 3)^2 + 4$

16. $-(3 \cdot 2)^2 - 5$ **17.** $(-2 \cdot 2)^3 - 15$ **18.** $(-2 \cdot 2)^3 - 25$

Sample Problem

$$\frac{2 + 7}{5 - 2} - \frac{2^2 + 2^2}{4}$$

Simplify above and below the fraction bar.

$$\frac{9}{3} - \frac{8}{4}$$

Divide as indicated.

$3 - 2$

Subtract.

Ans. 1

19. $\dfrac{3^2 + 5}{2} + \dfrac{5^2 - 4}{-3}$ **20.** $\dfrac{-2^3}{4} + \dfrac{5 + 3^3}{8}$

21. $\dfrac{-5^2 + 1}{6} + \dfrac{2(-3)^2}{6}$ **22.** $\dfrac{-2^3 + 1}{7} + \dfrac{3(-2)^3}{-4}$

23. $\dfrac{(2^2)(-3^2)}{5 + 1} - \dfrac{7^2 - 6^2}{5 + 8}$ **24.** $\dfrac{4^2 - 2^2}{-(4 - 2)^2} + \dfrac{3^3 - 2^4}{4 + 7}$

25. $\dfrac{-4 + 2 \cdot 3^3}{6 - 1} - \dfrac{6(-2)^3}{3(4)}$ **26.** $\dfrac{3^3 - 3}{-4(2)} + \dfrac{2(-2^3)}{4 - 8} - \dfrac{8^2}{16}$

27. $\dfrac{4^2 - 3^2}{-7} + \dfrac{2(3^2) + 2}{2(5)} - \dfrac{26}{3(-5) + 2}$

28. $\dfrac{26 - 2(3)^2}{-4^2 + 3(4)} - \dfrac{(4 - 6)^3}{3(5) - 7} - \dfrac{5 + 5^2}{6 + 3^2}$

29. $3\left(\dfrac{5^3 - 100}{3 + 2}\right)\left(\dfrac{2^5 + 4}{15 - 3^2}\right)$ **30.** $4\left(\dfrac{8^2 - 2(-3)^2}{2 - 5^2}\right)\left(\dfrac{6^3 - 4(5)^2}{5^2 + 4}\right)$

■ *Find the value of each of the following expressions, given* $x = -2$.

Sample Problem

$2x^2 - 2x + 1$

Substitute -2 for x.

$2(-2)^2 - 2(-2) + 1$

Evaluate powers.

$2(4) - 2(-2) + 1$

Multiply.

$8 - (-4) + 1$

Add or subtract.

$8 + 4 + 1$

Ans. 13

31. $(4x)^3$ 32. $(-4x)^3$ 33. $3x^2$ 34. $-3x^2$

35. $-x^2$ 36. $(-x)^2$ 37. $1 - 4x^2$ 38. $5 + 3x^2$

39. $6x^3$ 40. $-x^3$ 41. x^4 42. x^5

43. $3x^2 + x$ 44. $x^2 - x$ 45. $-x^2 + 5x$ 46. $2x^2 + 4x$

47. $\dfrac{2x^3 + 2}{2} + 1$ 48. $\dfrac{2x^3 + 1}{3} - 2$ 49. $\dfrac{x^2}{2} - \dfrac{2}{x}$ 50. $\dfrac{6}{x} - \dfrac{4}{x^2}$

■ *Given* $x = 1, \; y = -2.$

Sample Problems

a. $2x - 3y^2$ b. $2x^2 - 3xy + y^2$

$\qquad$ Substitute 1 for x and -2 for y.

$\quad 2(1) - 3(-2)^2 \qquad 2(1)^2 - 3(1)(-2) + (-2)^2$

$\qquad$ Evaluate powers; then multiply.

$\quad 2 - (12) \qquad\qquad 2 - (-6) + 4$

$\qquad$ Add or subtract.

$\qquad\qquad\qquad\qquad 2 + 6 + 4$

Ans. $-10 \qquad\qquad$ *Ans.* 12

51. $-x^2y^2$ 52. $5x^2y^3$ 53. $-3x^3y^2$

54. $x^2 + y$ 55. $3x^2 + y$ 56. $x^2 - 2y^2$

57. $x^2 + xy + y^2$ 58. $2x^2 - 4xy + y^2$ 59. $-x^2 - y - y^2$

60. $3x^2 - (3y)^2$ 61. $\dfrac{(x - y)^2}{3} + y^2$ 62. $\dfrac{(4x - 2y)^2}{4} - 3y$

■ *Given* $a = 1, \; b = -2, \; c = -3, \; d = 0.$

63. $a^2b^2c^2d^2$ 64. $-abc^2$ 65. $-4a^2bc$

66. $-2bc^2d$ 67. $a^2 + b^2$ 68. $ab^2 - cd^2$

69. $a^2 - b^2 - c$ 70. $a^2 + ac - b^2$ 71. $3a^2 - 2bc + 2d^2$

72. $-2ab - a^2 + 2bc$ 73. $\dfrac{cb^2}{a} - \dfrac{cd}{b}$ 74. $\dfrac{bd}{ac} + \dfrac{bc^2}{a^2}$

4.3 POLYNOMIALS; SUMS AND DIFFERENCES

TERMS

As we have noted any single product of factors, such as

$$xyz, \qquad 2, \qquad \text{or} \qquad -2x^2y$$

is called a term. If the term does not contain variables, as in

$$2, \quad 10, \quad \text{and} \quad 7,$$

then the term is called a **constant.**

POLYNOMIALS

A **polynomial** is the sum or difference of terms, where the exponents on the variables are natural numbers. For example,

$$4x^5 + 2x^2 - 7x - 5, \quad 3xy^2 - 2x - 3x^4 + 4,$$

$$4xy^2, \quad 5x^2y - 3x - 7, \quad \text{and} \quad 2x + 2y$$

are polynomials. Notice that a polynomial can have any number of terms.

If a polynomial has only one term, we call it a **monomial** (*mono* is the Greek prefix for "one"). For example,

$$7, \quad 4x, \quad y^2z, \quad 3x^2y^3, \quad \text{and} \quad 2x^2y$$

are monomials.

If a polynomial has exactly two terms, we call it a **binomial** (*bi* is the Latin prefix for "two"). For example,

$$2x + y, \quad 3xy^2 + 4, \quad \text{and} \quad x^2 - y^2$$

are binomials.

If a polynomial has exactly three terms, we call it a **trinomial** (*tri* is the Greek prefix for "three"). For example,

$$x + y - z, \quad 2a + 3b + 5c, \quad \text{and} \quad x^2 - 3x + 4$$

are trinomials.

Polynomials in one variable are generally written in descending powers of the variable.

Exponents decrease from left to right.

$$3x^4 - 2x^2 + x - 5$$

Constant term is written last.

DEGREE OF A POLYNOMIAL

In a term containing only one variable, the exponent on the variable is the **degree** of the term. The degree of a constant term is 0. For example,

$3x^2$ is of second degree,

$2y^3$ is of third degree,

$4x$ is of first degree (the exponent on x is understood to be 1), and

5 is of zero degree.

The degree of a polynomial in one variable is the degree of its term of highest degree. For example,

$2x + 1$ is of first degree,

$3y^2 - 2y + 4$ is of second degree, and

$y^5 - 3y^2 + y$ is of fifth degree.

SUMS AND DIFFERENCES

Recall that *like terms* must be exactly alike in their variable factors. Thus

$$x^3 \quad \text{and} \quad 5x^2,$$

$$2x^2y \quad \text{and} \quad 2xy^2$$

are not like terms because the variable factors are different. In expressions that contain some like terms and some unlike terms, we can combine only the like terms. For example,

$$x^3 + 5x^2 - 3x - 2x^2 + x = x^3 + 3x^2 - 2x.$$

Note that, by the commutative property, expressions such as xy and yx are equivalent. Hence, $5yx$ and $3xy$ are like terms. Similarly, $2x^2y$ and $4yx^2$ are like terms, so

$$2x^2y - 5yx + 4yx^2 + 3xy = 2x^2y - 5xy + 4x^2y + 3xy$$

$$= 6x^2y - 2xy.$$

Common Errors

Notice that,

$$2x + 3y \neq 5xy,$$

since $2x$ and $3y$ are *not* like terms; and

$$2x^2 + 3x^2 \neq 5x^4,$$

since we add the *numerical coefficients only;* the exponents are *not* added.

To add two polynomials, we need only remove parentheses and combine like terms. Thus

$$(4a^3 - 2a^2 - 3a + 1) + (2a^3 + 4a - 5) = 4a^3 - 2a^2 - 3a + 1 + 2a^3 + 4a - 5$$

$$= 6a^3 - 2a^2 + a - 4$$

We can also use a vertical format as shown below to add polynomials.

$$\text{Add:} \quad \begin{array}{r} 4x^2 - 7x + 2 \\ 2x^2 + 5x - 5 \\ \hline 6x^2 - 2x - 3 \end{array}$$

To subtract two polynomials, we must change signs properly when removing parentheses. Recall that a minus sign before parentheses applies to *each term* inside. For example,

$$(4x^2 + 2x - 5) - (2x^2 - 3x - 2)$$

Each sign is changed.

$$= 4x^2 + 2x - 5 - 2x^2 + 3x + 2$$

$$= 2x^2 + 5x - 3.$$

When using a vertical format for subtraction, we change the sign of each term of the bottom polynomial, and then add.

Subtract. $\begin{array}{r} +5x - 2y \\ +7x - 4y \end{array}$ ⟶ Change signs of each term and add. $\qquad \begin{array}{r} +5x - 2y \\ -7x + 4y \\ \hline -2x + 2y \end{array}$

Subtract. $\begin{array}{r} 4x^2 + 2x - 5 \\ 2x^2 - 3x - 2 \end{array}$ ⟶ Change signs of each term and add. $\qquad \begin{array}{r} 4x^2 + 2x - 5 \\ -2x^2 + 3x + 2 \\ \hline 2x^2 + 5x - 3 \end{array}$

EXERCISES 4.3

■ *a. Identify each expression as monomial, binomial, or trinomial.*
b. Write each term separately and state its numerical coefficient.

Sample Problem

$4y^3 + 2y^2 - y$

Ans.

a. trinomial

b. $4y^3$ — coefficient: 4 $2y^2$ — coefficient: 2 $-y$ — coefficient: -1

1. $2y^2$

2. $-3x^3$

3. $x - y^3$

4. $2y^3 - 4x$

5. $-3x^2 + 3y + 4z$

6. $3y^3 - 4y + x$

7. $-x^2y + yz^2$

8. $7xyz + 3x$

9. $7x^2y + x^2y^2$

10. $3xy + 2xz - yz$

11. $7x^2yz^3$

12. $x^5y^3z^2$

■ *Write each term separately and state the degree of the term.*

Sample Problem

$4x^2 - 2x^3 - x$

Ans. $4x^2$ $-2x^3$ $-x$
degree: 2   degree: 3   degree: 1

13. $-2x^2 + x - 4x$ **14.** $4y^2 - 2y + y$ **15.** $-y^4 - 2y^3 - y$
16. $4x^3 + 2x^2 + 2x$ **17.** $z^5 - 2z^2 + z$ **18.** $z^6 - 3z^3 + z$

■ *Give the degree of each polynomial.*

Sample Problem

$3y^4 - 2y^2 + 1$

Ans. Fourth degree (the term of highest degree, $3y^4$, is of fourth degree).

19. $2x^4 - x^2 + 4x$ **20.** $-4y^5 + 2y^2 - y$ **21.** $y^5 + 2y^3 + y$
22. $4x^3 - 2x^2 - 2x$ **23.** $z^4 + 2z^2 - z$ **24.** $-z^5 + 3z^3 + z$

■ *Simplify the following expressions.*

Sample Problem

$(5y^3 - y^3) + 2y^3$ Add or subtract numerical coefficients of like terms.

$4y^3 + 2y^3$

Ans. $6y^3$

25. $7x^2 - 4x^2$ **26.** $6y^3 - 2y^3$ **27.** $4y^2 - y^2$
28. $6b^3 - (2b^3 + b^3)$ **29.** $5b^4 - (3b^4 - b^4)$ **30.** $3a^6 + (2a^6 - a^6)$

Sample Problems

a. $(6x) - (-2x)$ b. $(2y^2) - (4y^2) - (-5y^2)$
              Remove parentheses; change signs.

$6x + 2x$ $2y^2 - 4y^2 + 5y^2$
           Add like terms.

Ans. $8x$ *Ans.* $3y^2$

31. $(6x) - (3x)$ **32.** $(3y) - (-3y)$ **33.** $(-2y) - (2y)$
34. $(-2xz) - (-3xz)$ **35.** $(5y^2) - (-y^2)$ **36.** $(-4y^2) - (y^2)$
37. $(2x) + (3x) - (-x)$ **38.** $(2y) - (-5y) - (3y)$
39. $(5ab^2) - (-ab^2) + (-3ab^2)$ **40.** $(7a) - (a) - (-2a)$

Sample Problem

$$2x^2y + 7x^2y + 4xy^2 - xy^2$$

Add or subtract numerical coefficients of like terms.

Ans. $9x^2y + 3xy^2$

41. $2x + 5x - 3y$ **42.** $2x + 6y - 2y$

43. $4ab^2 + 3a^2b - 2a^2b$ **44.** $ab^2 + a^2b + 5a^2b$

45. $8a^2b - ab^2 + a^2b$ **46.** $3ab^2 + 2a^2b - ab^2$

47. $6x^2y + 3xy^2 - 2xy^2 - x^2y$ **48.** $5x^2yz + 2xy^2z - 3xyz^2 - 2xy^2z$

49. $4x^3yz + 5x^2yz - 3x^3yz + 2xyz$ **50.** $15ab^2c + 6a^2bc - 3ab^2c + 5abc^2$

51. $25xz^2 + 17x^2z - 8xz^2 + 5xz^2$ **52.** $23ab + 35ac - 17ab + 18ab$

■ *Simplify.*

Sample Problem

$$3x - 2y - (4x + 2y) + x$$

Remove ( ); change signs.

$$3x - 2y - 4x - 2y + x$$

Add like terms.

Ans. $-4y$

53. $2x - y + (x + y)$ **54.** $3a - 2b + (2a + b)$

55. $2x + 3 - (x - 4)$ **56.** $(2x + 3) - x - 4$

57. $6a + 5b - (2a - 5b) - 2a$

58. $3c - 2d + 1 - (2 + 2c - d) + c - 1$

■ *Add or subtract the following polynomials.*

Sample Problem

$$(3x^2 - 2y + z) - (-2x^2 + 3y - z)$$

Remove ( ); change signs.

$$3x^2 - 2y + z + 2x^2 - 3y + z$$

Add like terms.

Ans. $5x^2 - 5y + 2z$

59. $(2y^2 - y + 1) - (6y^2 + 2y + 1)$

60. $(4x^2 - 3x - 1) - (5x^2 + x - 1)$

61. $(z^2 - 4z + 1) - (2z^2 + z + 1)$

62. $(y^2 - 3y + 5) - (y^2 + 4y - 3)$

63. $(2xy^2 + 6xy - x) - (2xy + x)$

64. $(5ax^2 + 3ax + 4) - (2ax^2 - 3)$

65. $(2x + y - z) + (x - 2y + z) - (x + y + 2z) - (x - 3y - 4z)$

66. $(a - b - c) + (a - b - c) - (a - b - c) + (a + b + c)$

67. $(2g + 3h - k) + (2g + h + k) - (2g + 2h + 2k) - (3g - h + k)$

68. $(2x + 2y - z) - (x + 2y - z) - (3x + 2y - z) + (x + 4y + 5z)$

■ *Add the polynomials.*

Sample Problem

$$\begin{array}{r} 3x^2 - 6x - 8 \\ \text{Add.} \quad \underline{x^2 + 2x + 10} \end{array} \qquad \text{Add like terms.}$$

Ans. $4x^2 - 4x + 2$

69.
$$\begin{array}{r} 8x^2 - 3x \\ \underline{-2x^2 + 5x} \end{array}$$

70.
$$\begin{array}{r} -3y^2 - 7 \\ \underline{5y^2 + 1} \end{array}$$

71.
$$\begin{array}{r} -4y^2 - y + 6 \\ \underline{2y - 3y^2 - 3} \end{array}$$

72.
$$\begin{array}{r} 6x^2 - 6x + 4 \\ \underline{-3x - 2x^2 + 4} \end{array}$$

■ *Subtract the bottom polynomial from the top polynomial.*

Sample Problem

$$\begin{array}{r} 4x^2 + 2x - 7 \\ \text{Subtract.} \quad \underline{-3x^2 - 5x + 1} \end{array} \quad \begin{array}{l} \text{Change signs of each} \\ \text{term and add.} \end{array} \quad \begin{array}{r} 4x^2 + 2x - 7 \\ \underline{+ 3x^2 + 5x - 1} \\ 7x^2 + 7x - 8 \end{array}$$

Ans. $7x^2 + 7x - 8$

73.
$$\begin{array}{r} 3x - 7 \\ \underline{2x + 4} \end{array}$$

74.
$$\begin{array}{r} 5x^2 + 6x \\ \underline{-2x^2 + x} \end{array}$$

75.
$$\begin{array}{r} -3x^2 + 4x - 2 \\ \underline{4x^2 - 3x - 1} \end{array}$$

76.
$$\begin{array}{r} 10y^2 + 2y + 1 \\ \underline{-4y^2 - 4y + 5} \end{array}$$

77.
$$\begin{array}{r} 3y^2 + 2y - 1 \\ \underline{1 - 2y - 2y^2} \end{array}$$

78.
$$\begin{array}{r} -4x^2 - 3x + 7 \\ \underline{2x^2 + 5 - 2x} \end{array}$$

■ *Use a numerical expression to show that the following pairs of expressions are not equivalent.*

79. $7x^2 + 5x^2$; $12x^4$

80. $8x^2 - 6x^2$; $2x^4$

81. $4x - 2y$; $2xy$

82. $17x^2 - 8y^2$; $9x^2y^2$

4.4 LAWS OF EXPONENTS

We have used exponents to indicate the number of times a given factor occurs in a product. For example, $x^3 = (x)(x)(x)$. Exponential notation provides us with a simple way to multiply expressions that contain powers with the same base.

Consider the product $(x^2)(x^3)$, which in factored form appears as

$$(x)(x) \cdot (x)(x)(x).$$

This, in turn, may be written x^5, since it contains x as a factor five times. Again, $(y^5)(y^2)$ is equivalent to

$$(y)(y)(y)(y)(y) \cdot (y)(y),$$

which may be written as y^7.

FIRST LAW OF EXPONENTS

In the examples above, we can obtain the product of two expressions with the same base by adding the exponents of the powers to be multiplied. We can make a more general statement by considering the product $(a^m)(a^n)$. In factored form, $(a^m)(a^n)$ appears as

$$\underbrace{(aaa \cdots a)}_{m \text{ factors}}\underbrace{(aaa \cdots a)}_{n \text{ factors}} = \underbrace{aaa \cdots a}_{(m + n) \text{ factors}},$$

which in exponential notation is written as a^{m+n}. Thus:

To multiply two powers with the same base, add the exponents.

This property is called the **first law of exponents.** In symbols:

$$a^m \cdot a^n = a^{m+n}. \qquad \mathbf{I}$$

For example,

Add exponents.

$$3^4 \cdot 3^3 = 3^{4+3} = 3^7,$$

Same base

Add exponents.

$$y^3 \cdot y^2 = y^{3+2} = y^5.$$

Same base

Common Error

Notice that

$$x^3 \cdot x^2 \neq x^{3 \cdot 2}, \quad \text{or } x^6.$$

If $x = 2$, then

$$2^3 \cdot 2^2 = 8 \cdot 4 = 32;$$

$$2^6 = 64.$$

(continued)

By the first law of exponents, we must *add* the exponents. Thus,

$$x^3 \cdot x^2 = x^{3+2} = x^5.$$

SECOND LAW OF EXPONENTS

Consider the expression $(x^4)^3$, which means the cube of x^4. We can simplify this expression as follows:

$$(x^4)^3 = (x^4)(x^4)(x^4) = x^{4+4+4} = x^{12}.$$

Similarly,

$$(y^2)^5 = (y^2)(y^2)(y^2)(y^2)(y^2) = y^{2+2+2+2+2} = y^{10}.$$

Because repeated addition is actually multiplication, we can obtain the same results in the previous examples by multiplying the exponents together. Thus in general:

> **To raise a power to a power, multiply the exponents.**

This is the **second law of exponents.** In symbols:

$$(a^m)^n = a^{mn} \qquad \textbf{II}$$

For example,

Multiply exponents.

$$(5^3)^6 = 5^{3 \cdot 6} = 5^{18},$$

Multiply exponents.

$$(y^4)^2 = y^{4 \cdot 2} = y^8,$$

Common Error

Notice the difference between the two expressions

$$(x^3)(x^4) = x^{3+4} = x^7$$

and

$$(x^3)^4 = x^{3 \cdot 4} = x^{12}.$$

The first of these is a multiplication, so the exponents are added. The second raises a power to a power, so the exponents are multiplied.

THIRD LAW OF EXPONENTS

By using the commutative and associative properties of multiplication, we may write

$$(2x)^3 = (2x)(2x)(2x)$$
$$= 2 \cdot 2 \cdot 2 \cdot x \cdot x \cdot x$$
$$= 2^3 x^3.$$

This example illustrates the **third law of exponents**:

A power of a product is equal to the product of the powers of each of its factors.

In symbols:

$$(ab)^n = a^n b^n \qquad \textbf{III}$$

For example,

$$(3xy^2)^4 = (3)^4 (x)^4 (y^2)^4$$
$$= 81 x^4 y^8,$$

where we used the second law of exponents to simplify $(y^2)^4$.

Note that $(-x)^n$ can be thought of as $(-1 \cdot x)^n$, so that in particular

$$(-x)^2 = (-1 \cdot x)^2 \qquad \text{and} \qquad (-x)^3 = (-1 \cdot x)^3$$
$$= (-1)^2 (x)^2 \qquad\qquad\qquad = (-1)^3 (x)^3$$
$$= x^2, \qquad\qquad\qquad\qquad = -x^3.$$

In general we have the following.

$$(-x)^n = x^n \qquad \text{if } n \text{ is an even integer,}$$

and

$$(-x)^n = -x^n \qquad \text{if } n \text{ is an odd integer.}$$

For example,

$$(-2)^4 = 2^4 = 16,$$
$$(-2)^5 = -2^5 = -32.$$

SIMPLIFYING ALGEBRAIC EXPRESSIONS INVOLVING PRODUCTS

By using the order of operations and the laws of exponents, we can simplify many algebraic expressions involving variables. Consider the product of two monomials

$$(2x^2 y)(5xy^3).$$

By applying the commutative and associative properties of multiplication, we can arrange the factors as

$$(2)(5)(x^2)(x)(y)(y^3),$$

which is equivalent to $10x^3y^4$. Thus, to multiply two or more monomials together, we first multiply their numerical coefficients, and then multiply together powers with the same base according to the first law of exponents. So

$$(-5x^2)(2x^3) = (-5)(2)(x^2)(x^3) = -10x^5,$$

and

$$(-6xy^2)(-3x^3y) = (-6)(-3)(x)(x^3)(y^2)(y) = 18x^4y^3.$$

To simplify an expression such as

$$3x^2y + 2x(xy),$$

we follow the proper order of operations. First we multiply to get

$$3x^2y + 2x^2y.$$

Then we combine like terms to get

$$5x^2y.$$

To simplify

$$(-x^2y^3)(-x)^4 - y(x^3y)^2,$$

we first perform the powers $(-x)^4$ and $(x^3y)^2$ to get

$$(-x^2y^3)(x^4) - y(x^6y^2),$$

then we multiply, and finally we add like terms, which gives

$$-x^6y^3 - x^6y^3 = -2x^6y^3.$$

EXERCISES 4.4

■ *Find each product by first writing in factored form, then simplifying.*

Sample Problems

a. $x^2 \cdot x^3$

$xx \cdot xxx$

Ans. x^5

b. $(a^4)(a^2)(a)$

$aaaa \cdot aa \cdot a$

Ans. a^7

1. $x^3 \cdot x^3$

2. $x^2 \cdot x^6$

3. $y^3 \cdot y^7$

4. $y \cdot y^5$

5. $(b^3)(b)(b^5)$

6. $(b^2)(b^3)(b^3)$

■ *Find the products by using the first law of exponents.*

Sample
Problems

a. $y^7 \cdot y^9$

y^{7+9}

Ans. y^{16}

b. $(c^5)(c^8)(c^3)$

c^{5+8+3}

Ans. c^{16}

7. $a^8 \cdot a^6$ **8.** $a^4 \cdot a^{13}$ **9.** $x \cdot x^5 \cdot x^7$

10. $x^4 \cdot x^5 \cdot x^4$ **11.** $(t^{10})(t^{20})(t^{30})$ **12.** $(t^{14})(t^{12})(t^7)$

■ *Find each power by first expressing as a repeated multiplication, then simplifying.*

Sample
Problems

a. $(y^2)^3$

$(y^2)(y^2)(y^2) = y^{2+2+2}$

Ans. y^6

b. $(c^8)^2$

$(c^8)(c^8) = c^{8+8}$

Ans. c^{16}

13. $(t^3)^5$ **14.** $(t^7)^3$ **15.** $(x^9)^4$ **16.** $(x^{10})^3$

■ *Find the powers by using the second law of exponents.*

Sample
Problems

a. $(x^7)^5$

$x^{7 \cdot 5}$

Ans. x^{35}

b. $(a^{13})^3$

$a^{13 \cdot 3}$

Ans. a^{39}

17. $(y^7)^3$ **18.** $(y^{12})^3$ **19.** $(b^{14})^5$ **20.** $(b^{11})^8$

21. $(x^{10})^{20}$ **22.** $(x^8)^{10}$ **23.** $(c^{18})^{30}$ **24.** $(c^{12})^{12}$

■ *Find the powers by first using the third law of exponents.*

Sample
Problems

a. $(5a^4)^2$

$(5)^2(a^4)^2$

Ans. $25a^8$

b. $(3xy^3)^4$

         Apply the third law of exponents.

$(3)^4(x)^4(y^3)^4$

         Apply the second law of exponents.

Ans. $81x^4y^{12}$

25. $(2x^3)^5$ **26.** $(6x^6)^2$ **27.** $(-4x^2y^4)^4$ **28.** $(-5xy^8)^3$

29. $(3a^3bc^6)^5$ **30.** $(10a^2b^4c^8)^4$ **31.** $(-3x^3yz^3)^3$ **32.** $(-2x^5y^3z^2)^6$

■ *Find the products.*

**Sample
Problems**

a. $(-3x^2y)(2x)$ b. $(-2x^2y)(4x)(-3xy^2)$

$\qquad\qquad\qquad\qquad\qquad\qquad\qquad\qquad$ Multiply numerical coefficients.

$\quad-6$ $\qquad\qquad\qquad\qquad$ 24

$\qquad\qquad\qquad\qquad\qquad\qquad\qquad\qquad$ Use the first law of exponents to
$\qquad\qquad\qquad\qquad\qquad\qquad\qquad\qquad$ multiply variable factors.

Ans. $-6x^3y$ $\qquad\qquad\qquad\qquad$ *Ans.* $24x^4y^3$

33. $(x)(4x^4)$ $\qquad\qquad$ **34.** $(2a^2b)(8ab^3)$ $\qquad\qquad$ **35.** $(a^3b^3)(b^2c)$

36. $(ac^2)(a^2b^3)(b^2c^2)$ $\qquad$ **37.** $(2abc)(5b^2c^2)(ab^2c)$ $\qquad$ **38.** $(7a^2bc)(2bc)(abc)$

39. $4x(3xy)$ $\qquad\qquad$ **40.** $3x(-2xy)$ $\qquad\qquad$ **41.** $5x(-x^2y)$

42. $-6xy^3(4xy^2)$ $\qquad$ **43.** $(-2x^3)(x^2y)(-4y^2)$ $\qquad$ **44.** $(3xy^3)(-x^4)(-2y^2)$

45. $(-a)(-4ab)(2a^2b^3)$ $\qquad\qquad\qquad$ **46.** $(2a^2b)(-a^2b)(8ab^3)$

47. $(-3abc^2)(-a^2b)(-6b^4c^2)(2b)$ $\qquad$ **48.** $(-5ab)(b^4)(-7a^3b^4)(b^2c)$

49. $(-9x^2y^3z^4)(3y^5z)(-x^3z^2)(6xy^2)$ $\qquad$ **50.** $(-4xy^3z^2)(-2x^5)(-y^2z^4)(3x^3z)$

■ *Simplify by using the laws of exponents.*

**Sample
Problems**

a. $(3x-4)^5(3x-4)^3$ $\qquad\qquad\qquad$ b. $2x^2y(3xy^3)^2$

$\qquad\qquad\qquad\qquad$ Apply the first $\qquad\qquad\qquad\qquad$ Apply the second and
$\qquad\qquad\qquad\qquad$ law of exponents. $\qquad\qquad\qquad\qquad$ third laws of exponents.

$\qquad\qquad\qquad\qquad\qquad\qquad\qquad\qquad\qquad$ $2x^2y(9x^2y^6)$

Ans. $(3x-4)^8$

$\qquad\qquad\qquad\qquad\qquad\qquad\qquad\qquad\qquad$ Apply the first law of
$\qquad\qquad\qquad\qquad\qquad\qquad\qquad\qquad\qquad$ exponents.

$\qquad\qquad\qquad\qquad\qquad\qquad\qquad\qquad$ *Ans.* $18x^4y^7$

51. $(a+b)^2(a+b)^4$ $\qquad\qquad\qquad$ **52.** $(1-3x)^6(1-3x)$

53. $[(a+b)^2]^4$ $\qquad\qquad\qquad\qquad$ **54.** $[(1-3x)^6]^2$

55. $[(x^2+2x)^3]^8$ $\qquad\qquad\qquad\qquad$ **56.** $[(-3-t)^{16}]^4$

57. $(x^2+2x)^3(x^2+2x)^8$ $\qquad\qquad$ **58.** $(-3-t)^{16}(-3-t)^4$

59. $x^3(x^2)^5$ $\qquad\qquad\qquad\qquad$ **60.** $x(x^4)^6$

61. $(x^2y)^3(xy^3)$ $\qquad\qquad\qquad\qquad$ **62.** $(x^3)^4(x^3y^4)$

63. $(2x^3y)^2(xy^3)^4$ $\qquad\qquad\qquad$ **64.** $(3xy^2)^3(2x^2y^2)^2$

65. $(ab^2)^3(bc^3)(-2abc)$ $\qquad\qquad$ **66.** $(3a^3c^2)(a^2b^4)^2(b^3c^2)$

67. $(5a^2b^5)(2ab^3c)^2(a^3c^3)^3$ $\qquad$ **68.** $(-4a^2c^3)(2b^5c^3)^2(a^2b)^5$

Sample Problems

a. $(-x)^2(-xy)^3$

b. $-x^2(x^2)^4$

Apply the third law of exponents.

$x^2(-x^3y^3)$

Apply the second law of exponents.

$-x^2(x^8)$

Apply the first law of exponents.

Ans. $-x^5y^3$

Ans. $-x^{10}$

69. $(-x)^3$

70. $(-a)^2(a)$

71. $-(a)^3$

72. $(-x)^3(-xy)$

73. $-(-xy)^2(xy^2)$

74. $(-a)^2(ab)(-b^3)^3$

75. $(-3)^2(-x^3)^2(-y^2)^2$

76. $-2(-x^3y)^3(y^2)$

77. $-4x(-x^2y^2)^2(-y^3)^2$

78. $3x(-xy^3)(-xy)^4$

79. $-5x(-3yz^2)^4(-z)^3$

80. $(-4x^2)^2(y^2)^3(y^2z)^3$

■ *Simplify by using the order of operations and the laws of exponents.*

Sample Problems

a. $a^6 + 3a^2(a^2)^2$

b. $(3x)(xy^3)^2 - xy + 4x^3(y^2)^3$

Apply the second and third laws of exponents.

$a^6 + 3a^2(a^4)$

$(3x)(x^2y^6) - xy + 4x^3(y^6)$

Apply the first law of exponents.

$a^6 + 3a^6$

$3x^3y^6 - xy + 4x^3y^6$

Combine like terms.

Ans. $4a^6$

Ans. $7x^3y^6 - xy$

81. $6a^2(b^2)^3 - b^2$

82. $a^2 + 4a^2(3b)^2$

83. $2y(y^3)^2 - 2y^4(3y)^3$

84. $6t^2(2t)^3 + 4t(2t^2)^2$

85. $2a(a^2)^4 + 3a^2(a)^6 - a^2(a^2)^3$

86. $b^2(b^3) + 3b(b^2)^2 - b^2(b)^3$

87. $5a(2b)^4 - 2a(3b)^3 + 3b^3$

88. $a(3b)^2 + 4a(2b^2) - 3ab(b)^2$

89. $4a^2(ab)^3 + a(a^2b)^3 + 3a^3b$

90. $8a(ab^2)^3 + 3ab(ab)^4 - 2(a^2b^2)^2ab$

■ *Use a numerical example to show that the following pairs of expressions are not equivalent.*

91. $x^2 \cdot x^4; \ x^8$

92. $x^3 \cdot x^3; \ x^9$

93. $x^2 + x^2; \ x^4$

94. $x^3 + x^2; \ x^5$

95. $a^2 \cdot b^3; \ (ab)^5$

96. $x^3 \cdot y^2; \ (xy)^5$

97. $(x^2)^3; \ x^5$

98. $(a^3)^3; \ a^6$

99. $(x^2)^3; \ x^8$

100. $(a^4)^2; \ a^{16}$

101. $(xy)^2; \ xy^2$

102. $(ab)^3; \ ab^3$

103. $-x^2; \ (-x)^2$

104. $-y^2; \ (-y)^2$

105. $-2x^2; \ (-2x)^2$

106. $-3x^2; \ (-3x)^2$

107. $-2y^3; \ -(2y)^3$

108. $-3y^3; \ (-3y)^3$

4.5 FORMULAS

Recall that equations that involve variables for the measures of two or more physical quantities are called **formulas.** We can solve for any one of the variables in a formula if the values of the other variables are known. We substitute the known values in the formula and solve for the unknown variable by the methods we used in preceding sections. For example, in the formula $d = rt$, if $d = 24$ and $r = 3$, we can solve for t by substituting 24 for d and 3 for r. That is,

$$d = rt,$$

$$(24) = (3)t,$$

$$8 = t.$$

It is often necessary to solve formulas, or equations involving more than one variable, for one of the variables in terms of the others. We use the same methods used in the preceding sections, treating all the other variables as if they were constants. For example, in the formula $d = rt$, we may solve for t in terms of r and d by dividing both members by r to yield

$$\frac{d}{r} = \frac{\cancel{r}t}{\cancel{r}},$$

$$\frac{d}{r} = t,$$

from which, by the symmetric law,

$$t = \frac{d}{r}.$$

EXERCISES 4.5

■ *Evaluate each formula for the specified symbol.*

Sample Problem

$f = ma$ Find m if $f = -64$ and $a = -32$.

Substitute for f and a.

$(-64) = m(-32)$

Divide each member by -32.

$$\frac{-64}{-32} = \frac{m(\cancel{-32})}{\cancel{-32}}$$

Ans. $m = 2$

1. $f = ma$ Find a if $m = 3$ and $f = -27$.

2. $f = ma$ Find m if $a = -32$ and $f = -96$.

3. $d = rt$ Find r if $d = 80$ and $t = 16$.
4. $d = rt$ Find t if $r = 60$ and $d = 240$.
5. $v = lwh$ Find h if $v = 60$, $l = 15$ and $w = 2$.
6. $v = lwh$ Find w if $v = 45$, $l = 5$ and $h = 3$.
7. $I = prt$ Find t if $I = 100$, $p = 25$ and $r = 2$.
8. $I = prt$ Find t if $I = 70$, $p = 7$ and $r = 2$.
9. $s = \dfrac{at^2}{2}$ Find a if $s = -144$ and $t = 3$.
10. $s = \dfrac{at^2}{2}$ Find a if $s = -64$ and $t = 2$.
11. $v = k + gt$ Find g if $v = 32$, $k = 20$ and $t = 4$.
12. $v = k + gt$ Find t if $v = 35$, $k = 15$ and $g = 4$.
13. $F = \dfrac{kmM}{d^2}$ Find M if $F = 10$, $k = 5$, $m = 4$ and $d = 2$.
14. $F = \dfrac{kmM}{d^2}$ Find k if $F = 8$, $m = 2$, $M = 12$ and $d = 3$.
15. $P = 2l + 2w$.   Find l if $P = 264$ and $w = 40$.
16. $P = 2l + 2w$.   Find w if $P = 258$ and $l = 74$.
17. $A = P + Prt$ Find P if $A = 840$, $r = 2$, and $t = 3$.
18. $A = P + Prt$ Find t if $A = 880$, $P = 80$, and $r = 5$.

■ *Solve each of the following formulas for the symbol in color.*

Sample Problem

$c = 2\pi r$

Exchange members. Divide each member by 2π.

$$\frac{2\pi r}{2\pi} = \frac{c}{2\pi}$$

Ans. $r = \dfrac{c}{2\pi}$

19. $d = rt$ **20.** $v = k + gt$ **21.** $v = lwh$ **22.** $f = ma$
23. $c = \pi d$ **24.** $I = prt$ **25.** $d = rt$ **26.** $v = lwh$
27. $v = lwh$ **28.** $f = ma$ **29.** $I = prt$ **30.** $I = prt$
31. $v = k + gt$ **32.** $s = \dfrac{at^2}{2}$ **33.** $F = \dfrac{kmM}{d^2}$ **34.** $F = \dfrac{kmM}{d^2}$
35. $P = 2l + 2w$ **36.** $P = 2l + 2w$ **37.** $A = P + Prt$ **38.** $A = P + Prt$

■ *Using the given symbols, write a formula to represent each of the following relationships.*

39. The area A of a square is equal to the square of the length s of a side.

40. The perimeter P of a square is equal to four times the length s of a side.

41. The area A of a rectangle is equal to the product of its length l and its width w.

42. The area A of a triangle is equal to one half the product of its base b and altitude h.

43. The volume V of a rectangular prism is found by multiplying its length l times its width w times its height h.

44. The perimeter P of a rectangle is equal to the sum of twice its length l and twice its width w.

45. The interest I for one year on an investment is found by multiplying the amount invested P by the rate of interest r.

46. The force F exerted by an object is equal to its mass m times its acceleration a.

47. The current I in a wire is equal to the voltage E divided by the resistance R of the wire.

48. The voltage E across a wire is equal to the product of the current I and the resistance R of the wire.

49. The volume V of a cylinder is equal to the product of π and the square of its radius r and its height h.

50. The amount A of an investment is equal to the sum of the principal P and the product of the principal and the interest rate r and the duration t of the investment.

51. The area A of a trapezoid is equal to one-half the product of the height h and the sum of the bases, B and b.

52. The pressure P exerted by a gas is equal to a constant k times its temperature T divided by its volume V.

4.6 APPLICATIONS FROM GEOMETRY

The study of geometry gives rise to a number of formulas relating geometric quantities. In this section we use the formulas to solve certain applied problems. The suggestions given in Section 3.4 to help you solve word problems apply to many types of problems, including applications from geometry. In a problem that involves a geometric figure, a sketch can be very helpful as you try to write an equation relating known and unknown quantities.

The following relationships, which may be familiar to you from your studies in arithmetic, are stated here for reference.

1. Square

Perimeter: $P = 4s$
Area: $A = s^2$

s = side

2. Rectangle

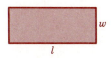

Perimeter: $P = l + l + w + w$
$P = 2l + 2w$
Area: $A = lw$

l = length
w = width

3. Triangle

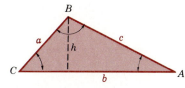

Perimeter: $P = a + b + c$
Area: $A = \dfrac{hb}{2}$
Sum of interior angles:
$\angle A + \angle B + \angle C = 180°$

b = base
h = height or altitude

(a) Isosceles triangle

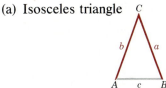

2 equal sides: $a = b$
2 equal angles: $\angle A = \angle B$

(b) Equilateral triangle

3 equal sides: $a = b = c$
3 equal angles: $\angle A = \angle B = \angle C$

(c) Right triangle

a = leg, b = leg, c = hypotenuse;
angle $C = 90°$

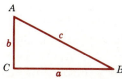

4. Angles

Acute angle: less than 90°

Right angle: equals 90°

Obtuse angle: greater than 90° and less than 180°

Straight angle: equals 180°

5. Circle

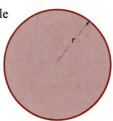

Diameter: $d = 2r$
Circumference: $C = 2\pi r$ or πd
Area: $A = \pi r^2$
(π is approximately equal to 3.14)

EXERCISES 4.6

■ *Evaluate each formula for the specified symbol. Use 3.14 for π.*

Sample Problem

$A = \dfrac{hb}{2}$.   Find h if $A = 16$ and $b = 4$.

> Substitute 16 for A and 4 for b.

$$16 = \frac{h(4)}{2}$$

> Divide each member by 4.

$$\frac{32}{4} = \frac{4h}{4}$$

Ans. $h = 8$

1. $C = 2\pi r$. Find C if $r = 3$.

2. $A = \pi r^2$. Find A if $r = 3$.

3. $A = lw$. Find A if $l = 2$ and $w = 6$.

4. $P = 2l + 2w$. Find P if $l = 2$ and $w = 6$.

5. $A = \dfrac{hb}{2}$. Find A if $h = 4$ and $b = 8$.

6. $\angle A + \angle B + \angle C = 180°$. Find $\angle A$ if $\angle B = 40°$ and $\angle C = 20°$.

7. $P = 4s$. Find s if $P = 12$.

8. $A = lw$. Find l if $A = 24$ and $w = 4$.

9. $P = 2l + 2w$. Find l if $P = 24$ and $w = 4$.

10. $A = \dfrac{hb}{2}$. Find h if $A = 40$ and $b = 10$.

11. $A = \dfrac{hb}{2}$. Find b if $A = 40$ and $h = 20$.

12. $C = 2\pi r$. Find r if $C = 20$.

■ *Solve. Use 3.14 for π.*

Sample Problem

Find the area of a circle with radius equal to 2 centimeters.

Use the formula relating the area and radius of a circle.

$A = \pi r^2$

Substitute 2 for r and 3.14 for π, and solve for A.

$A = 3.14(2)^2$

Ans. $A = 12.56$ square centimeters

13. What is the area of a triangle with a height of 6 centimeters and a base of 3 centimeters?

14. What is the area of a circle with a radius of length 5 centimeters?

15. What is the area of a rectangle with a length of 12 meters and a width of 10 meters?

16. What is the perimeter of a triangle if the sides are 3, 4, and 5 meters?

■ *Find the area of the shaded portion of each figure. All curves shown are parts of circles, and all horizontal and vertical lines meet at right angles.*

Sample Problem

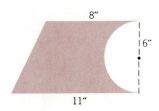

Consider the area of the three separate regions shown:

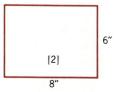

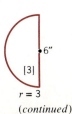

(*continued*)

$$A_{[1]} = \frac{(3)(6)}{2} = 9$$

$$A_{[2]} = (6)(8) = 48$$

$$A_{[3]} = \frac{3.14(3)^2}{2} = 14.13$$

The area of the shaded region is given by

$$A = A_{[1]} + A_{[2]} - A_{[3]}$$
$$= 9 + 48 - 14.13 = 42.87$$

Ans. 42.87 square inches

17.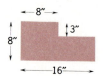

18.

19.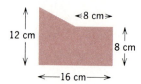

20.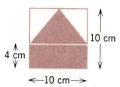

21.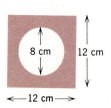

22.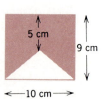

23.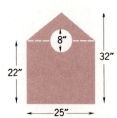

24.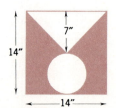

■ *Solve each of the following problems completely. Follow the six steps outlined on page 84.*

Sample Problem

The perimeter of a rectangle is 120 feet. The length is 10 feet greater than the width. What are the dimensions of the rectangle?

Steps 1–2   The width: x

The length: $x + 10$

Step 3

Step 4 $2l + 2w = P$

$2(x + 10) + 2x = 120$

Step 5 $2x + 20 + 2x = 120$

$4x + 20 = 120$

$4x = 100$

$x = 25$

Step 6   **Ans.** The width is 25 feet, and the length is $25 + 10$ or 35 feet.

25. A rectangle is 42 feet wide and has a perimeter of 210 feet. Find the length of the rectangle.

26. A rectangle is 26 meters long and has an area of 156 square meters. Find the width of the rectangle.

27. A triangle whose base is 16 centimeters has an area of 144 square centimeters. Find the altitude of the triangle.

28. A homeowner wishes to enclose a rectangular garden 18 meters longer than it is wide with 180 meters of wire fencing. What should be the dimensions of the garden?

29. A rectangle is 10 meters longer than it is wide, and its perimeter is 164 meters. What are its dimensions?

30. A tennis court for singles is 24 feet longer than twice its width, and its perimeter is 210 feet. Find its dimensions.

31. The perimeter of an isosceles triangle is 56 meters. The two equal sides are each 4 meters longer than the base. Find the length of each side.

32. One angle of a triangle is 10° larger than another, and the third angle is 29° larger than the smallest. How large is each angle?

33. One angle of a triangle is twice as large as another, and the third angle is 10° less than the larger of the other two. How large is each angle?

34. Two angles of a triangle are equal and the third is 20° less than the sum of the equal angles. How large is each angle?

35. The perimeter of a rectangle is 56 feet. What are its dimensions if the length is three times the width?

36. The length of a rectangle is five times its width. What are the dimensions of the rectangle if the perimeter is 36 feet?

37. How long is the side of a square whose perimeter is 24 feet?

38. The perimeter of a triangle is 104 inches. The second side is twice the first side, and the third side is 4 inches more than the second. How long is each side?

39. Two angles of a triangle are 40° and 70°. How large is the third angle?

40. Two angles of a triangle are equal, and the third angle is 40°. How large is each of the equal angles?

41. One angle of a triangle is 10° more than another, and the third is 20° more than the smallest. How large is each angle?

42. The largest angle of a triangle is three times the smallest, and the third angle is 30° smaller than the largest. How large is each angle?

43. One acute angle of a right triangle is twice the other acute angle. How large is each acute angle?

44. One acute angle of a right triangle is 10° less than three times the other acute angle. How large is each acute angle?

CHAPTER SUMMARY

[4.1] In the **power** a^n, where

$$a^n = a \cdot a \cdot a \cdot \ \cdot \ \cdot \ \cdot a \ (n \text{ factors}),$$

a is called the **base** and n is called the **exponent.**

[4.2] When a mathematical expression contains more than one operation, the operations *must be performed in a specified order* (see page 113).

We can *evaluate* algebraic expressions by replacing the variables with numbers and simplifying the resulting expression.

[4.3] A **polynomial** is the sum or difference of terms, where the exponents on the variables are natural numbers. Polynomials of one term, two terms, and three terms are called **monomials, binomials,** and **trinomials,** respectively. Any product of factors in a term is called the **coefficient** of the remaining factors. The **degree** of a term containing only one variable is the exponent on the variable.

To add or subtract polynomials, we add or subtract their corresponding like terms.

[4.4]   Products of powers with the same base can be rewritten in accordance with the **first law of exponents:**

$$a^m \cdot a^n \;\; = \;\; a^{m+n}$$

Powers of powers can be rewritten in accordance with the **second law of exponents:**

$$(a^m)^n \;=\; a^{mn}$$

Powers of products can be rewritten in accordance with the **third law of exponents:**

$$(ab)^n \;=\; a^n b^n$$

Powers of negative numbers may be simplified as follows:

$$(-x)^n \,=\, x^n \qquad \text{if } \mathbf{n} \text{ is an even integer}$$

$$(-x)^n \,=\, -x^n \qquad \text{if } \mathbf{n} \text{ is an odd integer.}$$

Many algebraic expressions involving variables can be simplified with the laws of exponents and order of operations.

[4.5]   We can solve a formula by substituting known values into the formula and solving for the unknown variable. We can also solve for a given variable in terms of the other variables.

[4.6]   The techniques for solving word problems can be applied to problems in geometry.

■ *The symbols introduced in this chapter appear on the inside front cover.*

CHAPTER REVIEW

1. Write in exponential form.

   a. $4aabbb$ b. $xyyzzz$ c. $(-3)(-3)ccd$

2. Write in completely factored form without exponents.

   a. $6xy^2$ b. a^3b^2 c. $-27cd^2$

3. Simplify:

   a. $\dfrac{3^2 - 1}{4} + \dfrac{2^3 + 1}{3}$ b. $\dfrac{6 - 2(-3)^2}{-2^2} + \dfrac{3 \cdot 4^2}{16 - 2^3}$

   c. $1 - (1 - 2^2)^2(2^2 - 2)^2$

4. If $a = 2$ and $b = 3$, find the value of each expression.

   a. $2(a + b)^2$ b. $2a^2 + 2b^2$ c. $(2a)^2 + (2b)^2$

5. If $x = -1$, $y = -2$, $z = 1$, find the value of each expression.

 a. $\dfrac{-x^2y}{-z}$ b. $\dfrac{x^2 - z^2}{-2y}$ c. $\dfrac{x^2 - y}{z}$

6. What is an algebraic expression consisting of two terms called?

7. In the expression $4x^3$, the number 4 is called the ? of x^3.

8. What is the numerical coefficient in the expression x^2?

9. What is the degree of the trinomial $2x^4 + x^2 + 6x$?

■ *Simplify.*

10. a. $-2x + x + 3x$

 b. $6x^2 + 2xy + 2x^2 - 3y^2 - 3xy$

 c. $2xy^2 + 3xy + 3x^2y - 2x^2y + xy^2$

11. a. $(3x^2 - 2x + 1) - (2x^2 - x + 2)$

 b. $(x^2 - 3x + 4) - (2x - 9)$

 c. $(2x^2 + 3y^2 - z^2) - (3x^2 + z^2)$

12. a. $(x + y + 2z) - (x - 2y + z)$

 b. $(2a + 3b - 4c) - (a + b + c)$

 c. $(x + y - 2z) + (2x + y - z) - (4x + 2y - 3z)$

13. a. $(xy^2)(-x^2y)$ b. $(3b^3)(-2a)(-2b)$ c. $(-r)^3(s^2)(-rs)$

14. a. $(2ab^2)^3(-3a^3b)^2$ b. $(-4x^2y^4)^2(3x^3)^3(-2xy)$ c. $(s^3t^5)^2(3st^4)^2(7t^5)$

15. a. $(4x - 1)^5(4x - 1)^4$ b. $[(4x - 1)^4]^5$ c. $(4x - 1)[(4x - 1)^2]^6$

16. a. $x^2(3x)^2 - x(4x^3) - (x^2)^2$

 b. $ab^2(2b^2) - a^2(ab)^2 - (3b^2)(-2a^3)$

 c. $2r(rs^2)^3 - (-3)^2(rs^3)(r^3s) - (r^2s^2)^2s^2$

17. Solve each of the following formulas for the symbol in color.

 a. $f = m\textcolor{red}{a}$ b. $v = k + g\textcolor{red}{t}$ c. $M = \dfrac{\textcolor{red}{a} + b}{2}$

18. Write a formula that states that the volume V of a sphere is equal to four thirds times the product of π and the cube of the radius r.

19. It takes 144 meters of wire to enclose a rectangular garden of width 34 meters. What is the length of the garden?

20. One angle of a triangle is 15° larger than a second angle, and the third angle is 30° larger than the smaller of the other two. How large is each angle?

CUMULATIVE REVIEW

1. Which of the following are natural numbers?

$$3^2 \qquad \frac{2^3}{3} \qquad \frac{4^2}{2} \qquad \frac{4 + 2^3}{12} \qquad 5^2 - 3^2$$

2. a. In the expression $3b^4$, the number 3 is called the $\underline{\ ?\ }$ of b^4.
   b. In the expression $3b^4$, the number 4 is called an $\underline{\ ?\ }$.

3. a. Write in exponential form: $2yyy$.
   b. Write in completely factored form without exponents: $36xy^3$.

4. If two numbers have like signs, the sign of their product is $\underline{\ ?\ }$.

5. If $x = 1$, $y = 2$, $w = -2$, find the value of $x^2y^2w^2$.

6. Simplify: **a.** $(x^3)(x^2)$.   **b.** $(x^3)^2$.

7. What is the value of $(-3)^4$? of -3^4?

8. a. Write $-(a - b + c)$ without parentheses.
   b. Write $-(a + b - c)$ without parentheses.

■ *Simplify.*

9. a. $\dfrac{3 + 2 \cdot 6}{3} - 2^2 + (-2)^2$ **b.** $(-2 \cdot 3)^2 + (-2 \cdot 3^2) - (2 \cdot 3^2)$

10. a. $(3x)(x^2)(-x)$ **b.** $3m^2(3mn^2)^2$

11. a. $(a - 2b + c) + (2a - b + c) - (a + b + c)$
    b. $(-7x^2 + 2x - 3) - (4x^2 - 6) + (x^2 - 5x + 12)$

12. a. $(2x)(x^2) - 3y^3 + x^3 + y^3$
    b. $(3x - x)^2 - (x - 3x)^2$

13. a. $(p^2q)^3(pq^3) - q^6$ **b.** $(3p^2)^3q^4 - 2q^4(p^3q)^2$

14. The number of feet s that an object falls in t seconds is given by the formula
    $s = 16t^2 + 24t$. How far will an object fall in 5 seconds?

15. Solve $E = h + kmv^2$ for m.

16. Solve the inequalities and graph their solution sets.

    a. $7 - 5y \le 19 - 3y$ **b.** $\dfrac{-3y}{4} + 2 > -10$

17. Solve each equation.

    a. $2(3a + 4) - 5(2a - 1) = 3(5a - 2)$

    b. $\dfrac{3(5x - 3x)}{2} = 18 - 3x$

18. The sum of three consecutive even integers is -48. What are the integers?

19. A train travels at 40 miles per hour on the first leg of its journey, and at 60 miles per hour for the remainder. The whole trip of 880 miles takes 16 hours. How long does the first portion take?

20. Eliza mixes eau de cologne that costs $12 a quart with dish soap worth $2 a quart to make 20 quarts of bubble bath worth $4 a quart. How much of each ingredient does she use?

■ *True or false.*

21. $(3x)^2$ is equivalent to $9x^2$.
22. $-x^2$ is equivalent to $(-x)^2$.
23. $3 + 2x$ is equivalent to $5x$.
24. $x^2 \cdot x^3$ is equivalent to x^5.
25. $-(4 - x)$ is equivalent to $-4 - x$.

5 PRODUCTS AND FACTORS

In this chapter we learn to multiply polynomials together and to reverse the process.

5.1 MULTIPLYING BY MONOMIALS

In Section 3.1 we used the distributive law to perform products like

$$3(5x - 2) = 15x - 6.$$

The foregoing example is the product of a monomial, 3, and a binomial, $5x - 2$. The distributive law also applies to products of polynomials and variable monomials. For example,

$$-3xy^2(4x^2 - 2xy + 2) = -12x^3y^2 + 6x^2y^3 - 6xy^2$$

and

$$2x(4x - y - 2) - 3y(x + 5y - 1) = 8x^2 - 2xy - 4x - 3xy - 15y^2 + 3y$$
$$= 8x^2 - 4x - 5xy + 3y - 15y^2.$$

EXERCISES 5.1

■ *Multiply.*

Sample Problems

a. $2x(x - 3)$

Ans. $2x^2 - 6x$

b. $-y(y^2 + 3y - 4)$

Ans. $-y^3 - 3y^2 + 4y$

c. $xy(x - 2y + 1)$

Ans. $x^2y - 2xy^2 + xy$

1. $2a(5a + 3)$

2. $3a(3a - 1)$

3. $-b(b - 2)$

4. $-3b(2b + 1)$

5. $(x + y)(xy)$

6. $(x - 2y)(xy)$

143

7. $-x^2(2x + 3y)$ **8.** $-y^2(x - 2y^2)$ **9.** $(x^2 - 2x + 1)(x)$

10. $(y^2 + 3y + 4)(y)$ **11.** $-y(y^2 - y + 2)$ **12.** $-x(x^2 + 3x - 2)$

13. $4x^3(x^2 - 3x + 4)$ **14.** $2y^3(y^2 + y - 2)$ **15.** $-y^3(y^3 - y + 1)$

16. $-x^3(y^3 + 2y^2 - 1)$ **17.** $-xy(x^2 + xy + y^2)$ **18.** $-xy(2x^2 - xy + 3y^2)$

19. $2x^2(3x^3 - 6x - 5)$ **20.** $4x^5(x^4 + 4x^2 + 2)$

21. $5a^2(-3a^2 + 2ab - b^2)$ **22.** $3b^3(2a^2 - 6ab + b^2)$

23. $-4p^2q(2p^3q - 3p^2q^2 + pq^3)$ **24.** $-3s^2t^2(6 - 2st + 3s^2t^2)$

25. $(3 + 2x^5 - 7y^4)(-x^2y^3)$ **26.** $(4z^4 - 8 - 3y^2)(-y^3z^2)$

27. $(-6x^3)(-3x^5 + 5x^4 - x^2 + 2x - 1)$

28. $(5y^4)(8y^7 - 6y^5 + 3y^4 + y^3 - 7y + 2)$

29. $(a^2b^3 - 6ab^5 + 3a^3b^2 - 8b^9)(-6a^3b^2)$

30. $(2p^4q - 3p^3q^2 + 5 - pq^3 + 7p^2q^5)(4pq^4)$

Sample Problems

a. $\overset{\frown}{c(y} - 3) + 2cy$ b. $\overset{\frown}{a(3} - a) - \overset{\frown}{2(a} + a^2)$

                           Apply the distributive law.

$cy - 3c + 2cy$ $3a - a^2 - 2a - 2a^2$

                           Combine like terms.

Ans. $3cy - 3c$ *Ans.* $a - 3a^2$

31. $-a(x + 1) + ax$ **32.** $by - b(1 - y)$

33. $a(x + 1) + x(a + 1)$ **34.** $2a(x + 3) - 3a(x - 3)$

35. $a(x + y) - 2(ax + y)$ **36.** $4x(3 + 2y) - 3(2x + y)$

37. $3(x^2 + 2x - 1) - 2(x^2 + x - 2)$ **38.** $ax(x^2 + 2x - 3) - a(x^3 + 2x^2)$

39. $3(x - 2y) - 2(x + 3y) + 2x$ **40.** $2x(3 - x) + 2(x^2 + 1) - 2$

41. $3y(2y - 5) + 2y^2 - 5(y^2 + 2y)$ **42.** $3(y^2 - 2y + 1) + 3(1 - y^2)$

43. $3(ax^2 + ax - a) - 2a(x^2 - 1)$

44. $3x^2(a - b + c) - 2x(ax - bx + cx)$

45. $-3ab(x + y - 2) - 2a(bx - by + 2b) + b(ax + 1)$

46. $4xy(2a - b + 1) + 6y(xb + xa - 5x) - 3bx(a + y - 5ay)$

47. $3ab^2(2 + 3a) - 2ab(3ab + 2b) - 2b^2(a^2 - 2a)$

48. $8x^2y(y - 3) + 5y^2(x - 3x^2) - xy(2y - 6xy)$

49. $3p^2q^2(p^3q - 3pq^2) - 4pq^4(p^4 - 5p^2)$

50. $-2p^3q^2(-3pq^2 + p^2q^3) - 5p^2q^3(6p^2q^2 - 3p^2q)$

5.2 FACTORING MONOMIALS FROM POLYNOMIALS

From the symmetric property of equality, we know that if

$$a(b + c) = ab + ac, \quad \text{then} \quad ab + ac = a(b + c).$$

Thus, if there is a monomial factor common to all terms in a polynomial, we can write the polynomial as the product of this **common factor** and another polynomial. For instance, since each term in $x^2 + 3x$ contains x as a factor, we can write the expression as the product $x(x + 3)$. Rewriting a polynomial in this way is called **factoring,** and the number x is said to be factored "from" or "out of" the polynomial $x^2 + 3x$.

We can usually find the common factor by considering the numerical coefficients and each variable separately. For example, to factor

$$4x^3 - 10x^2 + 2x \qquad\qquad (1)$$

we first note that 2 is a factor of the numerical coefficient of each term. The largest power of x that is a factor of each term is x^1, or x. Thus $2x$ is a common factor for the polynomial, so we write

$$2x(\qquad).$$

Now we find the missing factor for each term of the polynomial:

$$2x \cdot ? \quad = \quad 4x^3; \quad 2x \cdot 2x^2 = \quad 4x^3,$$

$$2x \cdot ? \quad = -10x^2; \quad 2x(-5x) = -10x^2,$$

$$2x \cdot ? \quad = \quad 2x; \quad 2x \cdot 1 \quad = \quad 2x,$$

and write the missing factors inside the parentheses to get

$$2x(2x^2 - 5x + 1).$$

This product is called the *factored form* of polynomial (1) above.

To factor a monomial from a polynomial:

1. Write a set of parentheses preceded by the monomial common to each term in the polynomial.
2. Find the missing factor for each term of the polynomial, and write the missing factors in the parentheses.

We can check that we factored correctly by multiplying the factors and verifying that the product is the original polynomial. Using the previous example we note that

$$2x(2x^2 - 5x + 1) = 4x^3 - 10x^2 + 2x.$$

Common Error

The factored form of a polynomial has only *one term*; it is expressed as a *product* of simpler polynomials. Thus

$$2 \cdot 2 \cdot x \cdot x \cdot x - 2 \cdot 5 \cdot x \cdot x + 2 \cdot x$$

is *not* considered a factored form for polynomial (1) above, because it consists of three terms.

In this book, we will restrict the common factors to monomials consisting of integer coefficients and to positive integer powers of the variables. The choice of sign for the monomial factor is a matter of convenience. Thus,

$$-3x^2 - 6x$$

can be factored either as

$$-3x(x + 2) \qquad \text{or as} \qquad 3x(-x - 2).$$

The first form is usually more convenient.

EXERCISES 5.2

■ *Find the missing factor.*

Sample Problems

a. $6x^3y^2 = 2xy \cdot$?
   $6x^3y^2 = 2xy \cdot 3x^2y$

Ans. $3x^2y$

b. $-72a^2b^5 = -8b^3 \cdot$?
   $-72a^2b^5 = -8b^3 \cdot 9a^2b^2$

Ans. $9a^2b^2$

1. $9x^3 = 3x \cdot$?
2. $8x^4 = 4x \cdot$?
3. $12x^2y = 6xy \cdot$?
4. $15x^3y^2 = 5xy \cdot$?
5. $-16a^3b^3 = -8ab^2 \cdot$?
6. $-18a^4b^2 = -3ab \cdot$?
7. $-24a^2b^3 = 6a^2b \cdot$?
8. $-36a^3b^3 = 9a^3b \cdot$?

■ *Factor.*

Sample Problems

a. $4x + 4y$

$4(\qquad)$

Ans. $4(x + y)$

b. $3xy - 6y$

Factor out the common monomial.

$3y(\qquad)$

Find the missing factor for each term.

Ans. $3y(x - 2)$

Sample Problems

9. $3x + 6$
10. $4x - 8$
11. $2x - 6y$
12. $10x + 5y$
13. $2y^2 - 2y$
14. $3y^2 + 6y$
15. $ay^2 + y$
16. $by^2 - b$
17. $9ay^2 + 6y$
18. $4bx^2 - 12x$
19. $28x^3 + 35x^2$
20. $24y^4 - 18y^2$
21. $4x^3 + 20x^4$
22. $18y^3 - 6y^2$
23. $24xy^2 - 27x^2y$
24. $32a^2b^2 + 24a^2b$
25. $72t^6 + 40t^4$
26. $63q^7 - 36q^3$
27. $28a^2b^3 - 44a^4b^2$
28. $60x^4y^6 + 54xy^3$

Sample Problems

a. $2x^3 - 4x^2 + 8x$ b. $3y^3 - 6y^2 + 9y$

                                Factor out the common monomial.

$2x(\qquad)$ $3y(\qquad)$

                                Find the missing factor for each term.

Ans. $2x(x^2 - 2x + 4)$ *Ans.* $3y(y^2 - 2y + 3)$

29. $3y^2 - 3y + 3$ **30.** $2y^2 - 4y - 2$ **31.** $ax + ay - az$

32. $2bx - 6by + 4bz$ **33.** $x^2 - 3x + xy$ **34.** $y^2 + 2y - 4xy$

35. $4y^3 - 2y^2 + 2y$ **36.** $6x^3 + 6x^2 - 9x$

37. $6ax^2y - 18axy^2 + 24axy$ **38.** $3a^2x^2y - 12ax^2y^2 + 9ax^2y$

39. $14x^3y - 35x^2y^2 + 21xy^3$ **40.** $45x^2y^2 + 18x^2y^3 - 27x^3y^3$

41. $25a^2b^3 - 5ab - 30a^3b^2$ **42.** $18a^3b - 36a^3b^2 - 72a^4b^2$

43. $48t^7 + 27t^5 - 45t^{11}$ **44.** $30w^9 - 42w^4 + 54w^8$

45. $16p^2q^3 - p^2q + 11p^3q^2 - 4p^4q^2$ **46.** $p^5q^5 - p^4q^6 + 2p^3q^3 - p^6q^4$

47. $3a^2bc^3 - 12ab^2c^2 + 15a^5b^3c^2$ **48.** $8a^2b^7c^5 - 80a^2b^8c^3 - 40a^3b^3c^9$

■ *Factor out a negative monomial.*

Sample Problems

a. $-3x^2 - 3xy$ b. $-x^3 - x^2 + x$

                                Factor out the common monomial, including -1.

$-3x(\qquad)$ $-x(\qquad)$

                                Find the missing factor for each term.

Ans. $-3x(x + y)$ *Ans.* $-x(x^2 + x - 1)$

49. $-a^2 - ab$ **50.** $-a^2 - a$ **51.** $-x - x^2$

52. $-ab - ac$ **53.** $-abc - ab - bc$ **54.** $-b^2 - bc - ab$

55. $-6y^3 - 3y^2 - 3y$ **56.** $-2x^2 - 4x - 2$ **57.** $-x + x^2 - x^3$

58. $-3x^2 + 3xy - 3x$ **59.** $-xy^5 - xy^4 + xy^2$ **60.** $-x^2y + xy^2 - 3xy$

■ *Express in factored form.*

Sample Problems

a. $a(x - 2) + 3(x - 2)$ b. $x^2(2x + 1) - 3x(2x + 1)$

                                Factor out the common binomial.

$(x - 2)(\qquad)$ $(2x + 1)(\qquad)$

Ans. $(x - 2)(a + 3)$ *Ans.* $(2x + 1)(x^2 - 3x)$

61. $6(x - 4) + a(x - 4)$

62. $b(3 - x) - 5(3 - x)$

63. $2x(x + 6) - 3(x + 6)$

64. $8x(x - 1) - 3(x - 1)$

65. $4(3x - 2) - 3x(3x - 2)$

66. $9(2x + 5) - x(2x + 5)$

67. $3x^2(2x + 3) - (2x + 3)$

68. $4x^2(6x - 5) + (6x - 5)$

69. $x^2(4 - x) + 3x(4 - x) - 2(4 - x)$

70. $5(x + 2) - 6x(x + 2) - x^2(x + 2)$

■ *Factor the right-hand member of each of the following equations.*

Sample Problems

a. $A = P + PRT$

$A = P(\qquad)$

Ans. $A = P(1 + RT)$

b. $S = 4kR^2 - 4kr^2$

Factor out the common monomial.

$S = 4k(\qquad)$

Ans. $S = 4k(R^2 - r^2)$

71. $d = k + kat$

72. $A = kR^2 + 2kr^2$

73. $S = kr^2h + kr^2$

74. $R = r + rat$

75. $V = 2ga^2D - 2ga^2d$

76. $L = 2an + n^2 - nd$

77. $A = ar^2 + br^2 + cr^2$

78. $S = 2kr^2 + 2krh$

5.3   POLYNOMIAL PRODUCTS I

We can also use the distributive law to multiply two polynomials. We begin by considering the product of two binomials. Although we will be interested primarily in multiplying expressions containing variables, we first illustrate the method on an arithmetic problem.

$$14 \cdot 12 = (10 + 4) \ (10 + 2) = 10 \cdot 10 + 10 \cdot 2 + 4 \cdot 10 + 4 \cdot 2$$

$$= 100 + 20 + 40 + 8$$

$$= 168.$$

Notice that *each* term of the first binomial is multiplied by *each* term of the second binomial. Now consider this procedure for an expression containing variables.

$$(x - 2) \quad (x + 3) = x^2 + 3x - 2x - 6$$

$$= x^2 + x - 6.$$

With practice, you will be able to add the second and third products mentally. This process for multiplying binomials is sometimes called the FOIL method. F, O, I, and L stand for:

1. the product of the First terms;
2. the product of the Outer terms;
3. the product of the Inner terms;
4. the product of the Last terms.

Common Errors

Note that

$$(x + 3)^2 = (x + 3) \quad (x + 3)$$

$$= x^2 + 3x + 3x + 9$$

$$= x^2 + 6x + 9.$$

Thus,

$$(x + 3)^2 \neq x^2 + 3^2.$$

Similarly,

$$(x - 3)^2 \neq x^2 - 3^2.$$

In general,

$$(a + b)^2 \neq a^2 + b^2 \quad \text{and} \quad (a - b)^2 \neq a^2 - b^2.$$

The method described above can easily be generalized to other types of polynomial products. For example,

$$(x - 2)(x^2 + 2x + 1) = x^3 + 2x^2 + x - 2x^2 - 4x - 2$$

$$= x^3 - 3x - 2$$

Notice again that each term of the second polynomial is multiplied by each term of the first polynomial.

A product of three polynomials can be performed by first multiplying two of the factors together, and then multiplying that product by the third factor. For example,

$$(x - 2)(x + 1)(x - 3) = (x - 2)[(x + 1)\ (x - 3)]$$

$$= (x - 2)[x^2 - 3x + x - 3]$$

$$= (x - 2)[x^2 - 2x - 3]$$

$$= x^3 - 2x^2 - 3x - 2x^2 + 4x + 6$$

$$= x^3 - 4x^2 + x + 6.$$

EXERCISES 5.3

■ *Write each product as an expression without parentheses.*

Sample Problems

a.

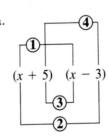

$(x + 5)\ (x - 3)$

$x^2 - 3x + 5x - 15$

Ans. $x^2 + 2x - 15$

b.

$(x - 3)\ (x + 3)$

$x^2 + 3x - 3x - 9$

Ans. $x^2 - 9$

Apply distributive law.

Simplify.

1. $(x + 3)(x + 4)$ 2. $(x - 2)(x - 3)$ 3. $(y - 3)(y + 1)$
4. $(y + 4)(y - 2)$ 5. $(a + 5)(a + 2)$ 6. $(a - 5)(a - 2)$
7. $(b - 4)(b + 2)$ 8. $(b + 5)(b - 3)$ 9. $(x + 1)(x + 8)$
10. $(x - 2)(x - 9)$ 11. $(y - 1)(y - 7)$ 12. $(y + 1)(y - 6)$
13. $(a + 4)(a + 4)$ 14. $(a - 3)(a - 3)$ 15. $(b - 5)(b + 5)$
16. $(b + 7)(b - 7)$ 17. $(x + 1)(x + 1)$ 18. $(x - 9)(x - 9)$

19. $(y - 1)(y + 1)$ **20.** $(y + 4)(y - 4)$ **21.** $(2 + x)(2 - x)$

22. $(5 - x)(5 - x)$ **23.** $(6 + y)(6 - y)$ **24.** $(9 - y)(9 + y)$

Sample Problems

a. $(x + 6)^2$ b. $(y - 3)^2$

    $(x + 6)(x + 6)$ $(y - 3)(y - 3)$ Rewrite expression.

      $x^2 + 6x + 6x + 36$ $y^2 - 3y - 3y + 9$ Apply distributive law.

                                                       Simplify.

Ans. $x^2 + 12x + 36$ *Ans.* $y^2 - 6y + 9$

25. $(x + 4)^2$ **26.** $(y - 5)^2$ **27.** $(x - 7)^2$ **28.** $(y + 1)^2$

29. $(x - 1)^2$ **30.** $(y + 8)^2$ **31.** $(x + 2)^2$ **32.** $(y - 10)^2$

Sample Problems

a. $(x - 2b)(x + 3b)$ b. $(y - a)(y + a)$

    $x^2 + 3bx - 2bx - 6b^2$ $y^2 + ay - ay - a^2$ Apply distributive law.

                                             Simplify.

Ans. $x^2 + bx - 6b^2$ *Ans.* $y^2 - a^2$

33. $(x - 3b)(x - b)$ **34.** $(x - a)(x - 2a)$ **35.** $(x + 2y)(x - y)$

36. $(x + b)(x + 2b)$ **37.** $(x + 2a)(x + 2a)$ **38.** $(x + 3b)(x + 3b)$

39. $(a - b)^2$ **40.** $(x - y)^2$ **41.** $(y - 6a)(y + 6a)$

42. $(x + 3z)(x - 3z)$ **43.** $(x - t)(x + t)$ **44.** $(y - c)(y + c)$

Sample Problems

a. $3(x - 2)(x + 3)$ b. $a(a + 1)(a + 3)$

    $3(x^2 + 3x - 2x - 6)$ $a(a^2 + 3a + a + 3)$ Multiply binomial factors.

    $3(x^2 + x - 6)$ $a(a^2 + 4a + 3)$ Simplify.

                                           Multiply by monomial.

Ans. $3x^2 + 3x - 18$ *Ans.* $a^3 + 4a^2 + 3a$

45. $2(x + 1)(x + 2)$ **46.** $4(x - 3)(x + 2)$ **47.** $6(y + 5)(y + 5)$

48. $3(y - 7)(y + 1)$ **49.** $6(x - 1)^2$ **50.** $3(y + 3)^2$

51. $a(a - 1)(a + 5)$ **52.** $b(b - 2)(b + 7)$ **53.** $a(a - 2)(a + 2)$

54. $b(b + 3)(b - 3)$ **55.** $x(y - 3)^2$ **56.** $x^2(y - 4)^2$

Sample Problems

a. $(x - 4)(x^2 - x + 3)$

Apply distributive law.

$x^3 - x^2 + 3x - 4x^2 + 4x - 12$

Add like terms.

Ans. $x^3 - 5x^2 + 7x - 12$

b. $(x + 1)[(x - 3) \ (x - 1)]$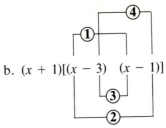

Multiply two binomials.

$(x + 1)[x^2 - 4x + 3]$

Apply distributive law.

$x^3 - 4x^2 + 3x + x^2 - 4x + 3$

Add like terms.

Ans. $x^3 - 3x^2 - x + 3$

57. $(x + 1)(x^2 + 3x + 2)$

58. $(x + 2)(x^2 + x + 4)$

59. $(x - 1)(x^2 + x + 1)$

60. $(x + 1)(x^2 - x + 1)$

61. $(x - 2)(x^2 - 3x + 2)$

62. $(x - 3)(x^2 + 2x - 1)$

63. $(x + 1)(x + 2)(x + 3)$

64. $(x + 2)(x + 3)(x + 4)$

65. $(x - 2)(x - 2)(x - 2)$

66. $(x - 1)(x - 1)(x - 1)$

67. $(x - 3)(x + 2)(x - 1)$

68. $(x - 2)(x - 4)(x + 1)$

69. Use a numerical example to show that $(x + 3)^2$ is not equivalent to $x^2 + 3^2$.

70. Use a numerical example to show that $(x - 5)^2$ is not equivalent to $x^2 - 5^2$.

5.4   FACTORING TRINOMIALS I

In Section 5.3, we saw how to find the product of two binomials. Now we will reverse this process. That is, given the product of two binomials, we will find the binomial factors. The process involved is another example of factoring. As before, we will only consider factors in which the terms have integral numerical coefficients. Such factors do not always exist, but we will study the cases where they do.

Consider the following product.

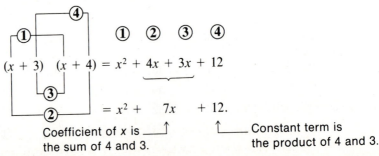

Coefficient of x is the sum of 4 and 3.

Constant term is the product of 4 and 3.

Notice that the first term in the trinomial, x^2, is product ①; the last term in the trinomial, 12, is product ④; and the middle term in the trinomial, $7x$, is the *sum* of products ② and ③.

In general,

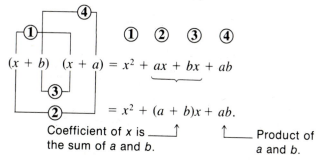

$$= x^2 + (a + b)x + ab.$$

Coefficient of x is ⌐‾‾↑   ↑‾‾⌐ Product of
the sum of a and b. a and b.

That is, the sum of a and b is the coefficient of x, and their product is the constant term.

We can use this information to factor any trinomial of the form $x^2 + Bx + C$. For example, to factor $x^2 + 7x + 12$ we look for two integers a and b whose *sum* is 7 and whose *product* is 12. If we find these integers we will then be able to factor the trinomial as

$$x^2 + 7x + 12 = (x + a)(x + b).$$

To find a and b we list all possible pairs of integers whose product is 12 and check to see which pair (if any) has a sum of 7.

Pairs of Factors	Product of Factors	Sum of Factors
12, 1	12	13
−12, −1	12	−13
3, 4	12	7
−3, −4	12	−7
2, 6	12	8
−2, −6	12	−8

We see that the only pair of factors whose product is 12 and whose sum is 7 is 3 and 4. Thus,

$$x^2 + 7x + 12 = (x + 3)(x + 4).$$

As another example, let us try to factor $x^2 + 5x + 12$. We look for two integers whose product is 12 and whose sum is 5. From the table in the previous example, we see that there is no pair of factors whose product is 12 and whose sum is 5. In this case, the trinomial is *not factorable*.

FURTHER TECHNIQUES

Note that when all terms of a trinomial are positive as in the previous example, we need only consider pairs of *positive* factors because we are looking for a pair of factors whose product and sum are positive. That is, the factored form of

$$x^2 + 7x + 12$$

would be of the form

$$(\quad + \quad)(\quad + \quad),$$

as in the previous examples.

When the first and third terms of a trinomial are positive but the middle term is negative, we need only consider pairs of *negative* factors because we are looking for a pair of factors whose product is positive but whose sum is negative. That is, the factored form of

$$x^2 - 7x + 12$$

would be of the form

$$(\quad - \quad)(\quad - \quad),$$

so we need only consider pairs of integers a and b that are both negative.

Pairs of Factors	Product of Factors	Sum of Factors
−12, −1	12	−13
−3, −4	12	−7
−2, −6	12	−8

From the table we see that

$$x^2 - 7x + 12 = (x - 3)(x - 4).$$

Finally, when the first term of a trinomial is positive and the third term is negative, the signs in the factored form are opposite. That is, the factored form of

$$x^2 - x - 12$$

would be of the form

$$(\quad + \quad)(\quad - \quad) \quad \text{or} \quad (\quad - \quad)(\quad + \quad),$$

so we need only consider pairs of integers a and b that are of opposite sign.

Pairs of Factors	Product of Factors	Sum of Factors
12, −1	−12	11
−12, 1	−12	−11
3, −4	−12	−1
−3, 4	−12	1
2, −6	−12	−4
−2, 6	−12	4

From the table we see that

$$x^2 - x - 12 = (x + 3)(x - 4).$$

With practice you will be able to list the pairs of factors and check their sums mentally and then write your answer directly.

It is easier to factor a trinomial completely if any common monomial factor is first factored out as in Section 5.2. For example, to factor

$$12x^2 + 36x + 24$$

we first factor out the common factor 12 to get

$$12(x^2 + 3x + 2).$$

Then we factor the trinomial by looking for two integers whose product is 2 and whose sum is 3. The integers are 2 and 1, so we may rewrite the trinomial as

$$12(x + 2)(x + 1),$$

which is said to be in **completely factored form.** In such cases it is not necessary to factor the numerical factor itself, that is, we do not write 12 as $2 \cdot 2 \cdot 3$.

We can check the results of a factorization by multiplying the binomial factors and verifying that the product is equal to the given trinomial. Thus, in the previous example, we would multiply the factors to obtain

$$12(x + 2)(x + 1) = 12(x^2 + 3x + 2)$$

$$= 12x^2 + 36x + 24.$$

EXERCISES 5.4

■ *Factor completely.*

Sample Problem

$x^2 - 5x + 6$

Because the third term is positive and the middle term is negative, find two negative integers whose product is 6 and whose sum is -5. List the possibilities.

Pairs of Factors	*Product*	*Sum*
$-6, -1$	6	-7
$-3, -2$	6	-5

The two integers are -3 and -2.

Ans. $(x - 3)(x - 2)$

Check by multiplying.

Check. $(x - 3)(x - 2) = x^2 - 2x - 3x + 6$
$= x^2 - 5x = 6$

1. $x^2 + 5x + 6$ **2.** $x^2 + 9x + 20$ **3.** $x^2 + 11x + 30$

4. $x^2 + 20x + 100$ **5.** $x^2 + 14x + 45$ **6.** $y^2 - 8y + 15$

7. $y^2 - 3y + 2$ **8.** $y^2 - 14y + 13$ **9.** $y^2 - 16y + 63$

10. $x^2 - 46x + 45$

Sample Problem

$x^2 + x - 2$

Find two integers whose product is -2 and whose sum is 1. The two integers are 2 and -1.

Ans. $(x + 2)(x - 1)$

Check by multiplying.

Check. $(x + 2)(x - 1) = x^2 - x + 2x - 2$
$$= x^2 + x - 2$$

11. $x^2 - x - 12$ **12.** $x^2 - 3x - 10$ **13.** $y^2 + y - 20$

14. $y^2 + 2y - 8$ **15.** $a^2 + 2a - 35$ **16.** $a^2 + 8a - 20$

17. $b^2 - 19b - 20$ **18.** $b^2 - 4b - 12$ **19.** $a^2 - 5a - 50$

20. $a^2 - a - 72$ **21.** $b^2 - 4b - 45$ **22.** $b^2 - 12b - 45$

23. $y^2 - 44y - 45$ **24.** $y^2 - 14y - 51$

Sample Problems

a. $3x^2 + 12x + 12$

$3(x^2 + 4x + 4)$

b. $2b^3 - 8b^2 - 10b$

Factor out the common monomial.

$2b(b^2 - 4b - 5)$

Factor the trinomial.

Ans. $3(x + 2)(x + 2)$

Ans. $2b(b - 5)(b + 1)$

Check by multiplying.

Check. $3(x + 2)(x + 2)$
$$= 3(x^2 + 4x + 4)$$
$$= 3x^2 + 12x + 12$$

Check. $2b(b - 5)(b + 1)$
$$= 2b(b^2 - 4b - 5)$$
$$= 2b^3 - 8b^2 - 10b$$

25. $2x^2 + 10x + 12$ **26.** $3a^2 - 3a - 18$

27. $y^3 - 2y^2 - 3y$ **28.** $b^3 + 2b^2 + b$

29. $5c^2 - 25c + 30$ **30.** $2a^2 - 38a - 40$

31. $4a^2b + 12ab - 72b$ **32.** $3x^2y - 6xy - 105y$

Sample Problem

$x^2 + 5xy + 6y^2$

Find two positive factors whose product is $6y^2$ and whose sum is $5y$. The two factors are $3y$ and $2y$.

Ans. $(x + 3y)(x + 2y)$

Check by multiplying.

Check. $(x + 3y)(x + 2y) = x^2 + 2xy + 3xy + 6y^2$
$$= x^2 + 5xy + 6y^2$$

33. $x^2 + 4ax + 4a^2$ **34.** $x^2 - 3xy + 2y^2$ **35.** $a^2 - 3ab + 2b^2$

36. $r^2 + 4rx + 3x^2$ **37.** $s^2 + 5as + 6a^2$ **38.** $x^2 + 15xy + 36y^2$

Sample Problem

$6 - 5x - x^2$

Write in decreasing powers of x.

$-x^2 - 5x + 6$

Since the coefficient of x^2 is -1, factor out -1 from each term.

$-1(x^2 + 5x - 6)$

Factor the trinomial.

Ans. $-(x + 6)(x - 1)$

Check by multiplying.

Check. $-(x + 6)(x - 1)$
$$= -(x^2 + 5x - 6)$$
$$= -x^2 - 5x + 6 \quad \text{or} \quad 6 - 5x - x^2$$

39. $10 + 7y + y^2$ **40.** $3 - 11y + y^2$ **41.** $8 - 9x + x^2$

42. $81 + 18x + x^2$ **43.** $32 - 12z + z^2$ **44.** $56 - 15y + y^2$

45. $21 - 4x - x^2$ **46.** $6 - x - x^2$ **47.** $24 + 10z - z^2$

48. $63 - 2y - y^2$ **49.** $18 + 7y - y^2$ **50.** $54 + 3y - y^2$

■ *Factor completely; if the trinomial is not factorable, so state.*

51. $x^2 + 4x + 2$ **52.** $x^2 + 3x + 1$ **53.** $y^2 - 4y + 2$

54. $y^2 - 2y - 2$ **55.** $x^2 + 12x + 30$ **56.** $x^2 + 11x + 20$

57. $x^2 + 5xy + 4y^2$ **58.** $x^2 - 2xy + 4y^2$ **59.** $7 - 3y + y^2$

60. $5 - 2y + y^2$

5.5 POLYNOMIAL PRODUCTS II

In this section we use the procedures developed in Section 5.3 to multiply polynomial factors whose highest-degree terms have numerical coefficients other than 1 or -1. Recall that the FOIL method can be used to find the product of two binomials. For example,

$$(3x + 4)\ (2x - 1) = 6x^2 - 3x + 8x - 4$$
$$= 6x^2 + 5x - 4.$$

For other types of products, we also use the distributive law. Thus

$$(2x + 1)(3x^2 - x + 2) = 6x^3 - 2x^2 + 4x + 3x^2 - x + 2$$
$$= 6x^3 + x^2 + 3x + 2,$$

and

$$(3x - 1)(x + 2)(2x - 1) = (3x - 1)[(x + 2)\ (2x - 1)]$$
$$= (3x - 1)[2x^2 + 3x - 2]$$
$$= 6x^3 + 9x^2 - 6x - 2x^2 - 3x + 2$$
$$= 6x^3 + 7x^2 - 9x + 2.$$

EXERCISES 5.5

■ *Write as a polynomial.*

Sample Problems

a. $(2x - 3)(x + 1)$

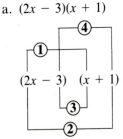

$$(2x - 3)\ (x + 1)$$

$$2x^2 + 2x - 3x - 3$$

Ans. $2x^2 - x - 3$

b. $(3x - 2y)(3x + y)$

$$(3x - 2y)\ (3x + y)$$

Apply the distributive law.

$$9x^2 + 3xy - 6xy - 2y^2$$

Combine like terms.

Ans. $9x^2 - 3xy - 2y^2$

1. $(2x + 1)(x + 3)$ **2.** $(4x - 2)(x - 1)$ **3.** $(3y - 2)(y + 1)$

4. $(y + 5)(5y - 2)$ **5.** $(3y + 1)(2y + 3)$ **6.** $(4y - 1)(3y + 2)$

7. $(5x - 2)(4x + 3)$ **8.** $(6x - 1)(2x + 5)$ **9.** $(2x + 3)(2x - 5)$

10. $(3x - 1)(2x + 1)$ **11.** $(4y - 3)(2y + 3)$ **12.** $(6y - 5)(6y + 3)$

Sample Problems

a. $(3x + 2)^2$ b. $(2x - y)^2$

$\qquad\qquad\qquad\qquad\qquad\qquad$ Rewrite as the product of binomials.

$\quad (3x + 2)(3x + 2)$ $(2x - y)(2x - y)$

$\qquad\qquad\qquad\qquad\qquad\qquad$ Multiply as indicated.

$\quad 9x^2 + 6x + 6x + 4$ $4x^2 - 2xy - 2xy + y^2$

$\qquad\qquad\qquad\qquad\qquad\qquad$ Combine like terms.

Ans. $9x^2 + 12x + 4$ *Ans.* $4x^2 - 4xy + y^2$

13. $(2x + 1)^2$ **14.** $(3x + 1)^2$ **15.** $(5x + 2)^2$ **16.** $(2x - 3)^2$

17. $(4y + 5)^2$ **18.** $(3y + 7)^2$ **19.** $(x - 2y)^2$ **20.** $(2x - 3y)^2$

21. $(3x - y)^2$ **22.** $(3x - 2y)^2$ **23.** $(8x + 3y)^2$ **24.** $(2x + y)^2$

25. $(2x + 3y)^2$ **26.** $(3x - 4y)^2$

Sample Problems

a. $(2x - 3)(2x + 3)$ b. $(3x - y)(3x + y)$

$\qquad\qquad\qquad\qquad\qquad\qquad\qquad$ Multiply as indicated.

$\quad 4x^2 + 6x - 6x - 9$ $9x^2 + 3xy - 3xy - y^2$

$\qquad\qquad\qquad\qquad\qquad\qquad\qquad$ Combine like terms.

Ans. $4x^2 - 9$ *Ans.* $9x^2 - y^2$

27. $(2x + 3)(2x - 3)$ **28.** $(3x + 1)(3x - 1)$

29. $(6y - 5)(6y + 5)$ **30.** $(4y - 3)(4y + 3)$

31. $(2x + a)(2x - a)$ **32.** $(3x - a)(3x + a)$

33. $(3x - 2y)(3x + 2y)$ **24.** $(2x - 5y)(2x + 5y)$

35. $(4x + 7y)(4x - 7y)$ **36.** $(5x + 9y)(5x - 9y)$

Sample Problems

a. $3(2x - 1)(x + 2)$ b. $x(x + 2)(3x - 5)$

$\qquad\qquad\qquad\qquad\qquad\qquad$ Multiply binomials.

$\quad 3(2x^2 + 4x - x - 2)$ $x(3x^2 - 5x + 6x - 10)$

$\qquad\qquad\qquad\qquad\qquad\qquad$ Simplify.

$\quad 3(2x^2 + 3x - 2)$ $x(3x^2 + x - 10)$

$\qquad\qquad\qquad\qquad\qquad\qquad$ Multiply by the monomial.

Ans. $6x^2 + 9x - 6$ *Ans.* $3x^3 + x^2 - 10x$

37. $2(3x + 1)(x - 3)$ **38.** $4(x - 2)(2x - 3)$ **39.** $-3(2y + 1)(2y - 1)$

40. $-6(3y + 2)(3y - 2)$ **41.** $3(2x - 5)^2$ **42.** $3(x + 1)^2$

43. $x(x - 2)(2x + 5)$ **44.** $y(y + 2)(y - 1)$ **45.** $-x(2x - 1)^2$

46. $-y(y + 1)^2$ **47.** $r(3r - 1)(3r + 1)$ **48.** $s(2s - 3)(2s + 3)$

49. $x^2y(x - 3)(2x + 1)$ **50.** $s^2t^2(2s + 1)(3s - 1)$ **51.** $-2x^3(3x + 2)^2$

52. $-4a^2(2a + 3)^2$ **53.** $xy(3x - y)(x + 2y)$ **54.** $a^2b(a - 3b)(2a + b)$

55. $-3p^2q^2(3p - q)(4p + q)$ **56.** $-6st^3(s - 5t)(s - 3t)$

57. $7xy(2x - y)(2x + y)$ **58.** $8x^2y(5x - 2y)(5x + 2y)$

59. $3x^2y^3(6x - 2y)^2$ **60.** $9x^3y(4x - 3y)^2$

a. $(5x - 2)(2x^2 - x + 3)$

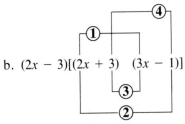

b. $(2x - 3)[(2x + 3)\ \ (3x - 1)]$

Apply the distributive law.
$$10x^3 - 5x^2 + 15x - 4x^2 + 2x - 6$$
Add like terms.

Ans. $10x^3 - 9x^2 + 17x - 6$

Multiply two binomials.
$$(2x - 3)[6x^2 + 7x - 3]$$
Apply the distributive law.
$$12x^3 + 14x^2 - 6x - 18x^2 - 21x + 9$$
Add like terms.

Ans. $12x^3 - 4x^2 - 27x + 9$

61. $(2x - 3)(3x^2 - 2x - 1)$ **62.** $(3x + 2)(3x^2 + 4x - 2)$

63. $(3x + 4)(-x^2 + 6x - 2)$ **64.** $(2x - 5)(-x^2 - 3x + 4)$

65. $(2x - 3)(4x^2 + 6x + 9)$ **66.** $(3x + 1)(9x^2 - 3x + 1)$

67. $(6x - 1)(2x + 1)(2x - 3)$ **68.** $(5x + 2)(3x - 2)(2x - 1)$

69. $(2 - 5x)(1 - 2x)(4 + 3x)$ **70.** $(3 + 2x)(3 - x)(2 - 3x)$

71. $(2x - 1)(2x - 1)(2x - 1)$ **72.** $(3x + 2)(3x + 2)(3x + 2)$

5.6   FACTORING TRINOMIALS II

In Section 5.4 we factored trinomials of the form $x^2 + Bx + C$ where the second-degree term had a coefficient of 1. Now we want to extend our factoring techniques to trinomials of the form $Ax^2 + Bx + C$, where the second-degree term has a coefficient other than 1 or -1.

First we consider a test to determine if a trinomial is factorable. A trinomial of the form $Ax^2 + Bx + C$ is factorable if we can find two integers whose product is $A \cdot C$ and whose sum is B. For example, to determine if $4x^2 + 8x + 3$

is factorable, we check to see if there are two integers whose product is $(4)(3) = 12$ and whose sum is 8 (the coefficient of x). Consider the following possibilities.

Pairs of Factors with Product 12	Sum of Factors
12, 1	13
−12, −1	−13
6, 2	8
−6, −2	−8
4, 3	7
−4, −3	−7

Since the factors 6 and 2 have a sum of 8, the value of B in the trinomial $Ax^2 + Bx + C$, the trinomial $4x^2 + 8x + 3$ is factorable.

On the other hand, the trinomial $4x^2 - 5x + 3$ is not factorable, since the table above shows that there is no pair of factors whose product is 12 and whose sum is −5. The test to see if a trinomial is factorable can usually be done mentally.

FACTORING BY "TRIAL AND ERROR"

Once we have determined that a trinomial of the form $Ax^2 + Bx + C$ is factorable, we proceed to find a pair of factors whose product is A, a pair of factors whose product is C, and an arrangement that yields the proper middle term. We illustrate by examples. Consider our earlier example

$$4x^2 + 8x + 3,$$

which we determined is factorable. First notice that there is no common monomial factor. If there were, we would factor out the common factor first. Since there isn't, we now proceed.

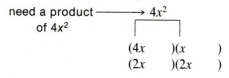

1. We consider all pairs of factors whose product is 4. Since 4 is positive, only positive integers need to be considered. The possibilities are 4, 1 and 2, 2.

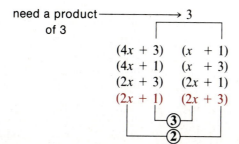

2. We consider all pairs of factors whose product is 3. Since the middle term is positive, consider positive pairs of factors only. The possibilities are 3, 1. We write all possible arrangements of the factors as shown.

(continued)

$(2x + 1)(2x + 3)$

3. We select the arrangement in which the sum of products ② and ③ yields a middle term of $8x$.

$(2x + 1)(2x + 3)$

$= 4x^2 + 6x + 2x + 3$

$= 4x^2 + 8x + 3$

4. We check by multiplying.

Before considering another example, we summarize the steps we used.

> **To Factor a Trinomial of the Form $Ax^2 + Bx + C$:**
> 1. Consider all pairs of integers whose product is A. These will be the coefficients of x in the two binomial factors. If A is positive, only positive integers need to be considered.
> 2. Consider all pairs of integers whose product is C. These will be the constants in the two binomial factors. Write down all possible different arrangements of the pairs of integers in the binomial factors.
> 3. Select the arrangement in which the sum of the outer and inner products yields the correct middle term Bx for the given trinomial.
> 4. Check by multiplying.

Now we consider the factorization of $6x^2 + x - 2$ where the constant term is negative. First notice that there is no common monomial factor. Next, we test to see if $6x^2 + x - 2$ is factorable. We look for two integers that have a product of $6(-2) = -12$ and a sum of 1 (the coefficient of x). The integers 4 and -3 have a product of -12 and a sum of 1, so the trinomial is factorable. We now proceed.

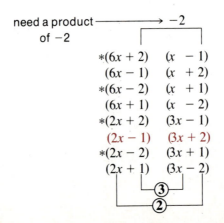

need a product ⟶ $6x^2$
of $6x^2$

(6x   )(x   )
(2x   )(3x   )

1. We consider all pairs of factors whose product is 6. Since 6 is positive, only positive integers need to be considered. The possibilities are 6, 1 and 2, 3.

need a product ⟶ -2
of -2

*(6x + 2)   (x − 1)
(6x − 1)   (x + 2)
*(6x − 2)   (x + 1)
(6x + 1)   (x − 2)
*(2x + 2)   (3x − 1)
(2x − 1)   (3x + 2)
*(2x − 2)   (3x + 1)
(2x + 1)   (3x − 2)

2. We consider all pairs of factors whose product is -2. The possibilities are 2, -1 and -2, 1. We write all possible arrangements of the factors as shown.
Because the original trinomial had no common monomial factor, neither can either of its binomial factors. Therefore we can eliminate all of the starred (∗) arrangements at left. We need only check the remaining possibilities.

$(2x - 1)(3x + 2)$

3. We select the arrangement in which the sum of products ② and ③ yields a middle term of x.

$(2x - 1)(3x + 2)$
$= 6x^2 + 4x - 3x - 2$
$= 6x^2 + x - 2$

4. We check by multiplying.

With practice, you will be able to check the combinations mentally and will not need to write out all the possibilities. Paying attention to the signs in the trinomial is particularly helpful for mentally eliminating possible combinations.

ALTERNATIVE METHOD

If the "trial and error" method of factoring does not yield quick results, an alternative method, which we will now demonstrate using the earlier example $4x^2 + 8x + 3$, may be helpful.

We know that the trinomial is factorable because we found two numbers whose product is 12 and whose sum is 8. Those numbers are 2 and 6. We use these numbers to rewrite $8x$ as $2x + 6x$.

Rewrite the middle term as $2x + 6x$.

$$4x^2 + 2x + 6x + 3$$

We now factor the first two terms, $4x^2 + 2x$, and the last two terms, $6x + 3$.

$$2x(2x + 1) + 3(2x + 1)$$

A common factor, $2x + 1$, is in each term, so we can factor again.

$$(2x + 1)(2x + 3)$$

This is the same result that we obtained before.

As usual, if a polynomial contains a common monomial factor in each of its terms, we should factor this monomial from the polynomial before looking for other factors. For example, to factor $24x^2 - 44x - 40$, we first factor 4 from each term to get

$$4(6x^2 - 11x - 10).$$

Next we look for two numbers whose product is $6(-10) = -60$ and whose sum is -11. Those numbers are -15 and 4. We use these numbers to rewrite $-11x$ as $-15x + 4x$.

Rewrite the middle term as $-15x + 4x$

$$4(6x^2 - 15x + 4x - 10)$$

We now factor the first two terms, $6x^2 - 15x$, and the last two terms, $4x - 10$.

$$4[3x(2x - 5) + 2(2x - 5)]$$

A common factor, $2x - 5$, is in each term, so we can factor again.

(continued)

$4(2x - 5)(3x + 2)$

Note that the monomial 4 which was initially factored out is retained at each step.

EXERCISES 5.6

■ *Factor. Assume the trinomials are factorable.*

Sample Problem

$6x^2 - 11x + 4$

Consider all pairs of positive factors whose product is $6x^2$.

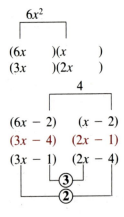

Since the middle term is negative, consider all pairs of negative factors whose product is 4.

Select the arrangement in which the sum of products ② and ③ yields a middle term of $-11x$.

Ans. $(3x - 4)(2x - 1)$

Check by multiplying.

$$(3x - 4)(2x - 1) = 6x^2 - 3x - 8x + 4$$
$$= 6x^2 - 11x + 4$$

1. $3a^2 + 4a + 1$
2. $2r^2 + 3r + 1$
3. $2x^2 - 3x + 1$
4. $2y^2 + 5y + 3$
5. $9b^2 - 6b + 1$
6. $4a^2 + 4a + 1$
7. $2x^2 - 7x + 3$
8. $4y^2 - 4y + 1$
9. $4y^2 - 5y + 1$
10. $4a^2 - 11a + 6$
11. $64x^2 + 64x + 15$
12. $4y^2 + 16y + 15$
13. $4y^2 - 3y - 1$
14. $4y^2 + 3y - 1$
15. $4a^2 + a - 5$
16. $16x^2 - 2x - 5$
17. $16x^2 - 38x - 5$
18. $16x^2 - 11x - 5$
19. $16x^2 + 79x - 5$
20. $9x^2 - 21x - 8$

Sample Problems

a. $9x^2 - 8 - 21x$

$9x^2 - 21x - 8$

b. $9 + 8x^2 - 18x$

Arrange in descending powers of x.

$8x^2 - 18x + 9$

Factor.

Ans. $(3x - 8)(3x + 1)$ *Ans.* $(4x - 3)(2x - 3)$
                                                    Check by multiplying.

Check.   *Check.*
$(3x - 8)(3x + 1)$ $(4x - 3)(2x - 3)$
$\quad = 9x^2 + 3x - 24x - 8$ $\quad = 8x^2 - 12x - 6x + 9$
$\quad = 9x^2 - 21x - 8.$ $\quad = 8x^2 - 18x + 9$

21. $2x^2 - 3 + x$ **22.** $2x^2 + 3 + 7x$ **23.** $2x^2 - 3 - x$
24. $6a^2 - 1 - a$ **25.** $1 + 6a^2 + 5a$ **26.** $23a + 4a^2 - 6$
27. $16x^2 - 5 - 16x$ **28.** $10y - 8 + 25y^2$

Sample Problems

a. $6x^2 + 15x - 9$ b. $4a^2b + 10ab + 6b$
                                                    Factor out monomial factor.
$\quad 3(2x^2 + 5x - 3)$ $\quad 2b(2a^2 + 5a + 3)$
                                                        Factor trinomial.

Ans. $3(2x - 1)(x + 3)$ *Ans.* $2b(2a + 3)(a + 1)$
                                                    Check by multiplying.

Check. $6x^2 + 15x - 9$ *Check.* $4a^2b + 10ab + 6b$

29. $6x^2 + 8x + 2$ **30.** $6x^2 + 21x + 9$ **31.** $8y^2 - 6y - 2$
32. $18x^2 - 3x - 3$ **33.** $18x^2 - 9x - 27$ **34.** $4x^3 - 10x^2 - 6x$
35. $27y^3 - 9y^2 - 6y$ **36.** $4y^3 + 10y^2 + 6y$ **37.** $12x^5 + 14x^4 - 10x^3$
38. $80x^4 - 28x^3 - 24x^2$ **39.** $36a^7 - 54a^6 + 20a^5$ **40.** $63a^6 - 102a^5 + 24a^4$

Sample Problem

$13xy + 3x^2 + 4y^2$
                        Arrange in descending powers of one variable.
$3x^2 + 13xy + 4y^2$
                        Factor.
Ans. $(3x + y)(x + 4y)$
                        Check by multiplying.
Check. $3x^2 + 13xy + 4y^2$

41. $2t^2 - 5st - 3s^2$ **42.** $2a^2 + 5ab - 3b^2$ **43.** $3x^2 - 7ax + 2a^2$
44. $9y^2 - 3yz - 2z^2$ **45.** $5by + 4y^2 + b^2$ **46.** $9ab + 9a^2 - 4b^2$
47. $4a^2 + 15b^2 + 16ab$ **48.** $9x^2 - 2y^2 + 3xy$
49. $12ab^2 + 15a^2b + 3a^3$ **50.** $27a^2b + 27ab^2 - 12b^3$
51. $50xy^3 + 20x^2y^2 - 16x^3y$ **52.** $4a^2bx^2 - 2abx - 12b$
53. $36x^5y^2 + 48x^4y^2 - 9x^3y^2$ **54.** $32x^4y - 48x^3y + 18x^2y$
55. $12x^4y^2 + 12x^2y^4 - 30x^3y^3$ **56.** $24x^4y^3 - 20x^2y^5 + 52x^3y^4$

■ *Factor completely; if the trinomial is not factorable, so state.*

Sample Problem

$3x^2 + 4x + 5$

We must find two numbers whose product is $(3)(5) = 15$ and whose sum is 4 (the coefficient of x). No such numbers exist.

Ans. The trinomial is not factorable.

57. $5x^2 - 4x + 3$ **58.** $2x^2 - 11x + 4$ **59.** $6y^2 - y + 3$

60. $4y^2 + 2y - 1$ **61.** $3x^2 + 5x - 1$ **62.** $4x^2 - 5x - 2$

63. $6x^2 + xy - 2y^2$ **64.** $5x^2 - 3xy + 2y^2$ **65.** $a^2 + ab + b^2$

66. $a^2 - ab - 2b^2$ **67.** $9a^2 + 21ab + 10b^2$ **68.** $20a^2 - 13ab + 2b^2$

5.7   FACTORING THE DIFFERENCE OF TWO SQUARES

Some polynomials occur so frequently that it is helpful to recognize these special forms, which in turn enables us to write their factored form directly. Observe that

$$(a + b)(a - b) = a^2 - ab + ab - b^2$$
$$= a^2 - b^2.$$

In this section we are interested in viewing this relationship from right to left, from the polynomial $a^2 - b^2$ to its factored form $(a + b)(a - b)$.

> **The difference of two squares, $a^2 - b^2$, equals the product of the sum $a + b$ and the difference $a - b$.**

For example,

$$x^2 - 9 = x^2 - 3^2 = (x + 3)(x - 3),$$

and

$$4y^2 - 25x^2 = (2y)^2 - (5x)^2 = (2y + 5x)(2y - 5x).$$

Common Error

Note that the *sum* of two squares, $a^2 + b^2$, *cannot* be factored. For example,

$$x^2 + 16 \quad \text{and} \quad 9x^2 + y^2$$

are not factorable.

EXERCISES 5.7

■ *Factor.*

Sample Problem

a. $x^2 - 16$

$x^2 - 4^2$

b. $25 - a^2b^2$

$5^2 - (ab)^2$

Write as the difference of squares.

Factor.

Ans. $(x + 4)(x - 4)$

Check. $x^2 - 16$

Ans. $(5 + ab)(5 - ab)$

Check. $25 - a^2b^2$

$x^2 - 9^2 = (x^2 - 9)(x^2 + 9)$

1. $x^2 - 9$ **2.** $y^2 - 25$ **3.** $x^2 - 1$ **4.** $x^2 - 81$

5. $x^2 - z^2$ **6.** $x^2 - 9y^2$ **7.** $x^2y^2 - 16$ **8.** $x^2y^2 - 36$

9. $a^2x^2 - 49b^2$ **10.** $x^2 - 100a^2b^2$ **11.** $36 - x^2$ **12.** $b^2 - y^2$

Sample Problems

a. $9x^2 - 4$

$(3x)^2 - 2^2$

b. $16a^2 - b^2y^2$

$(4a)^2 - (by)^2$

Write as the difference of squares.

Factor.

Ans. $(3x - 2)(3x + 2)$

Check. $9x^2 - 4$

Ans. $(4a - by)(4a + by)$

Check. $16a^2 - b^2y^2$

13. $4b^2 - 9$ **14.** $9b^2 - 1$ **15.** $25x^2 - 16$

16. $4y^2 - 25$ **17.** $9 - 4x^2$ **18.** $25 - 9y^2$

19. $81 - 4x^2$ **20.** $9 - 64y^2$ **21.** $4a^2 - 121b^2$

22. $64x^2 - 9y^2$ **23.** $25y^2 - 49x^2$ **24.** $100x^2 - 81y^2$

25. $49a^2x^2 - 144b^2y^2$ **26.** $49a^2x^2 - 36b^2y^2$ **27.** $4x^2y^2 - 81$

28. $121 - 49x^2y^2$ **29.** $36a^2b^2 - 1$ **30.** $1 - 100a^2b^2$

Sample Problems

a. $25x^6 - 16$

$(5x^3)^2 - 4^2$

b. $a^4 - b^4$

$(a^2)^2 - (b^2)^2$

Write as the difference of squares.

Factor.

$(a^2 - b^2)(a^2 + b^2)$

Factor.

Ans. $(5x^3 - 4)(5x^3 + 4)$

Check. $25x^6 - 16$

Ans. $(a - b)(a + b)(a^2 + b^2)$

Check. $(a^2 - b^2)(a^2 + b^2) = a^4 - b^4$

31. $x^4 - 9$ **32.** $x^6 - 25$ **33.** $x^4 - 81$

34. $x^8 - 16$　　　　　　**35.** $36y^8 - 49$　　　　　　**36.** $9y^{10} - 100$

37. $16x^6 - 9y^4$　　　　　**38.** $64x^{10} - 25y^2$　　　　**39.** $81x^8 - y^4$

40. $x^{12} - 16y^8$　　　　　**41.** $121a^4b^6 - 36$　　　　**42.** $49a^{10}b^{12} - 144$

Sample Problems

a. $x^3 - x^5$

$x^3(1 - x^2)$

b. $a^2x^2y - 16y$

Factor out monomial factor.

$y(a^2x^2 - 16)$
$y[(ax)^2 - 4^2]$

Factor the binomial.

Ans. $x^3(1 - x)(1 + x)$　　　　*Ans.* $y(ax - 4)(ax + 4)$

Check. $x^3 - x^5$　　　　　　　*Check.* $a^2x^2y - 16y$

43. $5x^2 - 5$　　　　　　**44.** $2x^2 - 8$　　　　　　**45.** $3x^3 - 3x$

46. $3a^2 - 75$　　　　　**47.** $2x^2 - 8y^2$　　　　　**48.** $3xy^2 - 12xb^2$

49. $3a^2b^2 - 12c^2d^2$　　**50.** $8x^2y^2z^2 - 18$　　　**51.** $4x^2y^2 - 16x^2$

52. $3x^2y^2 - 12y^2$　　　**53.** $3x^6 - 12x^2$　　　　　**54.** $9x^7 - 81x^3$

55. $4x^4y^2 - 16x^2y^4$　　**56.** $x^6y^8 - 81x^2y^2$　　　**57.** $32a^6b^2 - 2a^2b^6$

58. $2a^8 - 162a^4b^8$　　　**59.** $3x^5y - 48xy^5$　　　　**60.** $64x^5y^5 - 4xy$

CHAPTER SUMMARY

[5.1–5.2] Algebraic expressions containing parentheses can be written without parentheses by applying the distributive law in the form

$$a(b + c) = ab + ac.$$

A polynomial that contains a monomial factor common to all terms in the polynomial can be written as the product of the common factor and another polynomial by applying the distributive law in the form

$$ab + ac = a(b + c).$$

[5.3; 5.5] The distributive law can be used to multiply polynomials; the FOIL method applies to the product of two binomials.

$$(a + b)\ (c + d) = ac + ad + bc + bd$$

[5.4] Given a trinomial of the form $x^2 + Bx + C$, if there are two numbers, a and b, whose product is C and whose sum is B, then

$$x^2 + Bx + C = (x + a)(x + b);$$

otherwise; the trinomial is not factorable.

[5.6] Given a trinomial of the form $Ax^2 + Bx + C$, we determine if the trinomial is factorable by finding two numbers whose product is $A \cdot C$ and whose sum is B.

[5.7] We factor the difference of squares as

$$a^2 - b^2 = (a + b)(a - b).$$

■ *The symbols introduced in this chapter appear on the inside front cover.*

CHAPTER REVIEW

■ *Write as a polynomial.*

1. a. $3x(x^2 + x)$ b. $2xy(y - x)$ c. $-(x^2 - y + 1)$
2. a. $a(2 - a)$ b. $-b(a - b)$ c. $3b(a + b + c)$

■ *Factor.*

3. a. $3a^2 - 6a^2b$ b. $2x^3 + 4x^2 + 6x$ c. $-y^2 - y^3$
4. a. $a^2 + a^2b$ b. $4b - 4$ c. $b - b^2 - b^3$

■ *Write as a polynomial.*

5. a. $(x - 2)(x + 3)$ b. $(2a - 3)(3a - 4)$ c. $(2a - 3)^2$
6. a. $x(x + 1)(2x - 3)$ b. $3x^2(2x + 5)(x - 2)$ c. $4t^2(2t - 1)^2$
7. a. $(x - 3)(x^2 + 6x - 1)$
  b. $(3x - 4)(2x^2 - x + 2)$
  c. $(2x - 5)(x + 2)(3x - 1)$

■ *Factor. If the expression does not factor, so state.*

8. a. $x^2 - 4x - 21$ b. $10a^2 + 17a + 3$ c. $4x^2 + 9$
9. a. $a^2 - 10a - 21$ b. $3b^2 + 4b + 1$ c. $2b^2 + 3b - 2$
10. a. $2x^2 + 14x + 24$ b. $3y^2 + 24y - 60$ c. $4x^3 - 4x$

11. a. $x^2 - 3ax + 2a^2$ **b.** $x^4 - 25$ **c.** $4b^2 + 6bc - 4c^2$

12. a. $x^2 + x + 1$ **b.** $2x^2 + 4x + 2$ **c.** $4x^2 + 4x + 1$

13. If -3 is one of two factors of -12, what is the other factor?

14. If 2 and -3 are two factors of 30, what is the other factor?

15. The product of three factors is $48xy^2$. If two of the factors are -3 and $4x$, what is the third factor?

16. Factor the right-hand member of the formula $A = 2krh + 2kr^2$, and evaluate for $k = 3.14$, $r = 7$, and $h = 10$.

17. Factor the right-hand member of the formula $A = P + Prt$, and evaluate for $P = 400$, $r = 8$, and $t = 3$.

18. What factor is common to $x^2 + 4x + 3$ and $x^2 + x - 6$?

19. What factor is common to $x^2 - 4$ and $x^2 - 2x$?

20. What factors are common to $2x^3 - 18x$ and $6x^3 - 18x^2$?

CUMULATIVE REVIEW

1. Graph on a number line all natural numbers between 6 and 26 that are exactly divisible by 4.

2. Write $81x^3y^3$ in completely factored form.

3. If $a = 2$, $b = 3$, and $c = -2$, find the value of $a^2c^2 - 8abc$.

■ *Simplify.*

4. $x^2 - 2x + 3 - 4x + 2 - 2x^2$ **5.** $(x^2 - 3) - (2x^2 + x)$

6. $x(x - 3) - 2x(x + 4) + 3x^2$ **7.** $(x - 3)^2 - 2(x + 1)^2 + 2x^2 - 1$

8. Solve for x: $x - b = 2x + 3b$.   **9.** Solve for x: $-4x - 3 \geq 9$.

10. Write an algebraic expression for the cost of n articles at c cents each.

11. The temperature dropped $y°$ from a maximum value of 80°. Express the new temperature in terms of y.

12. The temperature dropped 6° from a maximum of $y°$. Express the new temperature in terms of y.

13. The minimum temperature of the day was y degrees; the maximum temperature was x degrees. Express the increase of temperature during the day in terms of x and y.

14. The sum of two numbers is 47. If x represents the smaller number, what expression, in terms of x, represents the larger number?

15. Find two consecutive odd integers whose sum if 56.

16. Find three consecutive integers whose sum is 111.

17. One integer is five less than a second integer. The larger integer plus four times the smaller equals 40. Find the integers.

18. A man had $1.45 in change consisting of nickels, dimes, and quarters. He had one more dime than quarters and three more nickels than quarters. How many of each coin did he have?

19. A 35-centimeter stick is cut into three pieces so that the two end pieces are each equal in length to one-third of the middle piece. How long is each piece?   (*Hint:* the length of the middle piece is three times the length of each end piece.)

20. A girl earns $4 per hour, and her older brother earns $5 per hour. One week the boy worked 10 hours longer than his sister. How long did each work if their total income was $185?

■ *True or false.*

21. $2x - x$ is equivalent to x.

22. $(9 - x)^2$ is equivalent to $81 - x^2$.

23. $(a + b)(a - b)$ is equivalent to $a^2 - b^2$.

24. $-(5 - y)$ is equivalent to $-5 - y$.

25. $x^4 \cdot x^3$ is equivalent to x^{12}.

6 PROPERTIES OF FRACTIONS

6.1 FORMS OF FRACTIONS; GRAPHICAL REPRESENTATIONS

As noted in Section 1.2, we often indicate the quotient of two numbers $a \div b$, $b \neq 0$, by the fraction $\frac{a}{b}$. We call a the **numerator** (or dividend) and b the **denominator** (or divisor).

In Section 1.2 we defined a quotient $\frac{a}{b}$ to be a number q such that

$$b \cdot q = a.$$

For example,

$$\frac{6}{2} = 3 \quad \text{because} \quad 2 \cdot 3 = 6, \quad \text{and} \quad \frac{12}{3} = 4 \quad \text{because} \quad 3 \cdot 4 = 12.$$

We can also view a quotient in terms of a product. For example,

$$\frac{6}{2} = 6 \cdot \frac{1}{2}, \quad \frac{12}{3} = 12 \cdot \frac{1}{3}, \quad \text{and} \quad \frac{3}{4} = 3 \cdot \frac{1}{4}.$$

In general,

$$\frac{a}{b} = a \cdot \frac{1}{b} \quad (b \neq 0).$$

We can use this property to help us graph arithmetic fractions on a number line. For example, to graph the fraction $\frac{3}{4}$, which is equivalent to $3 \cdot \frac{1}{4}$, we divide a unit on the number line into four equal parts and then mark a point at the third quarter, as shown in Figure 6.1. In general, we can locate the graph of any fraction by dividing a unit on the number line into a number of equal parts

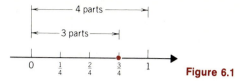

Figure 6.1

corresponding to the denominator of the fraction, and then counting off the number of parts corresponding to the numerator. For example, the graphs of

$$\frac{-3}{2}, \qquad \frac{-1}{4}, \qquad \text{and} \qquad \frac{5}{4}$$

are shown in Figure 6.2.

Figure 6.2

Fractions can involve algebraic expressions. In such cases, since division by 0 is undefined, we must restrict variables so that a divisor is never 0. In our work, *we will assume that no denominator is 0 unless otherwise specified.* For example,

$$\text{for } \frac{3}{x}, \text{ we assume that } x \neq 0;$$

$$\text{for } \frac{3x}{y + 5}, \text{ we assume that } y \neq -5.$$

SIGNS OF FRACTIONS

There are three signs associated with a fraction: the sign of the numerator, the sign of the denominator, and the sign of the fraction itself.

$$\text{sign of fraction} \rightarrow + \frac{-6}{+3}$$

(with "sign of numerator" pointing to the -6 and "sign of denominator" pointing to the $+3$)

Fractions that have different signs may have the same value. For example:

$$+ \frac{+6}{+3} = +2 = 2, \qquad - \frac{-6}{+3} = -(-2) = 2,$$

$$+ \frac{-6}{-3} = +2 = 2, \qquad - \frac{+6}{-3} = -(-2) = 2.$$

Each fraction above names the number 2.

Also,

$$+\frac{-6}{+3} = +(-2) = -2, \qquad -\frac{+6}{+3} = -(+2) = -2,$$

$$+\frac{+6}{-3} = +(-2) = -2, \qquad -\frac{-6}{-3} = -(+2) = -2.$$

Each fraction above names the number -2. The examples suggest the following.

Any two of the three signs of a fraction may be changed without changing the value of the fraction.

For example,

$$\frac{8}{4} = \begin{cases} \text{Change sign of numerator} \longrightarrow \dfrac{-8}{-4} \\ \text{and denominator.} \longrightarrow \\[4pt] \text{Change sign of numerator} \longrightarrow -\dfrac{-8}{4} \\ \text{and fraction.} \\[4pt] \text{Change sign of fraction} \longrightarrow -\dfrac{8}{-4} \\ \text{and denominator.} \end{cases}$$

and

$$\frac{-8}{4} = \begin{cases} \text{Change sign of numerator} \longrightarrow \dfrac{8}{-4} \\ \text{and denominator.} \longrightarrow \\[4pt] \text{Change sign of numerator} \longrightarrow -\dfrac{8}{4} \\ \text{and fraction.} \\[4pt] \text{Change sign of fraction} \longrightarrow -\dfrac{-8}{-4} \\ \text{and denominator.} \end{cases}$$

In general,

$$\frac{a}{b} = \frac{-a}{-b} = -\frac{-a}{b} = -\frac{a}{-b},$$

and

$$\frac{-a}{b} = \frac{a}{-b} = -\frac{a}{b} = -\frac{-a}{-b}.$$

In this book, the two forms $\dfrac{a}{b}$ and $\dfrac{-a}{b}$, which have positive signs on the denominator and on the fraction itself, will be considered *standard forms* for fractions. However, we shall sometimes leave answers in the form $-\dfrac{a}{b}$ instead of using the standard form $\dfrac{-a}{b}$.

If the numerator contains more than one term, there are alternative standard forms for a fraction. For example,

$$-\frac{x-3}{4} = \frac{-(x-3)}{4}$$

$$= \frac{-x+3}{4}$$

$$= \frac{3-x}{4},$$

and any of the three forms on the right-hand side of the equals sign may be taken as standard form.

Common Error

When we write a standard form of a fraction such as $-\frac{x-3}{4}$, we must be careful how we change the signs in the numerator. The use of parentheses helps us avoid errors. Note that in the previous example,

$$-\frac{x-3}{4} = \frac{-(x-3)}{4} = \frac{-x+3}{4}.$$

In particular note that

$$-\frac{x-3}{4} \neq \frac{-x-3}{4}.$$

EXERCISES 6.1

■ *Represent each quotient in fractional form.*

Sample Problems

a. $5 \div 8$

Ans. $\frac{5}{8}$

b. $x \div 2y$

Ans. $\frac{x}{2y}$

c. $(3x + 1) \div (y - 2)$

Ans. $\frac{3x+1}{y-2}$

1. $4 \div 7$ **2.** $7 \div 2$ **3.** $3x \div y$
4. $x \div 3y$ **5.** $7 \div (x - y)$ **6.** $(2x + y) \div 3$
7. $(x - 3) \div (4x + 1)$ **8.** $(4y - 2) \div (y + 3)$

■ *Write each quotient in the form* $a \cdot \frac{1}{b}$.

Sample Problems

a. $\frac{3}{7}$

Ans. $3 \cdot \frac{1}{7}$

b. $\frac{x+2}{5}$

Ans. $(x + 2) \cdot \frac{1}{5}$

c. $\frac{3}{y-1}$

Ans. $3 \cdot \frac{1}{y-1}$

9. $\dfrac{4}{7}$ **10.** $\dfrac{7}{9}$ **11.** $\dfrac{9}{5}$ **12.** $\dfrac{3}{7}$

13. $\dfrac{x-3}{4}$ **14.** $\dfrac{y+1}{6}$ **15.** $\dfrac{2}{x+3}$ **16.** $\dfrac{5}{y-2}$

■ *Graph each set of numbers on a number line. Use a separate graph for each exercise.*

Sample Problem $\dfrac{-4}{3}, \dfrac{2}{3}, \dfrac{5}{3}$

Ans.

17. $\dfrac{1}{4}, \dfrac{3}{4}$ **18.** $\dfrac{1}{3}, \dfrac{2}{3}$ **19.** $\dfrac{1}{2}, \dfrac{5}{2}$ **20.** $\dfrac{-1}{4}, \dfrac{-3}{4}$

21. $\dfrac{-5}{6}, \dfrac{1}{6}$ **22.** $\dfrac{-1}{2}, \dfrac{1}{2}$ **23.** $\dfrac{-5}{2}, \dfrac{5}{4}$ **24.** $3, \dfrac{3}{4}, \dfrac{3}{2}$

25. $-3, \dfrac{-3}{4}, \dfrac{3}{2}$ **26.** $\dfrac{-2}{3}, \dfrac{1}{3}, 0$ **27.** $\dfrac{2}{5}, \dfrac{3}{5}, \dfrac{4}{5}$ **28.** $3, \dfrac{-5}{3}, 0$

■ *Rewrite each fraction in standard form.*

Sample Problems   a. $-\dfrac{1}{2}$ b. $\dfrac{-x}{-y}$ c. $\dfrac{x-1}{-3}$

Ans. $\dfrac{-1}{2}$ *Ans.* $\dfrac{x}{y}$ *Ans.* $\dfrac{-(x-1)}{3}, \dfrac{-x+1}{3}$ or $\dfrac{1-x}{3}$

29. $\dfrac{-3}{-5}$ **30.** $-\dfrac{-1}{2}$ **31.** $-\dfrac{2}{-7}$ **32.** $-\dfrac{-1}{-3}$

33. $\dfrac{2}{-5}$ **34.** $-\dfrac{-2}{5}$ **35.** $-\dfrac{-a}{-b}$ **36.** $-\dfrac{-a}{b}$

37. $\dfrac{a}{-b}$ **38.** $-\dfrac{a}{-b}$ **39.** $\dfrac{-x}{y}$ **40.** $-\dfrac{3y}{x}$

41. $-\dfrac{7x}{-8y}$ **42.** $-\dfrac{2c}{-1}$ **43.** $-\dfrac{-c}{-1}$ **44.** $\dfrac{c}{-1}$

45. $-\dfrac{x+2}{4}$ **46.** $\dfrac{x+3}{-3}$ **47.** $-\dfrac{x+5}{-4}$ **48.** $-\dfrac{x-3}{2}$

49. $-\dfrac{2x-1}{x+2}$ **50.** $-\dfrac{4x-3}{x-5}$ **51.** $-\dfrac{-x+3}{2}$ **52.** $-\dfrac{-x-4}{3}$

53. For what value of x is $\dfrac{5}{x}$ undefined?

54. For what value of y is $\dfrac{3}{y}$ undefined?

55. For what value of x is $\dfrac{7}{x-3}$ undefined?

56. For what value of y is $\dfrac{5}{y+2}$ undefined?

57. Use a numerical example to show that $-\dfrac{x-1}{2}$ is not equivalent to $\dfrac{-x-1}{2}$.

58. Use a numerical example to show that $-\dfrac{-x+2}{3}$ is not equivalent to $\dfrac{x+2}{3}$.

6.2 REDUCING FRACTIONS TO LOWEST TERMS

FUNDAMENTAL PRINCIPLE OF FRACTIONS

Recall from your study of arithmetic that a fraction can be *reduced* if both numerator and denominator can be divided by a common factor. For example,

$$\frac{6}{8} = \frac{3 \cdot 2}{4 \cdot 2} = \frac{3}{4},$$

where we have used the following **fundamental principle of fractions.**

If both the numerator and the denominator of a given fraction are divided by the same nonzero number, the resulting fraction is equivalent to the given fraction.

In symbols,

$$\frac{a \cdot c}{b \cdot c} = \frac{a}{b}, \qquad c \neq 0.$$

For example,

$$\frac{xy}{3x} = \frac{y \cdot \overset{1}{\cancel{x}}}{3 \cdot \underset{1}{\cancel{x}}} = \frac{y}{3} \qquad \text{and} \qquad \frac{\overset{1}{\cancel{(x+2)}}(x-4)}{\underset{1}{\cancel{(x+2)}}(x+1)} = \frac{x-4}{x+1}.$$

We use the diagonal slash here to illustrate the process of dividing numerator and denominator by the same number.

Thus, to reduce an algebraic fraction, we *first* write the numerator and denominator in factored form and *then* divide each by their common factors. For example,

$$\frac{4x^2}{x^3} = \frac{4 \cdot \cancel{x} \cancel{x}}{\cancel{x} \cancel{x} x}, ^{*} \qquad\qquad \frac{10y^2}{4y} = \frac{5 \cdot \cancel{2} \cancel{y} y}{2 \cdot \cancel{2} \cancel{y}}, ^{*}$$

$$= \frac{4}{x} \qquad\qquad\qquad = \frac{5y}{2}$$

$$\frac{xy^2}{-x^2y} = \frac{\cancel{x} \cancel{y} y}{-1 \cancel{x} x \cancel{y}}, \qquad \text{and} \qquad \frac{-6a^3b}{15ab^2} = \frac{-2 \cdot \cancel{3} \cancel{a} aa \cancel{b}}{\cancel{3} \cdot 5 \cancel{a} \cancel{b} b}.$$

$$= \frac{-y}{x} \qquad\qquad\qquad = \frac{-2a^2}{5b}$$

It is *critical* that when we reduce a fraction in which the numerator and/or the denominator contains more than one term, we must factor wherever possible *before* applying the fundamental principle. Thus, to reduce

$$\frac{3x + 12y}{6x + 24y}$$

we first factor the numerator and denominator and then divide out common factors:

$$\frac{3x + 12y}{6x + 24y} = \frac{\cancel{3(x + 4y)}}{2 \cdot \cancel{3(x + 4y)}} = \frac{1}{2}.$$

Similarly,

$$\frac{x^2 - x - 6}{2x^2 + 5x + 2} = \frac{\cancel{(x + 2)}(x - 3)}{\cancel{(x + 2)}(2x + 1)} = \frac{x - 3}{2x + 1}.$$

Common Error

It is incorrect to "divide" out expressions which are not *factors* of the numerator and denominator. Thus

$$\frac{x^2 - x - 6}{2x^2 + 5x + 2} = \frac{\cancel{x}\cancel{x} - \cancel{x} - \cancel{2} \cdot 3}{2\cancel{x}\cancel{x} + 5\cancel{x} + \cancel{2}} = \frac{1 - 1 - 3}{2 + 5 + 1} = \frac{-3}{8}$$

is an *incorrect* attempt to reduce the fraction in the previous example, since the numerator and denominator have not been written in factored form.

Note that the fraction $\dfrac{a - b}{b - a}$ can be reduced by factoring -1 out of either the

* In cases such as this, where no quotient is indicated above or below the factors "divided out," the quotient 1 is understood.

numerator or the denominator. Since $a - b = -(b - a)$, we have

$$\frac{a - b}{b - a} = \frac{-(b - a)}{(b - a)} = -1.$$

We sometimes encounter expressions of the form $\dfrac{a - b}{b - a}$ when we factor the numerator and denominator of a fraction we wish to reduce. For example,

$$\frac{2x - 2y}{3y - 3x} = \frac{2(x - y)}{3(y - x)} = \frac{-2(y - x)}{3(y - x)} = \frac{-2}{3},$$

and

$$\frac{2x^2 - x - 6}{4 - x^2} = \frac{(x - 2)(2x + 3)}{(2 - x)(2 + x)} = \frac{-(2 - x)(2x + 3)}{(2 - x)(2 + x)} = \frac{-(2x + 3)}{2 + x}.$$

AN ALTERNATE METHOD OF REDUCING FRACTIONS

We have been using the fundamental principle of fractions to reduce fractions. Sometimes, we can reduce fractions by a more direct method. Consider the quotient

$$\frac{x^5}{x^2} = \frac{x \cdot x \cdot x \cdot x \cdot x}{x \cdot x} = x^3.$$

Note that $x^3 = x^{5-2}$, and hence

$$\frac{x^5}{x^2} = x^{5-2} = x^3.$$

Thus we can obtain the quotient above by subtracting the exponent of the denominator from the exponent of the numerator. This example suggests the following rule.

$$\frac{a^m}{a^n} = a^{m-n} \qquad (n < m). \qquad\qquad \textbf{IV}$$

If the greater exponent is in the denominator, that is, if n is greater than m, then

$$\frac{a^m}{a^n} = \frac{1}{a^{n-m}} \qquad (n > m). \qquad\qquad \textbf{IVa}$$

For example,

$$\frac{x^2}{x^5} = \frac{x \cdot x}{x \cdot x \cdot x \cdot x \cdot x} = \frac{1}{x^3}$$

or

$$\frac{x^2}{x^5} = \frac{1}{x^{5-2}} = \frac{1}{x^3}.$$

In Section 4.4 we considered *three* laws of exponents governing powers and products of monomials. For convenient reference we list the laws above as IV and IVa.

We can now reduce fractions either by factoring each power and dividing out common factors according to the fundamental principal of fractions, or by using laws IV and IVa above. For example,

$$\frac{6x^3y^2}{2xy^4} = \frac{2 \cdot 3xxxyy}{2xyyyy} = \frac{3x^2}{y^2};$$

or

$$\frac{6x^3y^2}{2xy^4} = \frac{3x^{3-1}}{y^{4-2}} = \frac{3x^2}{y^2}.$$

Common Errors

Many errors can be made in working with fractions. Careful attention to the following kinds of errors may help you to avoid them. Note that

$$\frac{1+3}{4+3} \neq \frac{1}{4}.$$

From the fundamental principles of fractions,

$$\frac{1 \cdot 3}{4 \cdot 3} = \frac{1}{4};$$

however,

$$\frac{1+3}{4+3} = \frac{4}{7}.$$

Thus, while we can divide out common factors, *we cannot divide out common terms.* That is,

$$\frac{2+4}{2} \neq \frac{2+4}{2},$$ Terms *cannot* be divided out.

$$\frac{x+3}{x} \neq \frac{x+3}{x},$$ Terms *cannot* be divided out.

As another example of a common error, note that

$$\frac{2x+4}{2} \neq x+4.$$

In the expression

$$\frac{2x+4}{2},$$

we cannot "divide out" the 2's until we write 2 as a *factor* of the entire numerator. Thus,

$$\frac{2x + 4}{2} = \frac{2(x + 2)}{2} = x + 2.$$

EXERCISES 6.2

■ *Reduce each fraction to an equivalent fraction in lowest terms by first completely factoring the numerator and the denominator and then dividing each by their common factors. Express your answer in standard form.*

Sample Problems

a. $\dfrac{6}{9}$

$\dfrac{3 \cdot 2}{3 \cdot 3}$

Ans. $\dfrac{2}{3}$

b. $\dfrac{10y^2}{4y}$

$\dfrac{5 \cdot 2\!\!\!/y\,y}{2 \cdot 2\!\!\!/y}$

Ans. $\dfrac{5y}{2}$

c. $\dfrac{xy^2}{-x^2y}$

$\dfrac{\not x\,\not y\,y}{-1x\,x\,\not y}$

Ans. $\dfrac{-y}{x}$

d. $\dfrac{-6a^3b}{15ab^2}$

$\dfrac{-2 \cdot 3\!\!\!/\,a\,a\,a\,\not b}{3 \cdot 5\!\!\!/\,a\!\!\!/\,b\,b}$

Ans. $\dfrac{-2a^2}{5b}$

1. $\dfrac{4}{6}$

2. $\dfrac{9}{12}$

3. $-\dfrac{18}{21}$

4. $-\dfrac{12}{30}$

5. $\dfrac{32}{20}$

6. $\dfrac{40}{24}$

7. $\dfrac{2x^3}{8x^4}$

8. $\dfrac{8x^2}{12x^5}$

9. $\dfrac{-6y}{9y^5}$

10. $\dfrac{-8y}{18y^3}$

11. $\dfrac{14}{-28x}$

12. $\dfrac{48}{-9x^3}$

13. $\dfrac{x^3y}{x}$

14. $\dfrac{xy^4}{y^2}$

15. $\dfrac{-x^2}{x^3y^2}$

16. $\dfrac{-y}{xy^3}$

17. $\dfrac{4ax^2}{2a}$

18. $\dfrac{6bx}{3b}$

19. $\dfrac{12bx^4}{8bx^2}$

20. $\dfrac{12ax^5}{20ax}$

21. $\dfrac{5ab^2c^3}{4abc}$

22. $\dfrac{12a^2b^3c}{10ab^2c}$

23. $\dfrac{26a^3b^2c}{6ab^2c^3}$

24. $\dfrac{24abc}{6a^2b^2c}$

Sample Problems

a. $\dfrac{(x - 2)(x + 3)}{(x - 5)(x + 3)}$

$\dfrac{(x - 2)(\not x \!+\! 3)}{(x - 5)(\not x \!+\! 3)}$

Ans. $\dfrac{x - 2}{x - 5}$

b. $\dfrac{6x - 4y}{12x - 8y}$

$\dfrac{2(3x - 2y)}{4(3x - 2y)}$ 2

Ans. $\dfrac{1}{2}$

c. $\dfrac{x^2 + x - 12}{9 - x^2}$

$\dfrac{(x + 4)(x - 3)^{-1}}{(3 + x)(3 - x)}$

Ans. $\dfrac{-(x + 4)}{x + 3}$ or $\dfrac{-x - 4}{x + 3}$

25. $\dfrac{3(a + b)}{4(a + b)}$ 26. $\dfrac{4(a + 2b)}{6(a + 2b)}$ 27. $\dfrac{12(x - y)}{-3}$ 28. $\dfrac{15(a + b)}{-5}$

29. $\dfrac{(a - b)}{(a - b)}$ 30. $\dfrac{(2x - y)}{(2x - y)}$ 31. $\dfrac{2x + 2y}{-(x + y)}$ 32. $\dfrac{3x - 3y}{-(3x + 3y)}$

33. $\dfrac{2a - 2x}{(x - a)^2}$ 34. $\dfrac{3a - 3x}{2(x - a)^2}$ 35. $\dfrac{-4x}{4x^2 + 16x}$ 36. $\dfrac{-3x}{6x^2 + 9x}$

37. $\dfrac{x + 1}{x^2 + 2x + 1}$ 38. $\dfrac{x - 4}{x^2 - 3x - 4}$ 39. $\dfrac{a - b}{a^2 - 2ab + b^2}$

40. $\dfrac{a - b}{a^2 - b^2}$ 41. $\dfrac{(a - b)^2}{a^2 - b^2}$ 42. $\dfrac{(x - 2y)^2}{x^2 - 4y^2}$

43. $\dfrac{3a - a^2}{a^2 - 2a - 3}$ 44. $\dfrac{a - a^2}{a^2 + a - 2}$ 45. $\dfrac{x^2 + x - 6}{x^2 - 9}$

46. $\dfrac{x^2 + 5x + 6}{x^2 - 4}$ 47. $\dfrac{a^2 + 6a + 9}{a^2 + 2a - 3}$ 48. $\dfrac{x^2 + 5x + 6}{x^2 + 6x + 9}$

49. $\dfrac{2x^2 + 5x - 12}{2x^2 + 9x + 4}$ 50. $\dfrac{6x^2 + x - 1}{3x^2 - 10x + 3}$ 51. $\dfrac{4a^2 - 9}{15 + 7a - 2a^2}$

52. $\dfrac{15a^2 - 22a - 5}{1 - 25a^2}$ 53. $\dfrac{7 - 23x + 6x^2}{4x^2 - 49}$ 54. $\dfrac{81x^2 - 25}{10 - 23x + 9x^2}$

55. $\dfrac{2a^2 - 4a + 2}{2a^2 + 2a - 4}$ 56. $\dfrac{3a^2 - 3a - 6}{3a^2 - 9a + 6}$ 57. $\dfrac{x^3 - x^2}{x^3 - x}$

58. $\dfrac{x^4 - x^2}{x^4 - 1}$ 59. $\dfrac{2x^2 + 5xy - 3y^2}{4x^2 + 4xy - 3y^2}$ 60. $\dfrac{4x^2 - 12xy + 9y^2}{2x^2 + 5xy - 12y^2}$

■ *Reduce each fraction if possible. Select the correct response,* a *or* b.

61. $\dfrac{x + 2}{y + 2}$;   a. $\dfrac{x}{y}$ b. Already in lowest terms

62. $\dfrac{2x + 3}{2y}$;   a. $\dfrac{x + 3}{y}$ b. Already in lowest terms

63. $\dfrac{2x + 4}{4}$;   a. $\dfrac{x + 2}{2}$ b. $2x$

64. $\dfrac{6 + 3y}{3y}$;   a. $\dfrac{2 + y}{y}$ b. 6

65. $\dfrac{3(x - 2y)}{3x}$;   a. $-2y$ b. $\dfrac{x - 2y}{x}$

66. $\dfrac{3a + a^2}{3a}$;   a. a^2 b. $\dfrac{3 + a}{3}$

67. $\dfrac{y^2 - 1}{y - 1}$;   a. $y + 1$ b. y

68. $\dfrac{a^3}{a^4 - a^3}$;　　　　a. $\dfrac{1}{a-1}$　　　　b. $\dfrac{1}{a^4}$

■ *Reduce each fraction:*
a. by using the fundamental principle of fractions.
b. by using laws IV and IVa on page 179.

Sample Problem　　$\dfrac{3x^2y^4}{6x^3y}$

a. $\dfrac{3x^2y^4}{6x^3y} = \dfrac{\cancel{3}xx\cancel{y}yyy}{2\cdot\cancel{3}\cancel{x}\cancel{x}x\cancel{y}}$

b. $\dfrac{3x^2y^4}{6x^3y} = \left(\dfrac{3}{6}\right)\left(\dfrac{x^2}{x^3}\right)\left(\dfrac{y^4}{y}\right) = \left(\dfrac{1}{2}\right)\left(\dfrac{y^{4-1}}{x^{3-2}}\right)$

Ans. $\dfrac{y^3}{2x}$　　　　　　*Ans.* $\dfrac{y^3}{2x}$

69. $\dfrac{12x^3y^2}{3xy}$　　**70.** $\dfrac{18x^4y^2}{6xy}$　　**71.** $\dfrac{15x^3y^3}{-3x^4y}$

72. $\dfrac{25x^2y^2}{-5x^5y}$　　**73.** $\dfrac{-2x^3y}{-30x^2y^3}$　　**74.** $\dfrac{-7x^3y^4}{-35x^5y}$

75. Use a numerical example to show that $\dfrac{x+y}{x}$ is not equivalent to y.

76. Use a numerical example to show that $\dfrac{2x+y}{2}$ is not equivalent to $x+y$.

6.3　QUOTIENTS OF POLYNOMIALS

MONOMIAL DIVISORS

In the last section, we simplified a quotient by reducing the fraction. In this section, we study two alternative methods of rewriting quotients in equivalent forms.

We use one method when the divisor is a monomial. As noted in Section 6.1,

$$\frac{a+b+c}{d} = (a+b+c)\cdot\frac{1}{d}$$

and by the distributive property,

$$(a+b+c)\cdot\frac{1}{d} = a\cdot\frac{1}{d} + b\cdot\frac{1}{d} + c\cdot\frac{1}{d}$$

$$= \frac{a}{d} + \frac{b}{d} + \frac{c}{d}.$$

Therefore,

$$\frac{a + b + c}{d} = \frac{a}{d} + \frac{b}{d} + \frac{c}{d}.$$

Thus, a fraction whose numerator is a sum or difference of several terms can be expressed as the sum or difference of fractions whose numerators are the terms of the original numerator and whose denominators are the same as the original denominator. For example,

$$\frac{2x^3 + 4x^2 + 2x}{2x} = \frac{2x^3}{2x} + \frac{4x^2}{2x} + \frac{2x}{2x}.$$

We can simplify each term of the right-hand member to get

$$x^2 + 2x + 1.$$

This approach can also be used when each term of the numerator is not exactly divisible by the denominator. For example,

$$\frac{3x^2 + 2x + 1}{x} = \frac{3x^2}{x} + \frac{2x}{x} + \frac{1}{x}$$

$$= 3x + 2 + \frac{1}{x}.$$

POLYNOMIAL DIVISORS

We use a second alternative method when the divisor has more than one term. In this case we use a process similar to arithmetic long division, as the following examples illustrate.

$21\overline{)674}$ $x + 3\overline{)x^2 + x - 7}$

Divide 2 into 6; the quotient is 3.   Divide x into x^2; the quotient is x.

$$\begin{array}{r} 3 \\ 21\overline{)674} \end{array}$$ $$\begin{array}{r} x \\ x + 3\overline{)x^2 + x - 7} \end{array}$$

Multiply 3 by 21: $3 \cdot 21 = 63$.   Multiply x by $x + 3$: $x(x + 3) = x^2 + 3x$.

$$\begin{array}{r} 3 \\ 21\overline{)674} \\ 63 \end{array}$$ $$\begin{array}{r} x \\ x + 3\overline{)x^2 + x - 7} \\ x^2 + 3x \end{array}$$

Subtract. (Change the sign and add.)   Subtract. (Change the sign and add.)

$$\begin{array}{r} 3 \\ 21\overline{)674} \\ -63 \\ \hline 4 \end{array}$$ $$\begin{array}{r} x \\ x + 3\overline{)x^2 + x - 7} \\ -x^2 - 3x \\ \hline - 2x \end{array}$$

"Bring down" 4.

$$3 \overline{)}$$
$$21\overline{)\,674}$$
$$\underline{-63}\downarrow$$
$$44$$

"Bring down" -7.

$$x$$
$$x + 3\overline{)\ x^2 + \ x - 7}$$
$$\underline{-x^2 - 3x}\qquad \downarrow$$
$$-\,2x - 7$$

Divide 2 into 4; the quotient is 2.

$$32 \leftarrow$$
$$21\overline{)\,674}$$
$$\underline{-63}$$
$$44$$

Divide x into $-2x$; the quotient is -2.

$$x - 2 \leftarrow$$
$$x + 3\overline{)\ x^2 + \ x - 7}$$
$$\underline{-x^2 - 3x}$$
$$-\,2x - 7$$

Multiply 2 by 21: $2 \cdot 21 = 42$.

$$32$$
$$21\overline{)\,674}$$
$$\underline{-63}$$
$$44$$
$$42 \leftarrow$$

Multiply -2 by $x + 3$:
$-2(x + 3) = -2x - 6.$

$$x - 2$$
$$x + 3\overline{)\ x^2 + \ x - 7}$$
$$\underline{-x^2 - 3x}$$
$$-\,2x - 7$$
$$-\,2x - 6 \leftarrow$$

Subtract. (Change the sign and add.)

$$32$$
$$21\overline{)\,674}$$
$$\underline{-63}$$
$$44$$
$$\underline{-42}$$
$$2 \leftarrow$$

Subtract. (Change the signs and add.)

$$x - 2$$
$$x + 3\overline{)\ x^2 + \ x - 7}$$
$$\underline{-x^2 - 3x}$$
$$-2x - 7$$
$$\underline{+2x + 6}$$
$$-\,1 \leftarrow$$

The remainder is 2.

$$32\tfrac{2}{21} \leftarrow$$
$$21\overline{)\,674}$$
$$\underline{-63}$$
$$44$$
$$\underline{-42}$$
$$2$$

The remainder is -1.

$$x - 2 + \tfrac{-1}{x + 3} \leftarrow$$
$$x + 3\overline{)\ x^2 + \ x - 7}$$
$$\underline{-x^2 - 3x}$$
$$-\,2x - 7$$
$$\underline{2x + 6}$$
$$-\,1$$

Ans. $32\dfrac{2}{21}$

Ans. $x - 2 + \dfrac{-1}{x + 3}$

As always, the division is not valid if the divisor is 0. Thus, in the example where the divisor is $x + 3$, we must restrict x from having a value of -3.

When using the long division process, it is convenient to write the dividend in descending powers of the variable. Furthermore, it is helpful to insert a term with a zero coefficient for all powers of the variable that are missing between the highest-degree term and the lowest-degree term. For example, to divide $3x - 1 + 4x^3$ by $2x - 1$, we would rewrite $3x - 1 + 4x^3$ as

descending powers

$$4x^3 + 0x^2 + 3x - 1$$

0 coefficient for the missing x^2 term

EXERCISES 6.3

■ *Divide.*

Sample Problems

a. $\dfrac{6x - 8}{2}$

$$\dfrac{\overset{3}{\cancel{6}x}}{\cancel{2}} - \dfrac{\overset{4}{\cancel{8}}}{\cancel{2}}$$

Ans. $3x - 4$

b. $\dfrac{12x^3 - 8x^2 + 4x}{2x}$

$$\dfrac{\overset{6x^2}{\cancel{12}x^{\cancel{3}}}}{\cancel{2}\cancel{x}} - \dfrac{\overset{4x}{\cancel{8}x^{\cancel{2}}}}{\cancel{2}\cancel{x}} + \dfrac{\overset{2}{\cancel{4}\cancel{x}}}{\cancel{2}\cancel{x}}$$

Ans. $6x^2 - 4x + 2$

1. $\dfrac{8x - 4}{4}$

2. $\dfrac{6x + 3}{3}$

3. $\dfrac{y^2 + 2y}{y}$

4. $\dfrac{y^2 - 4y}{y}$

5. $\dfrac{3x^2 + 9x}{3x}$

6. $\dfrac{4x^2 - 2x}{2x}$

7. $\dfrac{3y^3 - 2y^2 + y}{y}$

8. $\dfrac{y^4 + 2y^3 + y^2}{y^2}$

9. $\dfrac{6y^3 - 3y^2 + 9y}{3y}$

10. $\dfrac{12y^3 + 4y^2 - 8y}{4y}$

11. $\dfrac{4xy^2 - x^2y + xy}{xy}$

12. $\dfrac{9x^2y^2 + 3xy^2 - 3x^2y}{3xy}$

Sample Problems

a. $\dfrac{4x^2 + 2x + 1}{2}$

$$\dfrac{\overset{2}{\cancel{4}x^2}}{\cancel{2}} + \dfrac{\overset{}{\cancel{2}x}}{\cancel{2}} + \dfrac{1}{2}$$

Ans. $2x^2 + x + \dfrac{1}{2}$

b. $\dfrac{2x^3 - x^2 + 4}{-x}$

$$\dfrac{\overset{x^2}{\cancel{2}x^{\cancel{3}}}}{-\cancel{x}} + \dfrac{\overset{x}{-\cancel{x}^{\cancel{2}}}}{-\cancel{x}} + \dfrac{4}{-x}$$

Ans. $-2x^2 + x + \dfrac{-4}{x}$

13. $\dfrac{9x^2 + 6x - 1}{3}$

14. $\dfrac{8x^2 + 4x - 1}{4}$

15. $\dfrac{y^2 + 2y - 1}{y}$

16. $\dfrac{6y^2 + 4y - 3}{2y}$ **17.** $\dfrac{9x^4 - 6x^2 - 2}{3x^2}$ **18.** $\dfrac{6x^4 - 6x^2 - 4}{6x^2}$

19. $\dfrac{y^3 - 3y^2 + 2y - 1}{y}$ **20.** $\dfrac{2y^3 + 8y^2 + 2y - 1}{2y}$

21. $\dfrac{xy^2 + xy + x}{xy}$ **22.** $\dfrac{x^3y^2 + x^2y^3 + xy}{xy^2}$

23. $\dfrac{2x^2y^2 - 4xy^2 + 6xy}{2xy^2}$ **24.** $\dfrac{8x^3y + 4x^2y - 4xy}{4x^2y}$

Sample Problem $(x^2 - 3x - 4) \div (x + 1)$

$$x^2 \div x = x$$
$$-4x \div x = -4$$

$$x - 4$$
$$x + 1 \overline{)x^2 - 3x - 4}$$

Multiply: $x(x + 1) \longrightarrow x^2 + x$

Subtract; change signs of $x^2 + x$ and add $\rightarrow -4x - 4$

Multiply: $-4(x + 1) \longrightarrow -4x - 4$

Subtract; change signs of $-4x - 4$ and add $\rightarrow \quad 0 \longleftarrow$ remainder

Ans. $x - 4$

25. $(x^2 + 5x - 6) \div (x - 1)$ **26.** $(x^2 + x - 6) \div (x - 2)$

27. $(x^2 + 6x + 5) \div (x + 5)$ **28.** $(x^2 - 4x + 4) \div (x - 2)$

29. $(2x^2 - 7x - 4) \div (x - 4)$ **30.** $(2x^2 - x - 3) \div (x + 1)$

31. $(2x^2 + 5x - 3) \div (2x - 1)$ **32.** $(2x^2 - 9x - 5) \div (2x + 1)$

33. $(4x^2 + 4x - 3) \div (2x - 1)$ **34.** $(4x^2 - 8x - 5) \div (2x + 1)$

35. $(2x^3 + 3x^2 - x + 2) \div (x + 2)$ **36.** $(3x^3 - x^2 - 4x + 2) \div (x - 1)$

Sample Problem $(2x^2 + 3x - 3) \div (2x - 1)$

$$2x^2 \div 2x = x$$
$$4x \div 2x = 2$$

$$x + 2$$
$$2x - 1 \overline{)2x^2 + 3x - 3}$$

Multiply: $x(2x - 1) \longrightarrow 2x^2 - x$

Subtract $\longrightarrow 4x - 3$

Multiply: $2(2x - 1) \longrightarrow 4x - 2$

Subtract $\longrightarrow -1 \leftarrow$ remainder

Ans. $x + 2 + \dfrac{-1}{2x - 1}$

37. $(x^2 + 3x + 1) \div (x + 2)$ **38.** $(x^2 - x + 3) \div (x + 1)$

39. $(x^2 + 3x - 9) \div (x + 5)$ **40.** $(x^2 - 2x - 2) \div (x - 3)$

41. $(2x^2 + x - 2) \div (x + 1)$ **42.** $(3x^2 - 8x - 1) \div (x - 3)$

43. $(4x^2 - 4x - 5) \div (2x + 1)$ **44.** $(6x^2 + x + 2) \div (3x + 2)$

Sample Problem $(-1 + x^2) \div (x + 1)$

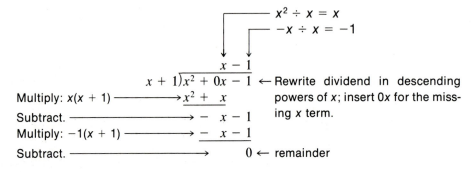

$$
\begin{array}{r}
x - 1 \\
x + 1 \overline{)x^2 + 0x - 1}
\end{array}
$$
$\leftarrow$ Rewrite dividend in descending powers of x; insert $0x$ for the missing x term.

Multiply: $x(x + 1) \longrightarrow x^2 + x$

Subtract. $\longrightarrow -x - 1$

Multiply: $-1(x + 1) \longrightarrow -x - 1$

Subtract. $\longrightarrow 0 \leftarrow$ remainder

Ans. $x - 1$

45. $(x^2 - 49) \div (x - 7)$ **46.** $(x^2 - 4) \div (x + 2)$

47. $(-7 + x^2) \div (x + 6)$ **48.** $(-10 + x^2) \div (x - 7)$

49. $(1 + 2x^3 - x^2) \div (x - 1)$ **50.** $(2x + 4x^3 - 3) \div (x + 2)$

6.4 BUILDING FRACTIONS

Just as it is often convenient to reduce fractions to lowest terms, it is also often convenient to build fractions to higher terms. In particular, we will have to build fractions when we add fractions with unlike denominators in the following chapter. In algebra, as in arithmetic, we use the fundamental principle in the following form in order to build fractions.

> **If both the numerator and denominator of a given fraction are multiplied by the same nonzero number, the resulting fraction is equivalent to the given fraction.**

In symbols:

$$\frac{a}{b} = \frac{a}{b}\left(\frac{c}{c}\right) = \frac{a \cdot c}{b \cdot c}, \qquad c \neq 0.$$

When we apply this principle, we are actually multiplying a quantity by 1 because

$$\frac{2}{2}, \frac{3}{3}, \frac{4}{4}, \ldots \quad \text{are equivalent to 1.}$$

To change $\dfrac{a}{b}$ to a fraction with a denominator bc:

1. Divide b, the denominator of the given fraction, into bc, the denominator to be obtained, to find the *building factor* c.

2. Multiply the numerator and denominator of the given fraction by the building factor c.

For example, to change

$$\frac{3}{x^2y} \qquad \text{to} \qquad \frac{?}{x^3y^2},$$

we divide x^2y into x^3y^2 to obtain the building factor xy. Then we multiply the numerator and denominator of the first fraction by this building factor to obtain

$$\frac{3(xy)}{x^2y(xy)} = \frac{3xy}{x^3y^2}.$$

If negative signs are attached to any part of the fraction, it is usually convenient to write the fraction in standard form before building it. For example, to change

$$-\frac{3x}{xy^2} \qquad \text{to} \qquad \frac{?}{x^2y^4},$$

we first write the fraction in standard form,

$$-\frac{3x}{xy^2} = \frac{-3x}{xy^2}.$$

Then we obtain the building factor, $(x^2y^4) \div (xy^2) = xy^2,$ and multiply the numerator and denominator by the building factor to get

$$\frac{-3x(xy^2)}{xy^2(xy^2)} = \frac{-3x^2y^2}{x^2y^4}.$$

If the denominator of either fraction contains more than one term, it is helpful to first write the denominators in factored form. For example, to change

$$\frac{6}{x+2} \qquad \text{to} \qquad \frac{?}{x^2 - 3x - 10},$$

we first factor $x^2 - 3x - 10$ to get

$$\frac{6}{x+2} = \frac{?}{(x+2)(x-5)}.$$

Next we obtain the building factor

$$(x + 2)(x - 5) \div (x + 2) = (x - 5).$$

Finally, we multiply the numerator and denominator of $\dfrac{6}{x + 2}$ by the building factor $(x - 5)$, to obtain

$$\frac{6(x - 5)}{(x + 2)(x - 5)}.$$

EXERCISES 6.4

■ *Build each fraction to an equivalent fraction with the given denominator.*

Sample Problem

$$\frac{3}{4} = \frac{?}{20}$$

Obtain building factor.

$(20 \div 4 = 5)$

Multiply numerator and denominator of given fraction by 5.

$$\frac{3(5)}{4(5)}$$

Ans. $\dfrac{15}{20}$

1. $\dfrac{2}{5} = \dfrac{?}{10}$ **2.** $\dfrac{3}{7} = \dfrac{?}{21}$ **3.** $\dfrac{2}{3} = \dfrac{?}{18}$ **4.** $\dfrac{4}{9} = \dfrac{?}{36}$

5. $\dfrac{11}{6} = \dfrac{?}{36}$ **6.** $\dfrac{4}{3} = \dfrac{?}{33}$ **7.** $\dfrac{7}{3} = \dfrac{?}{42}$ **8.** $\dfrac{9}{8} = \dfrac{?}{64}$

Sample Problem

$$\frac{2}{-5a} = \frac{?}{10a^2}$$

Write in standard form.

$$\frac{-2}{5a}$$

Obtain building factor.

$(10a^2 \div 5a = 2a)$

Multiply numerator and denominator of given fraction by building factor $2a$.

$$\frac{-2(2a)}{5a(2a)}$$

Ans. $\dfrac{-4a}{10a^2}$

9. $\dfrac{5}{3x} = \dfrac{?}{6x}$

10. $\dfrac{6}{-7a} = \dfrac{?}{14a^2}$

11. $\dfrac{-a}{b} = \dfrac{?}{12b^3}$

12. $-\dfrac{3a}{5b} = \dfrac{?}{15ab}$

13. $\dfrac{-x^2}{y^2} = \dfrac{?}{3y^3}$

14. $\dfrac{-ax}{by} = \dfrac{?}{ab^2y}$

Sample Problem

$ab = \dfrac{?}{ab^2}$

Obtain building factor.

$(ab^2 \div 1 = ab^2)$

Multiply numerator and denominator of given fraction by building factor ab^2

$\dfrac{ab(ab^2)}{1(ab^2)}$

Ans. $\dfrac{a^2b^3}{ab^2}$

15. $2 = \dfrac{?}{36}$

16. $-x = \dfrac{?}{y^2}$

17. $y = \dfrac{?}{xy}$

18. $3a = \dfrac{?}{9b^2}$

19. $x^2 = \dfrac{?}{3x^2y}$

20. $-2b^2 = \dfrac{?}{4a^2b}$

Sample Problem

$\dfrac{1}{3} = \dfrac{?}{3(x-a)}$

Obtain building factor.

$[3(x-a) \div 3 = (x-a)]$

Multiply numerator and denominator of given fraction by building factor $(x-a)$.

$\dfrac{1(x-a)}{3(x-a)}$

Ans. $\dfrac{x-a}{3(x-a)}$

21. $\dfrac{1}{2} = \dfrac{?}{2(x+y)}$

22. $\dfrac{2}{3} = \dfrac{?}{6(x-y)}$

23. $\dfrac{-2a}{5} = \dfrac{?}{5(a+4)}$

24. $\dfrac{3b}{-4} = \dfrac{?}{4(a+b)^2}$

25. $2a = \dfrac{?}{a+3}$

26. $3x = \dfrac{?}{6(x-2)}$

Sample Problem

$\dfrac{2x}{x-y} = \dfrac{?}{(x-y)(x+y)}$

Obtain building factor.

(continued)

$[(x - y)(x + y) \div (x - y) = (x + y)]$

Multiply numerator and denominator of given fraction by building factor $(x + y)$.

Ans. $\dfrac{2x(x + y)}{(x - y)(x + y)}$

27. $\dfrac{3}{x - y} = \dfrac{?}{(x - y)(x + y)}$

28. $\dfrac{2x}{-(x - y)} = \dfrac{?}{(x - y)(x - y)}$

29. $-\dfrac{3}{2x - 1} = \dfrac{?}{(x + 1)(2x - 1)}$

30. $\dfrac{-1}{a + b} = \dfrac{?}{(2a - b)(a + b)}$

31. $\dfrac{7a}{b + 2} = \dfrac{?}{(b - 3)(b + 2)}$

32. $\dfrac{6x^2}{3x - 4} = \dfrac{?}{(3x - 4)(2x + 5)}$

Sample Problem

$\dfrac{3}{x - 3} = \dfrac{?}{x^2 - 7x + 12}$

Factor denominator.

$\dfrac{3}{x - 3} = \dfrac{?}{(x - 3)(x - 4)}$

Obtain building factor.

$[(x - 3)(x - 4) \div (x - 3) = (x - 4)]$

Multiply numerator and denominator of given fraction by building factor $(x - 4)$.

Ans. $\dfrac{3(x - 4)}{(x - 3)(x - 4)}$

33. $\dfrac{a}{a - 3} = \dfrac{?}{a^2 - 3a}$

34. $\dfrac{2}{b} = \dfrac{?}{b + b^2}$

35. $\dfrac{-3}{x + y} = \dfrac{?}{x^2 - y^2}$

36. $\dfrac{-2}{a - b} = \dfrac{?}{a^2 - b^2}$

37. $\dfrac{y}{y - 1} = \dfrac{?}{y^2 + y - 2}$

38. $\dfrac{-1}{x - 1} = \dfrac{?}{x^2 - 2x + 1}$

39. $\dfrac{3x}{2x - 1} = \dfrac{?}{2x^2 - 9x + 4}$

40. $\dfrac{5x}{3x - 2} = \dfrac{?}{3x^2 - 8x + 4}$

41. $\dfrac{x - 2}{2x + 3} = \dfrac{?}{4x^2 + 4x - 3}$

42. $\dfrac{x + 3}{3x + 1} = \dfrac{?}{6x^2 + 5x + 1}$

43. $\dfrac{2a - b}{a - b} = \dfrac{?}{3a^2 - ab - 2b^2}$

44. $\dfrac{a + 4b}{3a - b} = \dfrac{?}{3a^2 + 8ab - 3b^2}$

45. $\dfrac{6x + 1}{x - 1} = \dfrac{?}{4 - 7x + 3x^2}$

46. $\dfrac{3x - 2}{2x - 3} = \dfrac{?}{3 - 8x + 4x^2}$

47. $\dfrac{x+1}{x^2-x} = \dfrac{?}{x^3-2x^2+x}$ **48.** $\dfrac{y+1}{y^2-1} = \dfrac{?}{(y-1)(y^2+2y+1)}$

49. $\dfrac{3x}{4x+2} = \dfrac{?}{4x^3-6x^2-4x}$ **50.** $\dfrac{x}{2x+1} = \dfrac{?}{12x^3-3x}$

6.5 INTEGER EXPONENTS; SCIENTIFIC NOTATION

In this section, we introduce another symbol for a fraction of the form $\dfrac{1}{a^n}$ and use this symbol to write certain numbers in simpler form.

INTEGER EXPONENTS

Recall that we have defined a power a^n (where n is a natural number) as follows:

$$a^n = a \cdot a \cdot a \cdot \cdots \cdot a \quad \textbf{(n factors).}$$

We now give meaning to powers where the exponent is 0 or a negative integer. First let us consider the quotient a^4/a^4. Using the property of quotients of powers on page 179, we have

$$\frac{a^4}{a^4} = a^{4-4} = a^0.$$

Note that for any a not equal to zero, the left-hand member equals 1 and the right-hand member equals a^0. In general, we define

$$a^0 = 1$$

for any number a except 0. For example,

$$3^0 = 1, \qquad 425^0 = 1, \qquad \text{and} \qquad (x^2 y)^0 = 1.$$

Now consider the quotient a^4/a^7. Using the two quotient laws for powers on page 179, we have

$$\frac{a^4}{a^7} = a^{4-7} = a^{-3},$$

and

$$\frac{a^4}{a^7} = \frac{1}{a^{7-4}} = \frac{1}{a^3},$$

Thus, for any a not equal to 0, we can view a^{-3} as equivalent to $\dfrac{1}{a^3}$. In general, we define

$$a^{-n} = \frac{1}{a^n}$$

for any number a except zero. For example,

$$3^{-2} = \frac{1}{3^2}, \qquad 10^{-4} = \frac{1}{10^4}, \qquad \text{and} \qquad y^{-5} = \frac{1}{y^5}.$$

**Common
Error**

Note that

$$3x^{-2} \neq \frac{1}{3x^2}.$$

The exponent only applies to the x, not the 3. Thus,

$$3x^{-2} = 3 \cdot \frac{1}{x^2} = \frac{3}{x^2}.$$

The laws introduced in Sections 4.4 and 6.2 for natural number exponents also apply to expressions involving integer exponents.

Laws for Integer Exponents

$$a^m \cdot a^n = a^{m+n} \qquad \qquad \text{I}$$

$$(a^m)^n = a^{mn} \qquad \qquad \text{II}$$

$$(ab)^n = a^n b^n \qquad \qquad \text{III}$$

$$\frac{a^m}{a^n} = a^{m-n}, \quad a \neq 0 \qquad \text{IV}$$

$$\frac{a^m}{a^n} = \frac{1}{a^{n-m}}, \quad a \neq 0 \qquad \text{IVa}$$

Thus

$$x^5 \cdot x^{-2} = x^{5-2} = x^3 \qquad \text{and} \qquad 3^{-5} \cdot 3^{-1} = 3^{-5-1} = 3^{-6}$$

by applying Law I;

$$(a^{-3})^{-2} = a^{-3(-2)} = a^6 \qquad \text{and} \qquad (2^{-3})^3 = 2^{-3(3)} = 2^{-9}$$

by applying Law II;

$$(3y)^{-2} = 3^{-2}y^{-2} \qquad \text{and} \qquad (ab)^{-3} = a^{-3}b^{-3}$$

by applying Law III.

If we use negative exponents, we may simplify a quotient by using either Law IV or Law IVa above. For example,

$$\frac{x^2}{x^5} = \frac{1}{x^{5-2}} = \frac{1}{x^3}$$

by applying Law IVa, and Law IV yields the equivalent result

$$\frac{x^2}{x^5} = x^{2-5} = x^{-3}.$$

SCIENTIFIC NOTATION

Very large numbers such as

$$5,980,000,000,000,000,000,000,000$$

and very small numbers such as

$$0.000\ 000\ 000\ 000\ 000\ 000\ 000\ 001\ 67$$

occur in many scientific areas. Large numbers can be rewritten in a more compact and useful form by using powers with positive exponents. We can also rewrite small numbers by using powers with negative exponents as introduced in this section.

First, let us consider some factored forms of 38,400 in which one of the factors is a power of 10.

$$38,400 = 3840 \times 10$$
$$= 384 \times 10^2$$
$$= 38.4 \times 10^3$$
$$= 3.84 \times 10^4$$

Although any one of such factored forms may be more useful than the original form of the number, a special name is given to the last form. A number expressed as the product of a number between 1 and 10 (including 1) and a power of 10 is said to be in **scientific form** or **scientific notation.** For example,

$$4.18 \times 10^4, \qquad 9.6 \times 10^2, \qquad \text{and} \qquad 4 \times 10^5$$

are in scientific form.

Now, let us consider some factored forms of 0.0057 in which one of the factors is a power of 10.

$$0.0057 = \frac{0.057}{10} = 0.057 \times \frac{1}{10} = 0.057 \times 10^{-1}$$
$$= \frac{0.57}{100} = 0.57 \times \frac{1}{10^2} = 0.57 \times 10^{-2}$$
$$= \frac{5.7}{1000} = 5.7 \times \frac{1}{10^3} = 5.7 \times 10^{-3}$$

In this case, 5.7×10^{-3} is the scientific form for 0.0057.

To Write a Number in Scientific Form:

1. Move the decimal point so that there is one nonzero digit to the left of the decimal point.
2. Multiply the result by a power of ten with an exponent equal to the number of places the decimal point was moved. The exponent is positive if the decimal point has been moved to the left, and it is negative if the decimal point has been moved to the right.

For example,

$$248. = 2.48 \times 10^2; \qquad 38.05 = 3.805 \times 10^1;$$
$$0.044 = 4.4 \times 10^{-2}; \qquad 0.00241 = 2.41 \times 10^{-3}.$$

If a number is written in scientific form and we want to rewrite it in standard form, we simply move the decimal point the number of places indicated by the power of 10. If the exponent is positive we move the decimal to the right; if it is negative we move the decimal to the left. For example,

$$3.84 \times 10^4 \qquad\qquad 5.6 \times 10^{-2} \qquad\qquad 7 \times 10^{-3}$$

$$= 3.8400 \times 10^4; \qquad = 05.6 \times 10^{-2}; \qquad = 007. \times 10^{-3}$$

$$= 38,400 \qquad\qquad = 0.056 \qquad\qquad = 0.007.$$

Factored forms other than scientific notation are also useful. Calculations involving very large or very small numbers can be simplified by first factoring each number as an *integer* times a power of 10. For example,

$$\frac{1,600,000 \times 0.00081}{0.0027 \times 64,000} = \frac{(16 \times 10^5) \times (81 \times 10^{-5})}{(27 \times 10^{-4}) \times (64 \times 10^3)}.$$

We next use the commutative and associative laws to group together the powers of ten and the integers, and then complete the calculations as follows.

$$\frac{(16 \times 10^5) \times (81 \times 10^{-5})}{(27 \times 10^{-4}) \times (64 \times 10^3)} = \frac{16}{64} \times \frac{81}{27} \times \frac{10^5 \times 10^{-5}}{10^{-4} \times 10^3}$$

$$= \frac{1}{4} \times 3 \times 10^1 = 0.75 \times 10 = 7.5.$$

EXERCISES 6.5

■ *Write each expression without using negative or zero exponents.*

Sample Problems

a. 6^{-2}

b. $3 \cdot 2^{-1}$

c. x^{-3}

Ans. $\dfrac{1}{6^2} = \dfrac{1}{36}$

Ans. $3 \cdot \dfrac{1}{2} = \dfrac{3}{2}$

Ans. $\dfrac{1}{x^3}$

1. 5^{-2} **2.** 4^{-3} **3.** x^{-6} **4.** y^{-2}

5. $(8x)^0$ **6.** $(3y)^0$ **7.** $3 \cdot 4^{-3}$ **8.** $4 \cdot 7^{-2}$

9. $3 \cdot 10^{-3}$ **10.** $2 \cdot 10^{-2}$ **11.** $4x^{-2}$ **12.** $7x^{-4}$

■ *Write each expression using negative exponents.*

Sample Problems

a. $\dfrac{1}{3^2}$ b. $\dfrac{5}{10^2} = 5 \cdot \dfrac{1}{10^2}$ c. $\dfrac{x}{8} = x \cdot \dfrac{1}{2^3}$

Ans. 3^{-2} *Ans.* $5 \cdot 10^{-2}$ *Ans.* $x \cdot 2^{-3}$

13. $\dfrac{1}{2^3}$ **14.** $\dfrac{1}{4^3}$ **15.** $\dfrac{1}{5^2}$ **16.** $\dfrac{1}{3^4}$

17. $\dfrac{1}{4}$ **18.** $\dfrac{1}{9}$ **19.** $\dfrac{x}{25}$ **20.** $\dfrac{y}{36}$

21. $\dfrac{2}{10^2}$ **22.** $\dfrac{3}{10^4}$ **23.** $\dfrac{x}{10^3}$ **24.** $\dfrac{y}{10^5}$

■ *Simplify by using the laws of exponents.*

Sample Problems

a. $x^{-7} \cdot x^3$ Apply law I.   b. $(2^{-5})^3$ Apply law II.

     x^{-7+3} $2^{-5(3)}$

Ans. x^{-4} *Ans.* 2^{-15}

25. $x^{-3} \cdot x^8$ **26.** $y^2 \cdot y^{-6}$ **27.** $(3x^{-4})(5x^{-2})$ **28.** $(2a^{-5})(3a^{-3})$

29. $10^{-7} \times 10^2$ **30.** $10^{-4} \times 10^6$ **31.** $(8^{-2})^5$ **32.** $(5^{-4})^3$

33. $(w^{-6})^{-3}$ **34.** $(t^{-1})^{-4}$ **35.** $(3x)^{-2}$ **36.** $(2y)^{-3}$

37. $(ab^2)^{-3}$ **38.** $(y^2z^4)^{-2}$ **39.** $\dfrac{x^{-8}}{x^{-3}}$ **40.** $\dfrac{a^{-7}}{a^{-6}}$

41. $\dfrac{b^{-4}}{b^{-8}}$ **42.** $\dfrac{x^{-3}}{x^{-9}}$ **43.** $\dfrac{6^6}{6^{-2}}$ **44.** $\dfrac{3^{10}}{3^{-5}}$

Sample Problems

a. $\dfrac{10^4 \times 10^{-5} \times 10^6}{10^{-3} \times 10^7}$ b. $\dfrac{10^{-3} \times 10^2}{10^4 \times 10^{-2} \times 10^3}$

$\dfrac{10^{(4-5+6)}}{10^{(-3+7)}}$ $\dfrac{10^{(-3+2)}}{10^{(4-2+3)}}$

$\dfrac{10^5}{10^4}$ $\dfrac{10^{-1}}{10^5}$

(continued)

$$10^{5-4}$$

$$10^{-1-5}$$

Ans. 10

Ans. 10^{-6}

45. $\dfrac{10^3 \times 10^{-7} \times 10^9}{10^{-2} \times 10^4}$

46. $\dfrac{10^2 \times 10^5 \times 10^{-3}}{10^2 \times 10^2}$

47. $\dfrac{10^{-5} \times 10^{-3} \times 10^2}{10^{-2} \times 10^{-4} \times 10^3}$

48. $\dfrac{10^{-6} \times 10^{-2} \times 10^4}{10^{-7} \times 10^{-3} \times 10^2}$

■ *Complete each factored form.*

49. $234 = 23.4 \times 10$

$\qquad = 2.34 \times$ _____

50. $4800 = 480 \times 10$

$\qquad = 48 \times 10^2$

$\qquad = 4.8 \times$ _____

51. $0.074 = 0.74 \times 10^{-1}$

$\qquad = 7.4 \times$ _____

52. $0.006 = 0.06 \times 10^{-1}$

$\qquad = 0.6 \times 10^{-2}$

$\qquad = 6 \times$ _____

■ *Express each of the following in scientific form.*

a. $62{,}000{,}000$

$\quad 62{,}000{,}000 \times 10^7$

b. 0.000431

$\quad 0.000431 \times 10^{-4}$

Ans. 6.2×10^7

Ans. 4.31×10^{-4}

53. 483 **54.** 5420 **55.** 0.072 **56.** 0.0063

57. 4000 **58.** $230{,}000$ **59.** 0.00063 **60.** 0.000007

■ *Express each of the following in standard decimal form.*

a. 1.47×10^5

$\quad 1.47000$

b. 4.2×10^{-3}

$\quad 004.2$

Ans. $147{,}000$

Ans. 0.0042

61. 4.3×10^4 **62.** 6.1×10^3 **63.** 5.7×10^{-4} **64.** 6.8×10^{-2}

65. 8.234×10^7 **66.** 1.413×10^4 **67.** 8×10^{-6} **68.** 2×10^{-5}

69. The mass of the earth is approximately

$$5,980,000,000,000,000,000,000,000,000 \text{ grams.}$$

Write this number in scientific form.

70. The mass of the hydrogen atom is approximately

$$0.000\ 000\ 000\ 000\ 000\ 000\ 000\ 000\ 001\ 67 \text{ gram.}$$

Write this number in scientific form.

71. Light travels at a speed of 300,000,000 meters per second. Write this number in scientific form.

72. Visible blue light has a wavelength of 0.000 000 45 meters. Write this number in scientific form.

73. The average body cell of an animal has a diameter of 0.000 015 meters. Write this number in scientific form.

74. The diameter of the earth is approximately 6,450,000 meters. Write this number in scientific form.

■ *Simplify by first writing in factored form.*

Sample Problems

a. $\dfrac{3200 \times 0.00015}{160 \times 0.000075}$

$\dfrac{(32 \times 10^2) \times (15 \times 10^{-5})}{(16 \times 10^1) \times (75 \times 10^{-6})}$

$\dfrac{32}{16} \times \dfrac{15}{75} \times \dfrac{10^2 \times 10^{-5}}{10^1 \times 10^{-6}}$

$2 \times \dfrac{1}{5} \times 10^2$

0.4×10^2

Ans. 40

b. $\dfrac{0.0014 \times 0.00009}{0.28 \times 0.03}$

$\dfrac{(14 \times 10^{-4}) \times (9 \times 10^{-5})}{(28 \times 10^{-2}) \times (3 \times 10^{-2})}$

$\dfrac{14}{28} \times \dfrac{9}{3} \times \dfrac{10^{-4} \times 10^{-5}}{10^{-2} \times 10^{-2}}$

$\dfrac{1}{2} \times 3 \times 10^{-5}$

1.5×10^{-5}

Ans. 0.000015

75. $\dfrac{48,000 \times 0.00042}{2400 \times 0.000021}$

76. $\dfrac{620,000 \times 0.0013}{31,000 \times 0.00026}$

77. $\dfrac{0.00084 \times 0.093}{0.00021 \times 0.0031}$

78. $\dfrac{0.00075 \times 0.00048}{0.0012 \times 0.0000025}$

79. $\dfrac{0.65 \times 500 \times 3.3}{1.3 \times 0.011 \times 0.5}$

80. $\dfrac{0.0625 \times 0.091 \times 0.042}{0.007 \times 0.00025 \times 1.3}$

CHAPTER SUMMARY

[6.1]   The quotient of two algebraic expressions is called a **fraction.** We can rewrite a quotient as a product by using the property

$$\frac{a}{b} = a \cdot \frac{1}{b}.$$

A fraction can be changed from one form to another equivalent form by any of the following properties:

$$\frac{a}{b} = \frac{-a}{-b} = -\frac{-a}{b} = -\frac{a}{-b},$$

$$\frac{-a}{b} = \frac{a}{-b} = -\frac{a}{b} = -\frac{-a}{-b}.$$

[6.2]   We can reduce fractions by using the following principles:

$$\frac{a \cdot c}{b \cdot c} = \frac{a}{b},$$

$$\frac{a^m}{a^n} = a^{m-n}, \quad \text{if } m \text{ is greater than } n,$$

$$\frac{a^m}{a^n} = \frac{1}{a^{n-m}}, \quad \text{if } n \text{ is greater than } m.$$

[6.3]   A fraction with a monomial denominator can be rewritten as follows:

$$\frac{a + b + c}{d} = \frac{a}{d} + \frac{b}{d} + \frac{c}{d}.$$

A fraction whose denominator is a polynomial with two or more terms can be rewritten by using a method of long division.

[6.4]   We can build fractions by using the fundamental principle:

$$\frac{a}{b} = \frac{a \cdot c}{b \cdot c}, \qquad c \neq 0.$$

To change a/b to a fraction with denominator bc:

1. Divide b into bc to obtain the building factor c.
2. Multiply numerator and denominator of the given fraction by the building factor c.

[6.5]   Powers a^0 and a^{-n} are defined as follows:

$$a^0 = 1 \quad (a \neq 0)$$

and

$$a^{-n} = \frac{1}{a^n} \quad (a \neq 0).$$

The laws of exponents that apply to natural number powers are also valid for integer powers. A number expressed as the product of a number between 1 and 10 (including 1) and a power of 10 is said to be in **scientific form** or **scientific notation.** We use scientific notation to rewrite very large and very small numbers.

■ *The symbols introduced in this chapter appear on the inside front cover.*

CHAPTER REVIEW

1. Graph the following numbers.

$$\frac{-27}{4}, \frac{-5}{2}, 2, \frac{11}{2}, \frac{37}{4}$$

2. Rewrite in the form $a \cdot \frac{1}{b}$.

   a. $\frac{4}{9}$ b. $\frac{x + 6}{3}$ c. $\frac{2y}{x + y^2}$

3. Rewrite in the form $\frac{ac}{b}$.

   a. $\frac{2}{3}(x - 3)$ b. $-\frac{1}{3}(x^2 + 1)$ c. $-\frac{3}{4}(2x + y)$

4. Change to equivalent fractions in standard form.

   a. $-\frac{3}{x + y}$ b. $-\frac{-a}{x}$ c. $-\frac{b - 2}{4}$

5. a. For what value of y will the fraction $\frac{3}{2y}$ be undefined?

   b. For what value of x will the fraction $\frac{4 + x}{6 + 2x}$ be undefined?

■ *Reduce to lowest terms.*

6. a. $\frac{96x^3y^2z^2}{84xy^3}$ b. $\frac{-21a^3(a - 3b)^5}{14a^5(a - 3b)^3}$

7. a. $\dfrac{b - 3}{b^2 - 2b - 3}$ **b.** $\dfrac{a^2 + a}{a - a^3}$

8. a. $\dfrac{2x + 4}{2x}$ **b.** $\dfrac{x^2 - x}{x}$

9. Simplify the quotient.

   a. $\dfrac{3x^3 + 6x^2 - 9x}{3x}$ **b.** $\dfrac{8x^4 - 4x^2 + 3}{2x^2}$

10. Simplify the quotient by using long division.

   a. $\dfrac{2x^2 - 5x + 1}{x - 1}$ **b.** $\dfrac{2x^3 + x - 3}{x + 2}$

■ *Build each fraction to an equivalent fraction with the given denominator.*

11. a. $\dfrac{4a}{b^2c} = \dfrac{?}{2a^2b^3c}$ **b.** $\dfrac{3}{x - y} = \dfrac{?}{2(x - y)}$

12. a. $\dfrac{x}{x - 2} = \dfrac{?}{x^2 - 3x + 2}$ **b.** $\dfrac{x - 1}{x^2 - 2x} = \dfrac{?}{x^4 - 4x^2}$

13. a. $\dfrac{5}{x - 3} = \dfrac{?}{3 - x}$ **b.** $\dfrac{3x}{x - 3} = \dfrac{?}{9 - x^2}$

14. Rewrite each expression without using zero or negative exponents.

   a. $2 \cdot 5^{-2}$ b. $3^0 \cdot x^{-2}$ c. $4y^{-3}$

15. Rewrite each expression without fractions by using negative exponents.

   a. $\dfrac{3}{x^3}$ b. $\dfrac{-6}{x}$ c. $\left(\dfrac{5x}{2y}\right)^4$

■ *Simplify by using the laws of exponents.*

16. a. $(5x^{-3})(2x^{-4})$ b. $(2a^{-3})^4$ c. $(xy^{-3})^{-5}$

17. a. $\dfrac{x^{-5}}{x^{-2}}$ b. $\dfrac{6x^{-6}}{3x^{-3}}$ c. $\dfrac{10^{-6} \times 10^2}{10^{-3} \times 10^{-4}}$

18. Express in scientific form.

   a. 34,700,000   b. 0.000873

19. Express in standard decimal form.

   a. 4.83×10^4 b. 3.81×10^{-4}

20. Simplify by first writing in factored form.

   a. $\dfrac{0.00045 \times 18,000}{72,000,000 \times 0.15}$ b. $\dfrac{630 \times 0.64}{5600 \times 0.024}$

CUMULATIVE REVIEW

1. If a and b represent numbers, then $a + b = b + a$. This statement is called the ? law of addition.

2. If $(24 \cdot 5) + (24 \cdot 8)$ is divided by 24, the quotient is ? .

3. The sum of the integers between -5 and $+5$ is ? .

4. Write $(300 \cdot 3) + (70 \cdot 3) + (5 \cdot 3)$ as $(? \cdot 3)$.

5. If $x = -2$ and $y = 4$, find the numerical value of $x^3 - 5y$.

6. Simplify: $(3 - x) + (2 + x) - (1 - x)$.

7. Factor completely: $x^3 - x$.

8. Factor completely: $5b^3 + 10b^2 + 5b$.

9. Divide $(x^2 - 6x + 8)$ by $(x - 2)$.

10. Fifty coins in nickels and dimes amount to $3.50. How many of each are there?

11. The sum of three consecutive even integers is 16 more than the next even integer. What are the integers?

12. Multiply $(3x + 2)$ by $(2x - 4)$.

13. Factor completely: $8x^2 - 10x - 3$.

14. Divide $(x^2 + 4x - 1)$ by $(x + 2)$.

15. Simplify the quotient: $\dfrac{2x^4 - x^2 + 1}{x^2}$

16. Write as a basic numeral: $(2x)^0 \cdot 3^{-3}$.

17. Write in scientific form: 4,810,000.

■ *In exercises 18–20, solve for x.*

18. $4x - 6 = 10$ **19.** $\dfrac{2}{3}x + 2 = 6$ **20.** $4 - 3x > 10$

■ *True or false.*

21. $\dfrac{3x + y}{3y}$ is equivalent to $\dfrac{x + y}{y}$.

22. $4x^{-1}$ is equivalent to $\dfrac{1}{4x}$.

23. $\dfrac{-x + 5}{3}$ is equivalent to $\dfrac{x + 5}{-3}$.

24. $(6 - a)^2$ is equivalent to $36 - a^2$.

25. $\dfrac{x^4}{x^{12}}$ is equivalent to $\dfrac{1}{x^3}$.

7 OPERATIONS WITH FRACTIONS

7.1 PRODUCTS OF FRACTIONS

The product of two fractions is defined as follows.

> **The product of two fractions is a fraction whose numerator is the product of the numerators and whose denominator is the product of the denominators of the given fractions.**

In symbols:

$$\frac{a}{b} \cdot \frac{c}{d} = \frac{ac}{bd}.$$

Any common factor occurring in both a numerator and a denominator of either fraction can be divided out either before or after multiplying. For example,

$$\frac{\overset{}{\cancel{3}}}{\underset{2}{\cancel{4}}} \cdot \frac{\cancel{2}}{\cancel{3}} = \frac{1}{2} \cdot \frac{1}{1} = \frac{1}{2},$$

or

$$\frac{3}{4} \cdot \frac{2}{3} = \frac{\cancel{6}}{\underset{2}{\cancel{12}}} = \frac{1}{2}.$$

The same procedure applies to fractions containing variables. For example,

$$\frac{12x^2}{5y} \cdot \frac{10y^3}{3x^2y} = \frac{\overset{4}{\cancel{12x^2}}}{\cancel{5y}} \cdot \frac{\overset{2y}{\cancel{10y^3}}}{\cancel{3x^2y}}$$

$$= \frac{4 \cdot 2y}{1} = 8y.$$

If a negative sign is attached to any of the factors, it is advisable to proceed as if all the factors were positive and then attach the appropriate sign to the result. A positive sign is attached if there are no negative signs or an even number of negative signs on the factors; a negative sign is attached if there is an odd number of negative signs on the factors. For example,

$$\frac{-3x}{4} \cdot \frac{2}{3y} = -\frac{3x}{4} \cdot \frac{2}{3y} = \frac{-x}{2y}.$$

When the fractions contain algebraic expressions, it is necessary to factor wherever possible and divide out common factors before multiplying. For example,

$$\frac{x^2 + 4x + 3}{x^2 - 9} \cdot \frac{x^2 - x - 6}{x^2 + 5x + 6} = \frac{(x+3)(x+1)}{(x+3)(x-3)} \cdot \frac{(x-3)(x+2)}{(x+3)(x+2)}$$

$$= \frac{x+1}{x+3},$$

and

$$\frac{x^2 - x}{2x + 6} \cdot \frac{x^2 + 5x + 6}{x^2 - 1} = \frac{x(x-1)}{2(x+3)} \cdot \frac{(x+3)(x+2)}{(x-1)(x+1)}$$

$$= \frac{x(x+2)}{2(x+1)} = \frac{x^2 + 2x}{2(x+1)}.$$

Note that when writing fractional answers, we will multiply out the numerator and leave the denominator in factored form. Very often, fractions are more useful in this form.

In algebra, we often rewrite an expression such as $a\left(\dfrac{b}{c}\right)$ as an equivalent expression $\dfrac{ab}{c}$. For example,

$$3\left(\frac{x}{4}\right) = \frac{3x}{4}, \qquad 4\left(\frac{1}{y}\right) = \frac{4}{y}, \qquad \text{and} \qquad -2\left(\frac{x}{x+y}\right) = \frac{-2x}{x+y}.$$

Use whichever form is most convenient for a particular problem.

Common Error

Recall that we can only divide out *common factors*, never common terms! For example,

$$\frac{1}{3+x} \cdot \frac{4+x}{y} \neq \frac{4}{3y}$$

because x is a term and *cannot* be divided out. Similarly,

$$\frac{1}{3x} \cdot \frac{3y + 2}{5} \neq \frac{y + 2}{5x}$$

because 3 is not a factor of the entire numerator $3y + 2$.

POWERS OF QUOTIENTS

Consider the following examples of products of fractions.

$$\left(\frac{x}{y}\right)^3 = \left(\frac{x}{y}\right)\left(\frac{x}{y}\right)\left(\frac{x}{y}\right)$$

$$= \frac{x \cdot x \cdot x}{y \cdot y \cdot y} = \frac{x^3}{y^3},$$

and

$$\left(\frac{a^2}{2}\right)^4 = \left(\frac{a^2}{2}\right)\left(\frac{a^2}{2}\right)\left(\frac{a^2}{2}\right)\left(\frac{a^2}{2}\right)$$

$$= \frac{a^2 \cdot a^2 \cdot a^2 \cdot a^2}{2 \cdot 2 \cdot 2 \cdot 2} = \frac{(a^2)^4}{2^4} = \frac{a^8}{16}.$$

These examples suggest the following rule.

> **The power of a quotient equals the quotient of the powers of the numerator and denominator.**

In symbols,

$$\left(\frac{a}{b}\right)^n = \frac{a^n}{b^n}, \qquad b \neq 0. \qquad\qquad \mathbf{V}$$

Thus, for example,

$$\left(\frac{-3x}{2y^5}\right)^3 = \frac{(-3x)^3}{(2y^5)^3} = \frac{-27x^3}{8y^{15}}.$$

EXERCISES 7.1

■ *Multiply.*

Sample Problem

$$\frac{3}{8} \cdot \frac{12}{27}$$

Factor numerators and denominators and divide numerators and denominators by common factors.

$$\frac{\cancel{3}}{2 \cdot 2 \cdot 2} \cdot \frac{2 \cdot 2 \cdot \cancel{3}}{3 \cdot 3 \cdot 3}$$

Multipy remaining factors of numerators and remaining factors of denominators.

Ans. $\dfrac{1}{6}$

1. $\dfrac{1}{2} \cdot \dfrac{3}{4}$ **2.** $\dfrac{16}{38} \cdot \dfrac{19}{12}$ **3.** $\dfrac{18}{121} \cdot \dfrac{99}{90}$

4. $\dfrac{3}{5} \cdot \dfrac{8}{12}$ **5.** $\dfrac{24}{20} \cdot \dfrac{20}{36} \cdot \dfrac{3}{4}$ **6.** $\dfrac{18}{30} \cdot \dfrac{6}{8} \cdot \dfrac{4}{20}$

Sample Problem

$$\frac{-5xy}{3} \cdot \frac{9y}{10x^2y}$$

Divide numerator and denominator by common factors.

$$\frac{-\cancel{5}xy}{\cancel{3}} \cdot \frac{\overset{3}{\cancel{9}y}}{\underset{2x}{\cancel{10}x^2\cancel{y}}}$$

Multiply remaining factors of numerators and remaining factors of denominators. Prefix appropriate sign.

Ans. $\dfrac{-3y}{2x}$

7. $\dfrac{1}{3} \cdot \dfrac{3y}{1}$ **8.** $\dfrac{2}{3} \cdot \dfrac{9x^2}{4}$ **9.** $\dfrac{6x^3}{5} \cdot \dfrac{2}{3x}$

10. $\dfrac{7a}{3} \cdot \dfrac{1}{a^3}$ **11.** $6x^2y \cdot \dfrac{2}{3x^2}$ **12.** $5x^2y^2 \cdot \dfrac{1}{x^3y^3}$

13. $\dfrac{-6xy}{3} \cdot \dfrac{4x}{8xy^2}$ **14.** $\dfrac{-24ab^2}{8a} \cdot \dfrac{21a^2b}{14b}$ **15.** $\dfrac{-21r^2s}{8t} \cdot \dfrac{-14t^2}{3rs}$

16. $\dfrac{-12a^2b}{5c} \cdot \dfrac{10bc^2}{24a^3b}$ **17.** $\dfrac{-6xyz}{4a^2b} \cdot \dfrac{10ab^2}{15xyz^2}$ **18.** $\dfrac{-56x^3yz^2}{24xy^2} \cdot \dfrac{-48z}{28x^2z^3}$

Sample Problem

$$\frac{2x - 4}{3x + 6} \cdot \frac{2x + 3}{x - 2}$$

Factor numerators and denominators and divide numerators and denominators by common factors.

$$\frac{2(\cancel{x-2})}{3(x+2)} \cdot \frac{(2x+3)}{(\cancel{x-2})}$$

Multiply remaining factors of numerators and remaining factors of denominators.

Ans. $\dfrac{2(2x+3)}{3(x+2)}$ or $\dfrac{4x+6}{3(x+2)}$

19. $\dfrac{3x - 9}{5x - 15} \cdot \dfrac{10x - 5}{8x - 4}$

20. $\dfrac{2x + 4}{3x - 9} \cdot \dfrac{x - 3}{x + 2}$

21. $\dfrac{5a + 25}{5a} \cdot \dfrac{10a}{2a + 10}$

22. $\dfrac{3b - 18}{3b} \cdot \dfrac{6b}{4b - 24}$

23. $\dfrac{2a - 4b}{8a + 24b} \cdot \dfrac{6b + 2a}{8b - 4a}$

24. $\dfrac{3x + 15y}{8x - 12y} \cdot \dfrac{18y - 12x}{10y + 2x}$

25. $\dfrac{2x + 3y}{x - 2y} \cdot \dfrac{3x - 6y}{x - 2y} \cdot \dfrac{x - 2y}{6x + 9y}$

26. $\dfrac{7x + 14}{14x - 28} \cdot \dfrac{2x - 4}{x + 2} \cdot \dfrac{x - 3}{x + 1}$

Sample Problem

$\dfrac{x^2 - 2x - 3}{x^2 - 9} \cdot \dfrac{x^2 + 5x + 6}{x^2 - 1}$

Factor numerators and denominators and divide numerators and denominators by common factors.

$\dfrac{(\cancel{x - 3})(\cancel{x + 1})}{(\cancel{x - 3})(\cancel{x + 3})} \cdot \dfrac{(\cancel{x + 3})(x + 2)}{(x - 1)(\cancel{x + 1})}$

Multiply remaining factors of numerators and remaining factors of denominators.

Ans. $\dfrac{x + 2}{x - 1}$

27. $\dfrac{x^2 - 3x - 10}{x^2 + 2x - 35} \cdot \dfrac{x^2 + 4x - 21}{x^2 + 9x + 14}$

28. $\dfrac{4y^2 - 1}{y^2 - 16} \cdot \dfrac{y^2 - 4y}{2y + 1}$

29. $\dfrac{6x^2 - x - 2}{12x^2 + 5x - 2} \cdot \dfrac{8x^2 - 6x + 1}{4x^2 - 1}$

30. $\dfrac{y^2 - y - 20}{y^2 + 7y + 12} \cdot \dfrac{y^2 + 9y + 18}{y^2 - 7y + 10}$

31. $\dfrac{x^2 + x - 2}{2 - 3x + x^2} \cdot \dfrac{x^2 - x - 2}{6 + 5x + x^2}$

32. $\dfrac{x^2 + x - 6}{6 - x - x^2} \cdot \dfrac{6 + x - x^2}{x^2 - 5x + 6}$

33. $\dfrac{x^2 - 4}{x^2 - 1} \cdot \dfrac{1 - x}{2x^2 + 4x}$

34. $\dfrac{a^2 - a}{4a^2 + 2a} \cdot \dfrac{2a + 1}{1 - a}$

35. $\dfrac{6x^2 - 11x - 10}{7x^2 - 29x + 4} \cdot \dfrac{6x^2 - 23x - 4}{12x^2 - 28x - 5}$

36. $\dfrac{12x^2 - 11x + 2}{12x^2 + 14x + 4} \cdot \dfrac{12x^2 - 2x - 4}{16x^2 - 8x + 1}$

37. $\dfrac{9x^2 - y^2}{4x^2 - 9y^2} \cdot \dfrac{15y^2 + 16xy + 4x^2}{5y^2 - 13xy - 6x^2}$

38. $\dfrac{4x^2 + 17xy + 15y^2}{25x^2 - 4y^2} \cdot \dfrac{6y^2 + 17xy + 5x^2}{25y^2 - 16x^2}$

39. $\dfrac{y^2 - y - 20}{y^2 - 6y + 5} \cdot \dfrac{y^2 + 5y - 6}{y^2 + 7y + 12} \cdot \dfrac{y^2 - 9}{y^2 - 36}$

40. $\dfrac{3x^3 - 3x^2}{x^2 - 3x + 2} \cdot \dfrac{4x^2 - 4}{2x^3 - x^2 - x} \cdot \dfrac{4x^2 - 1}{2x^2 - 5x + 2}$

41. $\dfrac{x^2 - xy}{xy + y^2} \cdot \dfrac{x^2 - 4y^2}{x^2 - y^2} \cdot \dfrac{x^2 - 2xy - 3y^2}{x^2 - 5xy + 6y^2}$

42. $\dfrac{x^3y - xy^3}{x^2y^2 - 1} \cdot \dfrac{x^2y^2 + xy}{x^3 + 2x^2y + xy^2} \cdot \dfrac{x^2y^2 + x^2y}{xy^3 + xy^2}$

■ *Change each of the following to the form* $\dfrac{ab}{c}$.

Sample Problems

a. $\dfrac{3}{4} a$

b. $\dfrac{5}{b} a$

c. $\dfrac{2}{3} (x - y)$

d. $-\dfrac{1}{3} x$

Ans. $\dfrac{3a}{4}$

Ans. $\dfrac{5a}{b}$

Ans. $\dfrac{2(x - y)}{3}$

Ans. $\dfrac{-x}{3}$

43. $\dfrac{2}{3} x$

44. $\dfrac{3}{4} y$

45. $-\dfrac{2}{5} a$

46. $-\dfrac{4}{7} b$

47. $\dfrac{3}{4} (a - b)$

48. $\dfrac{2}{3} (b - c)$

49. $-\dfrac{3}{5} (2x - y)$

50. $-\dfrac{4}{7} (x - 2y)$

■ *Change each of the following to the form* $\dfrac{a}{b} c$.

Sample Problems

a. $\dfrac{2a}{5}$

b. $\dfrac{b}{3}$

c. $\dfrac{2(x + 2y)}{5}$

Ans. $\dfrac{2}{5} a$

Ans. $\dfrac{1}{3} b$

Ans. $\dfrac{2}{5} (x + 2y)$

51. $\dfrac{3x}{7}$

52. $\dfrac{4y}{3}$

53. $\dfrac{-5a}{7}$

54. $\dfrac{-b}{5}$

55. $\dfrac{5(a - b)}{2}$

56. $\dfrac{-3(a + 2b)}{4}$

57. $\dfrac{x + y}{7}$

58. $\dfrac{-(x + y)}{5}$

■ *Simplify.*

Sample Problems

a. $\left(\dfrac{3x}{4yz}\right)^2$

$\dfrac{(3x)^2}{(4yz)^2}$

Ans. $\dfrac{9x^2}{16y^2z^2}$

b. $\left(\dfrac{-2ab^4}{c^5}\right)^3$

$\dfrac{(-2ab^4)^3}{(c^5)^3}$

Ans. $\dfrac{-8a^3b^{12}}{c^{15}}$

59. $\left(\dfrac{5a}{4b}\right)^3$ **60.** $\left(\dfrac{11a}{13b}\right)^2$ **61.** $\left(\dfrac{-3x^2y}{2z^3}\right)^4$ **62.** $\left(\dfrac{-4x^3y^3}{3z^2}\right)^3$

63. $\left(\dfrac{a^2b^3}{ab}\right)^4$ **64.** $\left(\dfrac{2a^5b^2}{a^3b}\right)^5$ **65.** $\left(\dfrac{-5x^2z^9}{3x^8yz^4}\right)^4$ **66.** $\left(\dfrac{-3x^6yz^5}{4x^{10}y^5}\right)^3$

67. Use a numerical example to show that $\dfrac{1}{x+y} \cdot \dfrac{y}{3}$ is not equivalent to $\dfrac{1}{3x}$.

68. Use a numerical example to show that $\dfrac{1}{x-2} \cdot \dfrac{2-x}{3}$ is not equivalent to $\dfrac{1}{3}$.

7.2 QUOTIENTS OF FRACTIONS

In dividing one fraction by another, we look for a number that, when multiplied by the divisor, yields the dividend. This is precisely the same notion as that of dividing one integer by another; $a \div b$ is a number q, the quotient, such that $bq = a$.

To find $\dfrac{1}{2} \div \dfrac{2}{3}$, we look for a number q such that $\dfrac{2}{3}q = \dfrac{1}{2}$. To solve this equation for q, we multiply each member of the equation by $\dfrac{3}{2}$. Thus,

$$\left(\dfrac{3}{2}\right)\dfrac{2}{3}q = \dfrac{1}{2}\left(\dfrac{3}{2}\right),$$

$$q = \dfrac{1}{2} \cdot \dfrac{3}{2},$$

and since $q = \dfrac{1}{2} \div \dfrac{2}{3}$,

$$\dfrac{1}{2} \div \dfrac{2}{3} = \dfrac{1}{2} \cdot \dfrac{3}{2},$$

which can be simplified to $\dfrac{3}{4}$.

In the example above, we call the number $\dfrac{3}{2}$ the **reciprocal** of the number $\dfrac{2}{3}$.

In general, the reciprocal of a fraction $\dfrac{a}{b}$ $(b \neq 0)$ is the fraction $\dfrac{b}{a}$ $(a \neq 0)$. That is, we obtain the reciprocal of a fraction by "inverting" the fraction. In general:

The quotient of two fractions equals the product of the dividend and the reciprocal of the divisor.

That is, to divide one fraction by another, we invert the divisor and multiply. In symbols:

$$\frac{a}{b} \div \frac{c}{d} = \frac{a}{b} \cdot \frac{d}{c}.$$

For example,

$$\frac{2}{x} \div \frac{3}{y} = \frac{2}{x} \cdot \frac{y}{3}$$

— Change to multiplication.

— Divisor inverted.

$$= \frac{2y}{3x}.$$

As in multiplication, when fractions in a quotient have signs attached, it is advisable to proceed with the problem as if all the factors were positive and then attach the appropriate sign to the solution. For example,

$$\frac{-3x}{2y} \div \frac{9x^2}{8y^2} = -\frac{3x}{2y} \cdot \frac{8y^2}{9x^2} = \frac{-4y}{3x}.$$

Some quotients occur so frequently that it is helpful to recognize equivalent forms directly. One case is

$$1 \div \frac{a}{b} = \frac{1}{1} \cdot \frac{b}{a} = \frac{b}{a}.$$

In other words, 1 divided by a fraction yields the reciprocal of the fraction. Thus,

$$1 \div \frac{2}{3} = \frac{3}{2}, \qquad 1 \div \frac{x}{2y} = \frac{2y}{x}, \qquad \text{and} \qquad 1 \div \frac{3}{x+y} = \frac{x+y}{3}.$$

In general:

$$a \div \frac{b}{c} = \frac{a}{1} \cdot \frac{c}{b} = \frac{ac}{b}$$

Thus,

$$4 \div \frac{3}{2} = \frac{8}{3}, \qquad x \div \frac{3}{y} = \frac{xy}{3}, \qquad \text{and} \qquad 2 \div \frac{3}{x+y} = \frac{2(x+y)}{3}.$$

When the fractions in a quotient involve algebraic expressions, it is easiest to factor wherever possible and divide out common factors before multiplying.

For example,

$$\frac{x^2 - 4}{3x} \div \frac{x + 2}{x} = \frac{(x + 2)(x - 2)}{3x} \cdot \frac{x}{x + 2}$$

$$= \frac{x - 2}{3},$$

and

$$\frac{a^2 - 3a + 2}{a^2 - 5a + 6} \div \frac{a^2 - a}{2a} = \frac{(a - 2)(a - 1)}{(a - 2)(a - 3)} \cdot \frac{2a}{a(a - 1)}$$

$$= \frac{2}{a - 3}.$$

EXERCISES 7.2

■ *Divide.*

Sample Problems

a. $\dfrac{1}{6} \div \dfrac{5}{8}$ b. $\dfrac{2}{3} \div 8$ c. $6 \div \dfrac{3}{4}$

Invert divisor and multiply.

$$\frac{1}{\overset{}{\underset{3}{6}}} \cdot \frac{\overset{4}{8}}{5}$$

$$\frac{\overset{}{2}}{3} \cdot \frac{1}{\overset{}{\underset{4}{8}}}$$

$$\frac{\overset{}{6}}{1} \cdot \frac{\overset{2}{4}}{\overset{}{\underset{}{3}}}$$

Ans. $\dfrac{4}{15}$ *Ans.* $\dfrac{1}{12}$ *Ans.* 8

1. $\dfrac{3}{4} \div \dfrac{5}{6}$ 2. $\dfrac{2}{7} \div \dfrac{4}{3}$ 3. $\dfrac{2}{3} \div \dfrac{4}{9}$ 4. $\dfrac{3}{5} \div \dfrac{7}{10}$

5. $4 \div \dfrac{6}{5}$ 6. $8 \div \dfrac{2}{3}$ 7. $10 \div \dfrac{5}{3}$ 8. $9 \div \dfrac{3}{2}$

9. $1 \div \dfrac{7}{8}$ 10. $1 \div \dfrac{5}{9}$ 11. $4 \div \dfrac{3}{7}$ 12. $6 \div \dfrac{2}{5}$

Sample Problems

a. $\dfrac{3a^2x}{2by} \div \dfrac{6ax^2}{b^2y^2}$ b. $\dfrac{-a^2}{b} \div a^3$ c. $a^3 \div \dfrac{a^2}{b^2}$

Invert divisor and multiply.

$$\frac{\overset{a}{3a^2x}}{2by} \cdot \frac{\overset{b\ y}{b^2y^2}}{6ax^2}$$

$$\frac{-a^2}{b} \cdot \frac{1}{a^3}$$

$$\frac{\overset{a}{a^3}}{1} \cdot \frac{b^2}{a^2}$$

$$2\ x$$

$$a$$

$Ans.\ \dfrac{aby}{4x}$ $Ans.\ \dfrac{-1}{ab}$ $Ans.\ ab^2$

13. $\dfrac{2c}{3d} \div \dfrac{4c}{6d}$ **14.** $\dfrac{c^2}{d} \div \dfrac{c^4}{d^2}$ **15.** $\dfrac{15}{27ab} \div \dfrac{16b}{9a}$

16. $\dfrac{a}{b^2} \div \dfrac{ab^2}{b^3}$ **17.** $\dfrac{-x^2y^2}{u^2v^2} \div \dfrac{xy^2}{u^2v}$ **18.** $\dfrac{14a^2b^3}{15x^2y} \div \dfrac{-21a^2b^2}{35xy}$

19. $\dfrac{3xy}{4} \div (-12y^2)$ **20.** $\dfrac{36x^3}{7y} \div (-3x^2)$ **21.** $\dfrac{2ab^2}{3} \div (4a^2b)$

22. $\dfrac{3a^2b^4}{7} \div (6ab^3)$ **23.** $y \div \dfrac{3x}{y^3}$ **24.** $x \div \dfrac{-2y}{x^2}$

25. $16y^2 \div \dfrac{4y}{3}$ **26.** $ax^2 \div \dfrac{x^2}{b}$ **27.** $1 \div \dfrac{x}{2y}$

28. $1 \div \dfrac{x^2}{3y}$ **29.** $1 \div \dfrac{2a}{a+b}$ **30.** $1 \div \dfrac{a^2}{a-b}$

Sample Problem $\dfrac{3xy + x}{y^2 - y} \div \dfrac{3y + 1}{y}$

Invert divisor.

$\dfrac{3xy + x}{y^2 - y} \cdot \dfrac{y}{3y + 1}$

Factor where possible and simplify.

$\dfrac{x(3y+1)}{y(y-1)} \cdot \dfrac{y}{3y+1}$

$Ans.\ \dfrac{x}{y - 1}$

31. $\dfrac{a^2 - ab}{ab} \div \dfrac{2a - 2b}{ab}$ **32.** $\dfrac{2x - 2y}{xy} \div \dfrac{4x - 4y}{xy}$

33. $\dfrac{6a - 12}{3a + 9} \div \dfrac{4a - 8}{5a + 15}$ **34.** $\dfrac{x^2 + xy}{x^2 - xy} \div \dfrac{x + y}{4x - 4y}$

35. $\dfrac{10x^2 - 5x}{12x^3 + 24x^2} \div \dfrac{x - 2x^2}{2x^2 + 4x}$ **36.** $\dfrac{ax - ay}{bx + by} \div \dfrac{cy - cx}{dx + dy}$

Sample Problem $\dfrac{x^2 - 2x - 8}{x^2 + x - 2} \div \dfrac{2x^2 - 5x + 2}{x^2 - 3x + 2}$

Invert divisor.

$$\frac{x^2 - 2x - 8}{x^2 + x - 2} \cdot \frac{x^2 - 3x + 2}{2x^2 - 5x + 2}$$

Factor where possible and simplify.

$$\frac{(x - 4)(x + 2)}{(x + 2)(x - 1)} \cdot \frac{(x - 2)(x - 1)}{(2x - 1)(x - 2)}$$

Ans. $\dfrac{x - 4}{2x - 1}$

37. $\dfrac{4x^2 - y^2}{x^2 - 4y^2} \div \dfrac{2x - y}{x - 2y}$

38. $\dfrac{x^2 - 9y^2}{16x^2 - y^2} \div \dfrac{x + 3y}{4x - y}$

39. $\dfrac{y^2 - 6y + 5}{y^2 + 8y + 7} \div \dfrac{y^2 - 3y - 10}{y^2 + 3y + 2}$

40. $\dfrac{x^2 - 8x + 15}{x^2 + 9x + 14} \div \dfrac{x^2 + 4x - 21}{x^2 - 6x - 16}$

41. $\dfrac{x^2 - x - 6}{x^2 + 2x - 15} \div \dfrac{x^2 - 4}{x^2 - 25}$

42. $\dfrac{6x^2 - x - 2}{12x^2 + 5x - 2} \div \dfrac{4x^2 - 1}{8x^2 - 6x + 1}$

43. $\dfrac{y^2 - y - 20}{y^2 - 6y + 5} \div \dfrac{y^2 - 7y + 10}{y^2 + y - 2}$

44. $\dfrac{x^2 - x - 2}{x^2 + 5x - 6} \div \dfrac{x^2 - 3x - 4}{x^2 - x - 12}$

45. $\dfrac{2x^2 - x - 28}{3x^2 - x - 2} \div \dfrac{4x^2 + 16x + 7}{3x^2 + 11x + 6}$

46. $\dfrac{y^2 + 7y + 10}{y^2 + 7y + 12} \div \dfrac{y^2 + 6y + 5}{y^2 + 8y + 16}$

47. $\dfrac{3y + 2}{5y^2 - y} \cdot \dfrac{2y^2 - y}{2y^2 - y - 1} \div \dfrac{6y^2 + y - 2}{10y^2 + 3y - 1}$

48. $\dfrac{x^2 + 4x + 3}{x^2 - 8x + 7} \cdot \dfrac{x^2 - 2x - 35}{x^2 - 7x - 8} \div \dfrac{x^2 + 8x + 15}{x^2 - 9x + 8}$

49. $\dfrac{3x^2 + 2x - 5}{2x^2 + x - 6} \cdot \dfrac{2x - 3}{3x + 5} \div \dfrac{x^2 + 2x - 3}{2x^2 + 3x - 2}$

50. $\dfrac{x^2 - x}{x^2 - 2x - 3} \cdot \dfrac{x^2 + 2x + 1}{x^2 + 4x} \div \dfrac{x^2 - 3x - 4}{x^2 - 16}$

51. $\dfrac{a^2 + 11a + 18}{a^2 + 4a - 5} \cdot \dfrac{a^2 - 6a - 7}{a^2 + 8a + 12} \div \dfrac{a^2 - 7a - 8}{a^2 + 2a - 15}$

52. $\dfrac{2y^2 - 5y - 3}{18y^2 + 3y - 1} \cdot \dfrac{6y^2 + 5y + 1}{2y + 1} \div \dfrac{2y^2 - 7y + 3}{36y^2 - 1}$

53. $\left(\dfrac{x + 3}{x} \cdot \dfrac{-2}{x + 3}\right) \div \dfrac{4x^2}{x^2 + 3x}$

54. $\left(\dfrac{x - 2}{3x} \cdot \dfrac{1}{2x - 4}\right) \div \dfrac{2x + 1}{5x^2}$

55. $\dfrac{x^2 - 1}{x + 3} \div \left(\dfrac{x + 4}{x^2} \cdot \dfrac{2x + 2}{x + 4}\right)$

56. $\dfrac{x - 4}{9 - x^2} \div \left(\dfrac{2x^2}{x + 1} \cdot \dfrac{2x - 8}{2x + 6}\right)$

57. $\dfrac{x^2 - 16}{4x - 16} \div \left(\dfrac{3x + 12}{3x + 6} \div \dfrac{4x - 8}{x^2 - 4}\right)$

58. $\dfrac{3x^3 - 3x^2}{9x^2 + 3x} \div \left(\dfrac{9x + 9}{9x - 27} \div \dfrac{3x^2 + x}{3x^2 - 9x}\right)$

59. $\dfrac{x^2 + 2x - 3}{x^2 + x - 2} \div \left(\dfrac{x^2 + 4x + 3}{x^2 - x - 6} \div \dfrac{x^2 - 9}{x^2 + x - 6}\right)$

60. $\dfrac{2x^2 + 3x - 2}{x^2 + 2x - 3} \div \left(\dfrac{x^2 + 4x + 4}{3x^2 + 7x + 2} \div \dfrac{2x^2 + x - 1}{6x^2 - x - 1}\right)$

7.3 SUMS AND DIFFERENCES OF FRACTIONS WITH LIKE DENOMINATORS

The sum of two or more arithmetic or algebraic fractions is defined as follows:

The sum of two or more fractions with common denominators is a fraction with the same denominator and a numerator equal to the sum of the numerators of the original fractions.

Thus,

$$\frac{4}{8} + \frac{9}{8} = \frac{13}{8},$$

$$\frac{2}{x} + \frac{5}{x} = \frac{7}{x},$$

$$\frac{3}{x+1} + \frac{x}{x+1} = \frac{3+x}{x+1}.$$

In general,

$$\frac{a}{c} + \frac{b}{c} = \frac{a+b}{c}.$$

When subtraction is involved, we note that

$$\frac{a}{c} - \frac{b}{c} = \frac{a}{c} + \left(-\frac{b}{c}\right)$$

$$= \frac{a}{c} + \frac{(-b)}{c}$$

$$= \frac{a-b}{c}.$$

Thus it is helpful to change to standard form before adding. For example,

$$\frac{3}{x} - \frac{5}{x} = \frac{3}{x} + \frac{-5}{x}$$

$$= \frac{3+(-5)}{x} = \frac{-2}{x}.$$

We must be especially careful with binomial numerators. For example, we should rewrite

$$\frac{5}{x+1} - \frac{x-2}{x+1}$$

as

$$\frac{5}{x+1} + \frac{-(x-2)}{x+1},$$

where *the entire numerator is enclosed within parentheses*. Now we can proceed to get

$$\frac{5}{x+1} + \frac{-(x-2)}{x+1} = \frac{5-(x-2)}{x+1}$$

$$= \frac{5-x+2}{x+1} \qquad \text{The sign of } each \\ term \text{ of the binomial} \\ \text{is changed.}$$

$$= \frac{7-x}{x+1}.$$

EXERCISES 7.3

■ *Write each sum or difference as a single term.*

Sample Problems

a. $\dfrac{x}{5} + \dfrac{y}{5}$

b. $\dfrac{2x}{7} - \dfrac{3y}{7}$

Write in standard form.

$$\frac{2x}{7} + \frac{-3y}{7}$$

Add numerators.

Ans. $\dfrac{x+y}{5}$

Ans. $\dfrac{2x-3y}{7}$

1. $\dfrac{2}{9} + \dfrac{5}{9}$

2. $\dfrac{4}{7} + \dfrac{2}{7}$

3. $\dfrac{5x}{11} - \dfrac{3}{11}$

4. $\dfrac{2}{13} - \dfrac{3y}{13}$

5. $\dfrac{2}{5} + \dfrac{1}{5} - \dfrac{x}{5}$

6. $\dfrac{4}{7} - \dfrac{x}{7} + \dfrac{y}{7}$

Sample Problems

a. $\dfrac{3}{2a} - \dfrac{1}{2a}$

b. $\dfrac{3x}{4y} - \dfrac{x}{4y}$

Write in standard form.

$$\frac{3}{2a} + \frac{-1}{2a}$$

$$\frac{3x}{4y} + \frac{-x}{4y}$$

Add numerators and simplify.

$$\frac{\overset{1}{2}}{2a}$$

$$\frac{\overset{x}{2x}}{\underset{2}{4y}}$$

Ans. $\dfrac{1}{a}$ *Ans.* $\dfrac{x}{2y}$

7. $\dfrac{5}{2a} + \dfrac{3}{2a}$ 8. $\dfrac{6}{5a} + \dfrac{4}{5a}$ 9. $\dfrac{7}{2b} - \dfrac{5}{2b}$

10. $\dfrac{5}{3a} - \dfrac{8}{3a}$ 11. $\dfrac{7}{3x} + \dfrac{7}{3x} - \dfrac{2}{3x}$ 12. $\dfrac{2}{7x} - \dfrac{10}{7x} - \dfrac{13}{7x}$

13. $\dfrac{4x}{5y} - \dfrac{x}{5y} + \dfrac{2x}{5y}$ 14. $\dfrac{7x}{3y} - \dfrac{5x}{3y} - \dfrac{14x}{3y}$

Sample Problems

a. $\dfrac{x-y}{a} + \dfrac{y}{a}$ b. $\dfrac{x+y}{x} + \dfrac{x-y}{x}$

                                            Add numerators.

$\dfrac{x-y+y}{a}$ $\dfrac{x+y+x-y}{x}$

                                          Simplify.

Ans. $\dfrac{x}{a}$ $\dfrac{2\cancel{x}}{\cancel{x}}$

                             Ans. 2

15. $\dfrac{x+1}{2} + \dfrac{3}{2}$ 16. $\dfrac{x-2}{5} + \dfrac{3}{5}$

17. $\dfrac{x-2y}{3x} + \dfrac{x+3y}{3x}$ 18. $\dfrac{3-x}{2y} + \dfrac{4-x}{2y}$

19. $\dfrac{x+1}{2a} + \dfrac{x-1}{2a}$ 20. $\dfrac{2x-y}{3y} + \dfrac{2x+2y}{3y}$

21. $\dfrac{x^2-x}{2} + \dfrac{x^2}{2} + \dfrac{3x}{2}$ 22. $\dfrac{2x-y}{3} + \dfrac{x-y}{3} + \dfrac{x+y}{3}$

23. $\dfrac{2x+y}{y} + \dfrac{x-2y}{y} + \dfrac{x+y}{y}$ 24. $\dfrac{x-2y}{2x} + \dfrac{x+y}{2x} + \dfrac{2x+y}{2x}$

Sample Problem

$\dfrac{x-2}{a+b} - \dfrac{2x+1}{a+b}$

                         Insert parentheses and write in standard form.

$\dfrac{(x-2)}{a+b} + \dfrac{-(2x+1)}{a+b}$

                         Add numerators.

$\dfrac{(x-2)-(2x+1)}{a+b}$

                         Remove parentheses. Simplify.

$$\frac{x - 2 - 2x - 1}{a + b}$$

Ans. $\dfrac{-x - 3}{a + b}$

25. $\dfrac{2x + 3}{2} - \dfrac{x - 3}{2}$

26. $\dfrac{2x - y}{3} - \dfrac{3x - y}{3}$

27. $\dfrac{2a + b}{a - b} - \dfrac{a - 2b}{a - b}$

28. $\dfrac{2a - b}{b} - \dfrac{a - 2b}{b}$

29. $\dfrac{3}{a + b} - \dfrac{a + 3}{a + b}$

30. $\dfrac{b - 1}{a} - \dfrac{b + 1}{a}$

31. $\dfrac{2x - y}{x + y} - \dfrac{x - 3y}{x + y} + \dfrac{2x}{x + y}$

32. $\dfrac{2u - 3v}{u + 2} - \dfrac{u + 2v}{u + 2} + \dfrac{u}{u + 2}$

33. $\dfrac{3}{x + 2y} - \dfrac{x + 3}{x + 2y} + \dfrac{x + 1}{x + 2y}$

34. $\dfrac{2x - y}{x - y} + \dfrac{x - 2y}{x - y} - \dfrac{3x - 3y}{x - y}$

Sample Problem

$$\frac{2x - y}{2a + 2b} + \frac{2x - 3y}{2a + 2b}$$

Add numerators.

$$\frac{(2x - y) + (2x - 3y)}{2a + 2b}$$

Remove parentheses and combine like terms.

$$\frac{4x - 4y}{2a + 2b}$$

Factor numerator and denominator and reduce.

$$\frac{\overset{2}{\cancel{4}}(x - y)}{\cancel{2}(a + b)}$$

Ans. $\dfrac{2x - 2y}{a + b}$

35. $\dfrac{2a + b}{3} + \dfrac{4a - 2b}{3}$

36. $\dfrac{6x - 6y}{5} + \dfrac{4x - 4y}{5}$

37. $\dfrac{2x + y}{2} + \dfrac{4x + y}{2}$

38. $\dfrac{x - y}{4} + \dfrac{3x - 7y}{4}$

39. $\dfrac{3u + 2v}{4u - 2v} - \dfrac{u + 2v}{4u - 2v}$

40. $\dfrac{u + 7}{2u - 4v} - \dfrac{5 - u}{2u - 4v}$

41. $\dfrac{x}{2x + 4} - \dfrac{2x - 3}{2x + 4} + \dfrac{3x - 5}{2x + 4}$

42. $\dfrac{x + y}{2(x - y)} - \dfrac{2y - 2x}{2(x - y)} + \dfrac{x - 3y}{2(x - y)}$

Sample Problem

$$\frac{3}{x^2 + 2x + 1} - \frac{2 - x}{x^2 + 2x + 1}$$

Insert parentheses and write in standard form.

$$\frac{3}{x^2 + 2x + 1} + \frac{-(2 - x)}{x^2 + 2x + 1}$$

Add numerators and simplify.

$$\frac{3 - (2 - x)}{x^2 + 2x + 1}$$

$$\frac{x + 1}{x^2 + 2x + 1}$$

Factor denominator and reduce to lowest terms.

$$\frac{(x + 1)}{(x + 1)(x + 1)}$$

$Ans.$ $\dfrac{1}{x + 1}$

43. $\dfrac{x + 1}{x^2 - 2x + 1} - \dfrac{5 - 3x}{x^2 - 2x + 1}$

44. $\dfrac{2x + 1}{x^2 - x - 6} - \dfrac{x - 1}{x^2 - x - 6}$

45. $\dfrac{2x - 3}{x^2 + 3x - 4} + \dfrac{7 - x}{x^2 + 3x - 4}$

46. $\dfrac{u + 3v}{u^2 - v^2} + \dfrac{3u + v}{u^2 - v^2}$

47. $\dfrac{x^2 - 2}{x^2 - x} - \dfrac{2 - 4x}{x^2 - x} - \dfrac{1}{x^2 - x}$

48. $\dfrac{3x^2 - 4}{x^2 - 4} - \dfrac{x^2}{x^2 - 4} - \dfrac{4}{x^2 - 4}$

7.4 SUMS OF FRACTIONS WITH UNLIKE DENOMINATORS

In Section 7.3, we added fractions with like denominators. In this section, we add fractions with unlike denominators.

LOWEST COMMON DENOMINATOR

In general, the smallest natural number that is a multiple of each of the denominators of a set of fractions is called the **lowest common denominator** (L.C.D.) of the set of fractions. Sometimes, we can obtain the L.C.D. by inspection. If the L.C.D. is not immediately evident, we can use a special procedure to find it. For example, consider the fractions $\frac{5}{12}$, $\frac{3}{10}$, and $\frac{1}{6}$. The lowest common denominator contains among its factors the factors of 12, 10, and 6.

$$\begin{array}{cc} \textit{Denominators} & \textit{Factors} \\ 12 & 2 \cdot 2 \cdot 3 \\ 10 & 2 \cdot \cdot 5 \\ 6 & 2 \cdot 3 \end{array}$$

The L.C.D. contains
the factors $\longrightarrow 2 \cdot 2 \cdot 3 \cdot 5$
whose product is $\longrightarrow 60.$

Thus the L.C.D. is 60. (This number is the smallest natural number that is divisible by 12, 10, and 6.)

The L.C.D. of a set of algebraic fractions is the simplest algebraic expression that is a multiple of each of the denominators in the set. Thus, the L.C.D. of the fractions

$$\frac{3}{x}, \qquad \frac{2}{x^2 - 1}, \qquad \text{and} \qquad \frac{1}{x^2(x - 1)}$$

is

$$x^2(x + 1)(x - 1)$$

because this is the simplest expression that is a multiple of each of the denominators.

If the L.C.D. is not evident by inspection, we can find it by the following procedure.

To Find the L.C.D.:

1. Completely factor each denominator, aligning common factors where possible.
2. Include in the L.C.D. each of these factors the greatest number of times it occurs in any single denominator.

For example, using the example above, we get

$$\begin{array}{rl} x = & x \\ x^2 - 1 = & (x + 1)(x - 1) \\ x^2(x - 1) = & x \qquad\qquad (x - 1)x \end{array}$$

The L.C.D. contains
the factors $\longrightarrow x\ (x + 1)(x - 1)x$
whose product is $\longrightarrow x^2(x + 1)(x - 1).$

Thus, the L.C.D. is $x^2(x + 1)(x - 1)$.

We can add fractions with unlike denominators by first building the fractions to *equivalent fractions with like denominators and then adding.*

To Add Fractions with Unlike Denominators:

1. Find the L.C.D. of the set of fractions.
2. Build each fraction to an equivalent fraction with the L.C.D. as the denominator.
3. Add the fractions using the property

$$\frac{a}{c} + \frac{b}{c} = \frac{a + b}{c}$$

For example, consider the sums $\frac{1}{2} + \frac{2}{5}$ and $\frac{a}{2} + \frac{b}{5}$. In each case, the L.C.D. is 10. We build each fraction to a fraction with 10 as the denominator. Thus,

$$\frac{(5)1}{(5)2} + \frac{(2)2}{(2)5} \quad \text{and} \quad \frac{(5)a}{(5)2} + \frac{(2)b}{(2)5}$$

are equivalent to

$$\frac{5}{10} + \frac{4}{10} \quad \text{and} \quad \frac{5a}{10} + \frac{2b}{10},$$

from which we obtain

$$\frac{9}{10} \quad \text{and} \quad \frac{5a + 2b}{10}.$$

Sometimes we want to add fractions whose denominators are binomials. For example, to write the sum

$$\frac{x}{x + 2} + \frac{x + 1}{x - 1}$$

as a single term, we also follow the three steps given above. By inspection, the L.C.D. is $(x + 2)(x - 1)$. Next we build each fraction to a fraction with denominator $(x + 2)(x - 1)$, inserting parentheses as needed, to get

$$\frac{x}{x + 2} + \frac{x + 1}{x - 1} = \frac{(x - 1)x}{(x - 1)(x + 2)} + \frac{(x + 1)(x + 2)}{(x - 1)(x + 2)}.$$

Now that we have like denominators, we can add the numerators and simplify to obtain

$$\frac{x(x - 1) + (x + 1)(x + 2)}{(x - 1)(x + 2)} = \frac{x^2 - x + x^2 + 3x + 2}{(x - 1)(x + 2)}$$

$$= \frac{2x^2 + 2x + 2}{(x - 1)(x + 2)}.$$

As another example, consider the sum

$$\frac{x}{x^2 - 4} + \frac{1}{x^2 - x - 2}.$$

In order to obtain the L.C.D., we first factor each denominator.

$$x^2 - 4 = (x + 2)(x - 2)$$

$$x^2 - x - 2 = \quad (x - 2)(x + 1)$$

The L.C.D. is $(x + 2)(x - 2)(x + 1)$.

Next we build each fraction to a fraction whose denominator is the L.C.D. to obtain

$$\frac{x}{x^2 - 4} + \frac{1}{x^2 - x - 2} = \frac{x}{(x - 2)(x + 2)} + \frac{1}{(x - 2)(x + 1)}$$

$$= \frac{(x + 1)x}{(x + 1)(x - 2)(x + 2)} + \frac{1(x + 2)}{(x - 2)(x + 1)(x + 2)}.$$

Finally we add the numerators and simplify to get

$$\frac{x(x + 1) + (x + 2)}{(x + 1)(x - 2)(x + 2)} = \frac{x^2 + x + x + 2}{(x + 1)(x - 2)(x + 2)}$$

$$= \frac{x^2 + 2x + 2}{(x + 1)(x - 2)(x + 2)}$$

Common Error

Note that we can only add fractions with like denominators. Thus,

$$\frac{2}{x} + \frac{3}{y} \neq \frac{5}{x + y}.$$

Also, we only add the *numerators* of fractions with like denominators. Thus,

$$\frac{3}{2x} + \frac{4}{2x} \neq \frac{7}{4x}.$$

EXERCISES 7.4

■ *Find the lowest common denominator for each set of fractions.*

Sample Problem

$$\frac{1}{6}, \frac{1}{8}, \frac{1}{36}$$

Completely factor each denominator and align common factors.

$$6 = 2 \cdot 3$$
$$8 = 2 \qquad \cdot 2 \cdot 2$$
$$36 = 2 \cdot 3 \qquad \cdot 2 \cdot 3$$

L.C.D. $2 \cdot 3 \cdot 2 \cdot 2 \cdot 3$

Include in the L.C.D. each of these factors the greatest number of times it occurs in any single denominator.

Ans. 72

1. $\dfrac{1}{4}, \dfrac{1}{6}, \dfrac{1}{12}$ **2.** $\dfrac{1}{3}, \dfrac{1}{15}, \dfrac{1}{5}$ **3.** $\dfrac{1}{6}, \dfrac{1}{9}, \dfrac{1}{18}$ **4.** $\dfrac{1}{5}, \dfrac{1}{10}, \dfrac{1}{15}$

5. $\dfrac{1}{3}, \dfrac{1}{14}, \dfrac{1}{28}$ **6.** $\dfrac{1}{2}, \dfrac{1}{16}, \dfrac{1}{20}$ **7.** $\dfrac{3}{4}, \dfrac{5}{8}, \dfrac{7}{18}$ **8.** $\dfrac{1}{6}, \dfrac{1}{5}, \dfrac{1}{14}$

9. $\dfrac{2}{3}, \dfrac{4}{8}, \dfrac{5}{6}$ **10.** $\dfrac{3}{4}, \dfrac{5}{8}, \dfrac{7}{18}$ **11.** $\dfrac{a}{15}, \dfrac{b}{20}, \dfrac{c}{6}$ **12.** $\dfrac{a}{27}, \dfrac{b}{12}, \dfrac{c}{18}$

Sample Problem $\dfrac{1}{x}, \dfrac{3}{x^2 y}, \dfrac{5}{x^2 y^3}$

Completely factor each denominator and align common factors.

$$x = x$$
$$x^2 y = x \cdot x \cdot y$$
$$x^2 y^3 = x \cdot x \cdot y \cdot y \cdot y$$

L.C.D. $x \cdot x \cdot y \cdot y \cdot y$

Include in the L.C.D. each of these factors the greatest number of times it occurs in any single denominator.

Ans. $x^2 y^3$

13. $\dfrac{2}{x}, \dfrac{3}{x^2}, \dfrac{1}{y}$ **14.** $\dfrac{a}{x^2}, \dfrac{a}{x^2 y}, \dfrac{1}{z}$ **15.** $\dfrac{1}{xy}, \dfrac{2}{yz}, \dfrac{3}{xz}$

16. $\dfrac{a}{x^2 y}, \dfrac{2}{xyz}, \dfrac{3}{yz^2}$ **17.** $\dfrac{1}{8xy}, \dfrac{2}{3x^2}, \dfrac{2}{4xy}$ **18.** $\dfrac{2}{4xy}, \dfrac{3}{6yz^2}, \dfrac{1}{3xy^2 z}$

Sample Problem $\dfrac{1}{x}, \dfrac{1}{x^2 - 1}, \dfrac{1}{x^2 + 2x + 1}$

Completely factor each denominator and align common factors.

$$x = x$$
$$x^2 - 1 = \begin{vmatrix} (x + 1)(x - 1) \end{vmatrix}$$
$$x^2 + 2x + 1 = (x + 1) \qquad (x + 1)$$

$$\text{L.C.D.} \quad x(x + 1)(x - 1)(x + 1)$$

Include in the L.C.D. each of these factors the greatest number of times it occurs in any single denominator.

Ans. $x(x - 1)(x + 1)^2$

19. $\dfrac{2}{x^2 - y^2}, \dfrac{1}{x - y}$

20. $\dfrac{3}{x^2 + 2x}, \dfrac{4}{x + 2}$

21. $\dfrac{3}{x^2}, \dfrac{4}{x^2 + 2x}$

22. $\dfrac{5}{x^2 + 2x + 1}, \dfrac{3}{x^2 + 4x + 3}$

23. $\dfrac{2}{x^2 + 3x - 4}, \dfrac{3}{(x - 1)^2}, \dfrac{2}{x + 4}$

24. $\dfrac{3}{a^2 - a - 6}, \dfrac{a + 2}{a^2 + 7a + 10}, \dfrac{3}{(a - 3)^2}$

Sample Problem $\dfrac{1}{4x - 6y}, \dfrac{1}{9y - 6x}$

Completely factor each denominator.

$$4x - 6y = \qquad\qquad 2 \quad (2x - 3y)$$
$$9y - 6x = 3(3y - 2x) = \begin{vmatrix} -3 (2x - 3y) \end{vmatrix}$$

$$\text{L.C.D.} \quad 2(-3)(2x - 3y)$$

Factor -1 from $3y - 2x$ to obtain $2x - 3y$.

Ans. $-2 \cdot 3(2x - 3y)$

25. $\dfrac{1}{5a - 10b}, \dfrac{1}{6b - 3a}$

26. $\dfrac{1}{4b - 12a}, \dfrac{1}{9a - 3b}$

27. $\dfrac{2}{x^2 - 3x}, \dfrac{3}{9 - x^2}$

28. $\dfrac{4}{x^2 - 36}, \dfrac{5}{6x - x^2}$

29. $\dfrac{x}{x^2 + x - 2}, \dfrac{2}{3x - 3x^2}, \dfrac{1}{4x^2 + 8x}$

30. $\dfrac{x}{x^2 + x - 6}, \dfrac{5}{4x - 2x^2}, \dfrac{5}{4x^2 + 12x}$

■ *Rewrite each sum as a single term.*

Sample Problem $\dfrac{1}{4x} + \dfrac{5}{6x}$

Find L.C.D. $12x$ and build each fraction to a fraction with denominator $12x$.

$$\frac{(3)1}{(3)4x} + \frac{5(2)}{6x(2)}$$

$$\frac{3}{12x} + \frac{10}{12x}$$

Add numerators.

Ans. $\dfrac{13}{12x}$

31. $\dfrac{3}{8} + \dfrac{1}{4}$ **32.** $\dfrac{1}{6} + \dfrac{1}{2}$ **33.** $\dfrac{1}{3} + \dfrac{1}{4}$ **34.** $\dfrac{1}{2} + \dfrac{2}{5}$

35. $\dfrac{3}{4} + \dfrac{1}{5}$ **36.** $\dfrac{2}{3} + \dfrac{3}{2}$ **37.** $\dfrac{x}{8} + \dfrac{5x}{2}$ **38.** $\dfrac{2y}{3} + \dfrac{5y}{12}$

39. $\dfrac{5}{ax} + \dfrac{3}{x}$ **40.** $\dfrac{2}{ax} + \dfrac{3}{a}$ **41.** $\dfrac{2}{x} + \dfrac{4}{3y}$ **42.** $\dfrac{2}{3x} + \dfrac{1}{2y}$

43. $\dfrac{1}{x} + \dfrac{1}{y} + \dfrac{1}{z}$ **44.** $\dfrac{1}{x} + \dfrac{2}{y} + \dfrac{3}{z}$

Sample Problem $\dfrac{x + 1}{3} + \dfrac{2x - 1}{2}$

Find L.C.D. 6 and build each fraction to a fraction with denominator 6. Insert parentheses as needed.

$$\frac{(2)(x + 1)}{(2)3} + \frac{(2x - 1)(3)}{2(3)}$$

Add numerators.

$$\frac{2(x + 1) + 3(2x - 1)}{6}$$

Remove parentheses and simplify.

$$\frac{2x + 2 + 6x - 3}{6}$$

Ans. $\dfrac{8x - 1}{6}$

45. $\dfrac{3y - 2}{3} + \dfrac{2y - 1}{6}$ **46.** $\dfrac{3x + 4}{2} + \dfrac{4x - 1}{4}$

47. $\dfrac{x - 2}{6} + \dfrac{x + 3}{2}$ **48.** $\dfrac{5x + 1}{6x} + \dfrac{3x - 2}{2x}$

49. $\dfrac{x - y}{2x} + \dfrac{x + y}{3x}$ **50.** $\dfrac{2x - y}{4y} + \dfrac{x - 2y}{3y}$

Sample Problem $\dfrac{x}{x + 1} + \dfrac{x - 2}{x + 2}$

Find L.C.D. $(x + 2)(x + 1)$ and build each fraction to a fraction with this denominator.

$$\dfrac{(x + 2)x}{(x + 2)(x + 1)} + \dfrac{(x - 2)(x + 1)}{(x + 2)(x + 1)}$$

Add numerators.

$$\dfrac{x(x + 2) + (x - 2)(x + 1)}{(x + 2)(x + 1)}$$

Remove parentheses in numerator.

$$\dfrac{x^2 + 2x + x^2 + x - 2x - 2}{(x + 2)(x + 1)}$$

Ans. $\dfrac{2x^2 + x - 2}{(x + 2)(x + 1)}$

51. $\dfrac{3}{x + 3} + \dfrac{4}{x - 3}$ **52.** $\dfrac{2}{x + 2} + \dfrac{3}{x - 2}$ **53.** $\dfrac{x}{x - y} + \dfrac{2y}{x + y}$

54. $\dfrac{y}{x - y} + \dfrac{x}{x + y}$ **55.** $\dfrac{x}{x - 2} + \dfrac{x + 1}{x + 2}$ **56.** $\dfrac{x - 3}{x + 1} + \dfrac{2x}{x - 1}$

57. $\dfrac{y + 1}{y - 3} + \dfrac{y - 4}{y + 2}$ **58.** $\dfrac{y - 2}{y + 1} + \dfrac{y + 3}{y + 2}$

Sample Problem $\dfrac{x}{x^2 - 9} + \dfrac{1}{x^2 - x - 6}$

Factor denominators.

$$\dfrac{x}{(x + 3)(x - 3)} + \dfrac{1}{(x + 2)(x - 3)}$$

Find L.C.D. $(x + 3)(x - 3)(x + 2)$ and build each fraction to a fraction with this denominator.

$$\dfrac{(x + 2)x}{(x + 2)(x + 3)(x - 3)} + \dfrac{1(x + 3)}{(x + 2)(x - 3)(x + 3)}$$

Add numerators.

$$\dfrac{x(x + 2) + (x + 3)}{(x + 2)(x + 3)(x - 3)}$$

Remove parentheses in numerator.

$$\dfrac{x^2 + 2x + x + 3}{(x + 2)(x + 3)(x - 3)}$$

Ans. $\dfrac{x^2 + 3x + 3}{(x + 2)(x + 3)(x - 3)}$

59. $\dfrac{y}{y^2 - 1} + \dfrac{3}{y^2 + 2y + 1}$

60. $\dfrac{4}{y^2 - 5y + 4} + \dfrac{y}{y^2 - 16}$

61. $\dfrac{2x}{x^2 + 3x + 2} + \dfrac{x}{x^2 - 4}$

62. $\dfrac{x}{x^2 - x - 6} + \dfrac{3x}{x^2 + 3x + 2}$

63. $\dfrac{x - 1}{x^2 + 3x} + \dfrac{x}{x^2 + 6x + 9}$

64. $\dfrac{x}{x^2 + 5x + 4} + \dfrac{x - 2}{x^2 + x}$

65. $\dfrac{2x + 1}{6x^2 - 7x - 3} + \dfrac{3x - 1}{8x^2 - 18x + 9}$

66. $\dfrac{2x - 3}{6x^2 - 17x - 3} + \dfrac{2x - 1}{5x^2 - 17x + 6}$

67. $\dfrac{x + y}{x^2 + xy - 2y^2} + \dfrac{x - y}{2x + 4y}$

68. $\dfrac{x + 3y}{x^2 + 2xy + y^2} + \dfrac{x - y}{x^2 + 4xy + 3y^2}$

69. Use a numerical example to show that $\dfrac{1}{x} + \dfrac{1}{y}$ is not equivalent to $\dfrac{1}{x + y}$.

70. Use a numerical example to show that $\dfrac{3}{x} + \dfrac{2}{y}$ is not equivalent to $\dfrac{3 + 2}{x + y}$.

7.5 DIFFERENCES OF FRACTIONS WITH UNLIKE DENOMINATORS

We subtract fractions with unlike denominators in a similar way that we add such fractions. However, we first write each fraction in standard form. Thus, any fraction in the form

$$\frac{a}{b} - \frac{c}{d}$$

is first written as

$$\frac{a}{b} + \frac{-c}{d}.$$

We can now add the fractions following the steps on page 221 in Section 7.4. For example, to write the difference

$$\frac{3}{5x} - \frac{2}{3x}$$

as a single term, we begin by writing $-\dfrac{2}{3x}$ in standard form as $\dfrac{-2}{3x}$. The L.C.D. is $15x$, so we build each fraction to an equivalent fraction with this denominator to get

$$\frac{(3)3}{(3)5x} + \frac{-2(5)}{3x(5)} = \frac{9}{15x} + \frac{-10}{15x}.$$

Adding the numerators yields

$$\frac{9 - 10}{15x} = \frac{-1}{15x}.$$

Again, special care must be taken with binomial numerators. For example, the difference

$$\frac{1}{2} - \frac{x - 1}{3}$$

should first be written as

$$\frac{1}{2} + \frac{-(x - 1)}{3},$$

where *the entire numerator is enclosed within parentheses* to indicate that the negative sign applies to *each* term of the numerator. Then we obtain the L.C.D. 6 and build each fraction to a fraction with denominator 6, add numerators, and simplify:

$$\frac{(3)1}{(3)2} + \frac{-(x - 1)(2)}{3(2)} = \frac{3}{6} + \frac{-2(x - 1)}{6}$$

$$= \frac{3 - 2x + 2}{6}$$

$$= \frac{-2x + 5}{6}.$$

As another example, we will write the difference

$$\frac{x}{x + 3} - \frac{1}{x - 2}$$

as a single term. We begin by writing $-\frac{1}{x - 2}$ in standard form as $\frac{-1}{x - 2}$. The L.C.D. is $(x + 3)(x - 2)$, so we build each fraction to an equivalent fraction with this denominator to get

$$\frac{x}{x + 3} - \frac{1}{x - 2} = \frac{(x - 2)x}{(x - 2)(x + 3)} + \frac{-1(x + 3)}{(x - 2)(x + 3)}.$$

Adding numerators and simplifying yields

$$\frac{x(x - 2) - 1(x + 3)}{(x - 2)(x + 3)} = \frac{x^2 - 2x - x - 3}{(x - 2)(x + 3)}$$

$$= \frac{x^2 - 3x - 3}{(x - 2)(x + 3)}.$$

Finally, consider the difference

$$\frac{x}{x^2 + 5x + 6} - \frac{1}{x^2 + 7x + 12}.$$

To find the L.C.D., we first factor the denominators and write the fractions in standard form to get

$$\frac{x}{x^2 + 5x + 6} - \frac{1}{x^2 + 7x + 12} = \frac{x}{(x + 3)(x + 2)} + \frac{-1}{(x + 4)(x + 3)}.$$

The L.C.D. is $(x + 3)(x + 2)(x + 4)$, so we build each fraction to an equivalent fraction with this denominator to get

$$\frac{(x + 4)x}{(x + 4)(x + 3)(x + 2)} + \frac{-1(x + 2)}{(x + 4)(x + 3)(x + 2)}.$$

Adding numerators and simplifying yields

$$\frac{x(x + 4) - (x + 2)}{(x + 4)(x + 3)(x + 2)} = \frac{x^2 + 4x - x - 2}{(x + 4)(x + 3)(x + 2)}$$

$$= \frac{x^2 + 3x - 2}{(x + 4)(x + 3)(x + 2)}.$$

EXERCISES 7.5

■ *Rewrite each difference as a single term.*

Sample Problem

$$\frac{2}{3x} - \frac{3}{4x}$$

Write in standard form; find L.C.D. $12x$ and build each fraction to a fraction with denominator $12x$.

$$\frac{(4)2}{(4)3x} + \frac{-3(3)}{4x(3)}$$

$$\frac{8}{12x} + \frac{-9}{12x}$$

Add numerators.

Ans. $\dfrac{-1}{12x}$

1. $\dfrac{4}{5} - \dfrac{1}{2}$ **2.** $\dfrac{7}{8} - \dfrac{1}{3}$ **3.** $\dfrac{1}{4} - \dfrac{5}{6}$ **4.** $\dfrac{1}{6} - \dfrac{5}{9}$

5. $\dfrac{x}{3} - \dfrac{y}{4}$ **6.** $\dfrac{y}{2} - \dfrac{x}{7}$ **7.** $\dfrac{1}{2x} - \dfrac{1}{6x}$ **8.** $\dfrac{2}{3y} - \dfrac{4}{9y}$

9. $\dfrac{2}{x} - \dfrac{3}{y}$ **10.** $\dfrac{4}{y} - \dfrac{1}{x}$ **11.** $\dfrac{3}{2x} - \dfrac{5}{6y}$ **12.** $\dfrac{1}{3y} - \dfrac{3}{4x}$

Sample Problem

$$\frac{x - 1}{4} - \frac{2x + 5}{2}$$

Insert parentheses and write in standard form; find L.C.D. 4 and build each fraction to a fraction with denominator 4.

$$\frac{(x-1)}{4} + \frac{-(2x+5)(2)}{2(2)}$$

Add numerators.

$$\frac{(x-1) - 2(2x+5)}{4}$$

Remove parentheses and simplify.

$$\frac{x - 1 - 4x - 10}{4}$$

$$Ans. \quad \frac{-3x - 11}{4}$$

13. $\dfrac{x-2}{6} - \dfrac{x+1}{3}$ **14.** $\dfrac{2x+1}{3} - \dfrac{x-1}{9}$ **15.** $\dfrac{3y-2}{3} - \dfrac{2y-1}{6}$

16. $\dfrac{3x+4}{2} - \dfrac{4x-1}{4}$ **17.** $\dfrac{2-x}{6} - \dfrac{3+x}{2}$ **18.** $\dfrac{y+2}{3} - \dfrac{y-4}{6}$

19. $\dfrac{5x+1}{6x} - \dfrac{3x-2}{2x}$ **20.** $\dfrac{2b-c}{2c} - \dfrac{c+a}{c}$ **21.** $\dfrac{x-y}{2x} - \dfrac{x+y}{3x}$

22. $\dfrac{4y-9}{3y} - \dfrac{3y-8}{4y}$ **23.** $\dfrac{2a-b}{4b} - \dfrac{a-3b}{6a}$ **24.** $\dfrac{a-b}{ab} - \dfrac{b-c}{bc}$

Sample Problem

$$\frac{1}{x+1} - \frac{1}{2x+2}$$

Write in standard form.

$$\frac{1}{x+1} + \frac{-1}{2x+2}$$

Factor denominators where possible.

$$\frac{1}{x+1} + \frac{-1}{2(x+1)}$$

Find L.C.D. $2(x+1)$ and build each fraction to a fraction with denominator $2(x+1)$.

$$\frac{(2)1}{(2)(x+1)} + \frac{-1}{2(x+1)}$$

Add numerators.

$$Ans. \quad \frac{1}{2(x+1)}$$

25. $\dfrac{2}{x+y} - \dfrac{1}{2x+2y}$ **26.** $\dfrac{2}{x+1} - \dfrac{3}{2x+2}$

27. $\dfrac{5}{6x + 6} - \dfrac{3}{2x + 2}$

28. $\dfrac{7}{5y - 10} - \dfrac{5}{3y - 6}$

29. $\dfrac{3}{2a + b} - \dfrac{2}{4a + 2b} + \dfrac{1}{8a + 4b}$

30. $\dfrac{3}{2x + 3y} - \dfrac{5}{4x + 6y} + \dfrac{1}{8x + 12y}$

Sample Problem

$\dfrac{1}{3x - 3} - \dfrac{3}{2 - 2x}$

Write in standard form.

$\dfrac{1}{3x - 3} + \dfrac{-3}{2 - 2x}$

Find L.C.D. $-2 \cdot 3(x - 1)$

$3x - 3 = \qquad\qquad 3(x - 1)$

$2 - 2x = 2(1 - x) = -2 \quad (x - 1)$

L.C.D. $\quad -2 \cdot 3(x - 1)$

Factor -1 from $1 - x$ to obtain $x - 1$.

Build each fraction to a fraction with denominator $-2 \cdot 3(x - 1)$.

$\dfrac{(-2) \cdot 1}{(-2) \cdot 3(x - 1)} + \dfrac{-3(3)}{-2(x - 1)(3)}$

Add numerators.

$\dfrac{-2(1) - 3(3)}{-2 \cdot 3(x - 1)}$

Simplify.

$\dfrac{-11}{-6(x - 1)}$

Ans. $\dfrac{11}{6(x - 1)}$

31. $\dfrac{5}{2x - 4} - \dfrac{2}{6 - 3x}$

32. $\dfrac{3}{4x - 12} - \dfrac{4}{3x - 9}$

33. $\dfrac{5}{6x - 2} - \dfrac{1}{15x - 5}$

34. $\dfrac{1}{8x - 4} - \dfrac{7}{6x - 3}$

35. $\dfrac{3}{2x - 3y} - \dfrac{2}{9y - 6x}$

36. $\dfrac{4}{3x - 9y} - \dfrac{2}{3y - x}$

Sample Problem

$\dfrac{x}{x + 2} - \dfrac{1}{x - 1}$

Write in standard form.

$\dfrac{x}{x + 2} + \dfrac{-1}{x - 1}$

Find L.C.D. $(x - 1)(x + 2)$ and build each fraction to a fraction with denominator $(x - 1)(x + 2)$.

$$\frac{(x-1)x}{(x-1)(x+2)} + \frac{-1(x+2)}{(x-1)(x+2)}$$

Add numerators.

$$\frac{x(x-1) - 1(x+2)}{(x-1)(x+2)}$$

Remove parentheses in numerator and simplify.

$$\frac{x^2 - x - x - 2}{(x-1)(x+2)}$$

$$Ans. \quad \frac{x^2 - 2x - 2}{(x-1)(x+2)}$$

37. $\dfrac{x}{x+3} - \dfrac{x}{x-3}$

38. $\dfrac{2}{x+2} - \dfrac{3}{x+3}$

39. $\dfrac{3}{3x-4} - \dfrac{5}{5x+6}$

40. $\dfrac{1}{x+y} - \dfrac{1}{x-y}$

41. $\dfrac{1}{2a+1} - \dfrac{3}{a-2} + \dfrac{2}{2a+1}$

42. $\dfrac{x}{2x-y} - \dfrac{y}{x-2y} + \dfrac{y}{2x-y}$

43. $\dfrac{x}{x-2y} + \dfrac{y}{2y-x} - \dfrac{y}{x+2y}$

44. $\dfrac{a}{2a+1} - \dfrac{2}{a-2} + \dfrac{2a}{2-a}$

Sample Problem

$$\frac{x+1}{x+2} - \frac{x-1}{x-2}$$

Enclose numerators and denominators in parentheses and write in standard form.

$$\frac{(x+1)}{(x+2)} + \frac{-(x-1)}{(x-2)}$$

Find L.C.D. $(x-2)(x+2)$ and build each fraction to a fraction with this denominator.

$$\frac{(x-2)(x+1)}{(x-2)(x+2)} + \frac{-(x-1)(x+2)}{(x-2)(x+2)}$$

Add numerators.

$$\frac{(x-2)(x+1) - (x-1)(x+2)}{(x-2)(x+2)}$$

Perform indicated multiplication. Write products in parentheses.

$$\frac{(x^2 - x - 2) - (x^2 + x - 2)}{(x-2)(x+2)}$$

Remove parentheses in numerator and simplify.

$$\frac{x^2 - x - 2 - x^2 - x + 2}{(x - 2)(x + 2)}$$

Ans. $\dfrac{-2x}{(x - 2)(x + 2)}$

45. $\dfrac{x - 2}{x + 2} - \dfrac{x + 2}{x - 2}$ **46.** $\dfrac{y - 4}{y - 2} - \dfrac{y - 7}{y - 5}$ **47.** $\dfrac{x + 1}{x + 2} - \dfrac{x + 2}{x + 3}$

48. $\dfrac{x + y}{x - y} - \dfrac{x - y}{x + y}$ **49.** $\dfrac{2a - 3b}{a + b} - \dfrac{a + b}{a - b}$ **50.** $\dfrac{a + 2b}{2a - b} - \dfrac{2a + b}{a - 2b}$

Sample Problem

$$\frac{x}{x^2 - 9} - \frac{1}{x^2 + 4x - 21}$$

Factor denominators and write in standard form.

$$\frac{x}{(x - 3)(x + 3)} + \frac{-1}{(x + 7)(x - 3)}$$

Find L.C.D. $(x + 7)(x - 3)(x + 3)$ and build each fraction to a fraction with this denominator.

$$\frac{(x + 7)x}{(x + 7)(x - 3)(x + 3)} + \frac{-1(x + 3)}{(x + 7)(x - 3)(x + 3)}$$

Add numerators.

$$\frac{x(x + 7) - (x + 3)}{(x + 7)(x - 3)(x + 3)}$$

Remove parentheses in numerator.

$$\frac{x^2 + 7x - x - 3}{(x + 7)(x - 3)(x + 3)}$$

Simplify.

Ans. $\dfrac{x^2 + 6x - 3}{(x + 7)(x - 3)(x + 3)}$

51. $\dfrac{1}{x^2 - x - 2} - \dfrac{1}{x^2 + 2x + 1}$ **52.** $\dfrac{2}{x^2 - 5x + 6} - \dfrac{5}{x^2 + 2x - 15}$

53. $\dfrac{2}{x^2 - x - 6} - \dfrac{3}{9 - x^2}$ **54.** $\dfrac{3}{y^2 - 5y + 4} - \dfrac{3}{1 - y^2}$

55. $\dfrac{3x}{x^2 + 3x - 10} - \dfrac{2x}{x^2 + x - 6}$ **56.** $\dfrac{3x}{x^2 - 2x - 3} - \dfrac{x}{x^2 + 3x + 2}$

57. $\dfrac{5x}{x^2 + 3x + 2} - \dfrac{3x - 6}{x^2 + 4x + 4}$ **58.** $\dfrac{2y}{y^2 - 16} - \dfrac{y + 1}{y^2 + 3y - 4}$

59. $\dfrac{x + 1}{x^2 - 2x - 8} - \dfrac{x - 3}{x^2 - x - 12}$ **60.** $\dfrac{y - 1}{y^2 - 2y - 15} - \dfrac{y + 3}{y^2 - y - 20}$

61. $\dfrac{t - 1}{t^2 - 4} - \dfrac{t + 1}{4t - 2t^2}$ **62.** $\dfrac{t + 2}{4t^2 - 6t} - \dfrac{2t}{9 - 4t^2}$

63. $\dfrac{2y + 3}{2y^2 - 7y - 4} - \dfrac{3y + 1}{3y^2 - 14y + 8}$ **64.** $\dfrac{3x + 1}{3x^2 + 11x - 4} - \dfrac{2x - 5}{2x^2 + 5x - 12}$

65. $\dfrac{8}{x^2 - 4y^2} - \dfrac{2}{x^2 - 5xy + 6y^2}$ **66.** $\dfrac{6}{2x^2 - 7xy - 4y^2} - \dfrac{4}{2x^2 - 5xy - 3y^2}$

67. Use a numerical example to show that $\dfrac{1}{x} - \dfrac{1}{y}$ is not equivalent to $\dfrac{1}{x - y}$.

68. Use a numerical example to show that $\dfrac{8}{x} - \dfrac{5}{y}$ is not equivalent to $\dfrac{8 - 5}{x - y}$.

7.6   COMPLEX FRACTIONS

A fraction that contains one or more fractions in either its numerator or denominator or both is called a **complex fraction.** For example,

$$\frac{\dfrac{1}{2}}{\dfrac{3}{4}} \qquad \text{and} \qquad \frac{a + \dfrac{1}{3}}{a - \dfrac{1}{2}}$$

are complex fractions. Like simple fractions, complex fractions represent quotients. For example,

$$\frac{\dfrac{1}{2}}{\dfrac{3}{4}} = \frac{1}{2} \div \frac{3}{4} \tag{1}$$

and

$$\frac{a + \dfrac{1}{3}}{a - \dfrac{1}{2}} = \left(a + \frac{1}{3}\right) \div \left(a - \frac{1}{2}\right). \tag{2}$$

In cases like Equation (1), where the numerator and denominator of the complex fraction do not contain sums or differences, we can simply invert the divisor and multiply. That is,

$$\frac{\frac{1}{2}}{\frac{3}{4}} = \frac{1}{2} \div \frac{3}{4} = \frac{1}{2} \cdot \frac{4}{3}$$

Divisor inverted.

Multiply.

$$= \frac{1}{2} \cdot \frac{\overset{2}{\cancel{4}}}{3} = \frac{2}{3}.$$

We can also use the fundamental principle of fractions to simplify this complex fraction. We multiply the numerator and denominator by the L.C.D. of all fractions in the numerator and denominator; the result is a simple fraction equivalent to the given complex fraction:

Multiply numerator and denominator by 4, the L.C.D. of 2 and 4.

$$\frac{\frac{1}{2}}{\frac{3}{4}} = \frac{4\left(\frac{1}{2}\right)}{4\left(\frac{3}{4}\right)}$$

$$= \frac{\overset{2}{\cancel{4}}\left(\frac{1}{\cancel{2}}\right)}{\cancel{4}\left(\frac{3}{\cancel{4}}\right)} = \frac{2}{3}.$$

Result is a simple fraction.

In cases like Equation (2), where the numerator or denominator of the complex fraction contains sums or differences, we *cannot* simply invert the divisor and multiply. However, we can use the fundamental principle of fractions to simplify the complex fraction. Thus,

Multiply numerator and denominator by 6, the L.C.D. of 3 and 2.

$$\frac{a + \frac{1}{3}}{a - \frac{1}{2}} = \frac{6\left(a + \frac{1}{3}\right)}{6\left(a - \frac{1}{2}\right)} = \frac{6a + 6\left(\frac{1}{3}\right)}{6a - 6\left(\frac{1}{2}\right)}$$

$$= \frac{6a + 2}{6a - 3}.$$

In Exercises 7.6, we will simplify all complex fractions using the fundamental principle of fractions.

Some expressions that contain negative exponents can be rewritten as complex fractions and simplified by the methods of this section. For example,

$$\frac{1 - 4x^{-2}}{1 + 2x^{-1}} = \frac{1 - \dfrac{4}{x^2}}{1 + \dfrac{2}{x}}.$$

We multiply the numerator and denominator by x^2, the L.C.D. of $\dfrac{4}{x^2}$ and $\dfrac{2}{x}$, to obtain

$$\frac{x^2\left(1 - \dfrac{4}{x^2}\right)}{x^2\left(1 + \dfrac{2}{x}\right)} = \frac{x^2 \cdot 1 - x^2\left(\dfrac{4}{x^2}\right)}{x^2 \cdot 1 + x^2\left(\dfrac{2}{x}\right)}$$

$$= \frac{x^2 - 4}{x^2 + 2x},$$

and finally we reduce the fraction to get

$$\frac{x^2 - 4}{x^2 + 2x} = \frac{(x - 2)(x + 2)}{x(x + 2)} = \frac{x - 2}{x}.$$

EXERCISES 7.6

■ *Simplify each complex fraction by using the fundamental principle of fractions.*

Sample Problems

a. $\dfrac{\dfrac{2}{3}}{\dfrac{4}{9}}$

b. $\dfrac{\dfrac{3c^2}{4d^4}}{\dfrac{5c^3}{12d^4}}$

Find the L.C.D. for all fractions in numerator and denominator; 9 in problem a and $12d^4$ in problem b; multiply numerator and denominator by the L.C.D.

$$\frac{(9)\,\dfrac{2}{3}}{(9)\,\dfrac{4}{9}}, \qquad \frac{(12d^4)\,\dfrac{3c^2}{4d^4}}{(12d^4)\,\dfrac{5c^3}{12d^4}}$$

Divide numerator and denominator by common factors.

$$\frac{\dfrac{3}{6}}{\dfrac{4}{2}} \qquad\qquad \frac{9c^2}{5c^3}$$

$$c$$

Ans. $\dfrac{3}{2}$ \qquad\qquad Ans. $\dfrac{9}{5c}$

1. $\dfrac{\dfrac{3}{4}}{\dfrac{1}{2}}$ 2. $\dfrac{\dfrac{2}{3}}{\dfrac{4}{5}}$ 3. $\dfrac{\dfrac{5}{6}}{\dfrac{2}{3}}$ 4. $\dfrac{\dfrac{4}{9}}{\dfrac{2}{3}}$

5. $\dfrac{\dfrac{4}{5}}{\dfrac{7}{10}}$ 6. $\dfrac{\dfrac{8}{5}}{\dfrac{16}{7}}$ 7. $\dfrac{\dfrac{3x}{y}}{\dfrac{x}{2y}}$ 8. $\dfrac{\dfrac{2x}{y}}{\dfrac{x}{3y}}$

9. $\dfrac{\dfrac{3}{ax}}{\dfrac{9}{bx}}$ 10. $\dfrac{\dfrac{2a}{bx}}{\dfrac{3a}{cx}}$ 11. $\dfrac{\dfrac{2x}{y^2}}{\dfrac{6x^2}{5y}}$ 12. $\dfrac{\dfrac{4x^2}{3y}}{\dfrac{2x}{9y^2}}$

13. $\dfrac{\dfrac{1}{3}}{5}$ 14. $\dfrac{\dfrac{1}{2}}{7}$ 15. $\dfrac{1}{\dfrac{2x}{5}}$ 16. $\dfrac{1}{\dfrac{3y}{2}}$

17. $\dfrac{\dfrac{5}{10}}{3}$ 18. $\dfrac{\dfrac{4}{12}}{7}$ 19. $\dfrac{3x}{\dfrac{x}{2}}$ 20. $\dfrac{4y}{\dfrac{y}{6}}$

Sample Problem

$$\dfrac{1 - \dfrac{1}{3}}{2 + \dfrac{5}{6}}$$

Find the L.C.D. for all fractions in numerator and denominator. Multiply each term in numerator and each term in denominator by the L.C.D. 6.

$$\dfrac{(6)1 - (\cancel{6})\dfrac{1}{\cancel{3}}}{(6)2 + (\cancel{6})\dfrac{5}{\cancel{6}}}$$

Simplify.

$$\dfrac{6 - 2}{12 + 5}$$

Ans. $\dfrac{4}{17}$

21. $\dfrac{\dfrac{2}{3}}{3 - \dfrac{1}{3}}$ 22. $\dfrac{1 + \dfrac{1}{5}}{\dfrac{2}{5}}$ 23. $\dfrac{1 - \dfrac{1}{3}}{2 + \dfrac{2}{3}}$ 24. $\dfrac{3 + \dfrac{1}{10}}{2 + \dfrac{3}{5}}$

25. $\dfrac{\dfrac{1}{2} - \dfrac{3}{8}}{\dfrac{5}{4} + \dfrac{1}{2}}$ **26.** $\dfrac{\dfrac{2}{3} - \dfrac{1}{6}}{\dfrac{1}{3} + \dfrac{5}{6}}$ **27.** $\dfrac{\dfrac{1}{2} + \dfrac{1}{3}}{\dfrac{1}{3} - \dfrac{1}{6}}$ **28.** $\dfrac{\dfrac{3}{4} - \dfrac{1}{2}}{\dfrac{1}{6} + \dfrac{1}{3}}$

Sample Problem

$$\dfrac{x + \dfrac{y}{z}}{x - \dfrac{z}{y}}$$

Find the L.C.D. for all fraction in numerator and denominator. Multiply each term in numerator and each term in denominator by the L.C.D. yz.

$$\dfrac{(yz)x + (yz)\dfrac{y}{z}}{(yz)x - (yz)\dfrac{z}{y}}$$

Simplify.

Ans. $\dfrac{xyz + y^2}{xyz - z^2}$

29. $\dfrac{2 - \dfrac{a}{b}}{2 - \dfrac{b}{a}}$ **30.** $\dfrac{y - \dfrac{1}{y}}{y + \dfrac{1}{y}}$ **31.** $\dfrac{4 - \dfrac{1}{x^2}}{2 - \dfrac{1}{x}}$ **32.** $\dfrac{a + \dfrac{a}{b}}{1 + \dfrac{1}{b}}$

33. $\dfrac{\dfrac{x}{3y} - \dfrac{1}{2}}{\dfrac{4}{3y} - \dfrac{2}{x}}$ **34.** $\dfrac{\dfrac{3}{2b} - \dfrac{1}{b}}{\dfrac{4}{a} + \dfrac{3}{2a}}$ **35.** $\dfrac{\dfrac{2}{y} + \dfrac{1}{2y}}{y + \dfrac{y}{2}}$ **36.** $\dfrac{\dfrac{1}{ab} - \dfrac{1}{b}}{\dfrac{1}{a} - \dfrac{1}{ab}}$

37. $\dfrac{x + y}{\dfrac{1}{x} + \dfrac{1}{y}}$ **38.** $\dfrac{x - y}{\dfrac{x}{y} - \dfrac{y}{x}}$ **39.** $\dfrac{\dfrac{4}{x^2} - \dfrac{4}{z^2}}{\dfrac{2}{z} - \dfrac{2}{x}}$ **40.** $\dfrac{\dfrac{6}{b} - \dfrac{6}{a}}{\dfrac{3}{a^2} - \dfrac{3}{b^2}}$

41. $\dfrac{3x - \dfrac{y^2}{3x}}{1 - \dfrac{y}{3x}}$ **42.** $\dfrac{1 - \dfrac{2b}{a}}{a - \dfrac{4b^2}{a}}$ **43.** $\dfrac{\dfrac{x}{z} + 2}{\dfrac{2}{x} + z}$ **44.** $\dfrac{\dfrac{1}{b} + \dfrac{9}{bc}}{\dfrac{9}{bc} + \dfrac{1}{c}}$

45. $\dfrac{\dfrac{3}{a} - \dfrac{9}{a^2b}}{\dfrac{6}{ab^2} - \dfrac{2}{b}}$ **46.** $\dfrac{\dfrac{4}{x^2y} - \dfrac{5}{xy^2}}{\dfrac{10}{x} - \dfrac{8}{y}}$

47. $\dfrac{8 - \dfrac{2}{x}}{4 - \dfrac{13}{x} + \dfrac{3}{x^2}}$

48. $\dfrac{6 + \dfrac{1}{x} - \dfrac{2}{x^2}}{9 + \dfrac{6}{x}}$

49. $\left(a - \dfrac{a}{b}\right) \div \left(2 + \dfrac{3}{b}\right)$

50. $\left(y - \dfrac{2}{y}\right) \div \left(y + \dfrac{2}{y}\right)$

51. $\left(4 - \dfrac{1}{x^2}\right) \div \left(2 + \dfrac{1}{x}\right)$

52. $\left(9 - \dfrac{1}{x^2}\right) \div \left(3 + \dfrac{1}{x}\right)$

Sample Problem

$$\dfrac{\dfrac{a}{a - b} - \dfrac{b}{a + b}}{\dfrac{b}{a - b} + \dfrac{a}{a + b}}$$

Find the L.C.D. for all fractions in numerator and denominator. Multiply each term in numerator and denominator by the L.C.D. $(a + b)(a - b)$.

$$\dfrac{(a + b)(a - b)\,\dfrac{a}{a - b} - (a + b)(a - b)\,\dfrac{b}{a + b}}{(a + b)(a - b)\,\dfrac{b}{a - b} + (a + b)(a - b)\,\dfrac{a}{a + b}}$$

Simplify.

$$\dfrac{a(a + b) - b(a - b)}{b(a + b) + a(a - b)}$$

$$\dfrac{a^2 + ab - ab + b^2}{ab + b^2 + a^2 - ab}$$

$$\dfrac{a^2 + b^2}{b^2 + a^2}$$

Ans. 1

53. $\dfrac{x + 3 - \dfrac{8}{x + 1}}{\dfrac{-6}{x + 1} + x + 2}$

54. $\dfrac{x + 1 - \dfrac{12}{x - 2}}{\dfrac{-2}{x - 2} + x - 3}$

55. $\dfrac{\dfrac{1}{a + 2} - \dfrac{2}{a - 1}}{\dfrac{2}{a + 1} - \dfrac{1}{a + 2}}$

56. $\dfrac{\dfrac{3}{a - 2} - \dfrac{1}{a + 2}}{\dfrac{1}{a - 2} - \dfrac{3}{a - 3}}$

57. $\dfrac{x - 1 - \dfrac{1}{x - 1}}{x + 1 - \dfrac{1}{x - 1}}$

58. $\dfrac{\dfrac{1}{x - 1} - \dfrac{x}{x + 1}}{1 - \dfrac{x}{x + 1}}$

Sample Problem

$$\frac{1 + x^{-1}}{1 - x^{-1}}$$

Write without negative exponents.

$$\frac{1 + \dfrac{1}{x}}{1 - \dfrac{1}{x}}$$

Multiply each term of numerator and denominator by the L.C.D. x.

$$\frac{x \cdot 1 + \cancel{x} \cdot \dfrac{1}{\cancel{x}}}{x \cdot 1 - \cancel{x} \cdot \dfrac{1}{\cancel{x}}}$$

Simplify.

Ans. $\dfrac{x + 1}{x - 1}$

59. $\dfrac{y^{-1} + 2}{y^{-1} - 2}$ **60.** $\dfrac{3 + z^{-1}}{2 + z^{-1}}$ **61.** $\dfrac{x^{-2} + 1}{x^{-1} + 1}$ **62.** $\dfrac{4 - x^{-2}}{2 + x^{-1}}$

63. $\dfrac{x^{-2} - y^{-2}}{x^{-1} - y^{-1}}$ **64.** $\dfrac{1 + x^{-2}}{2 + x^{-1}}$ **65.** $\dfrac{2 - xy^{-1}}{1 + xy^{-1}}$ **66.** $\dfrac{x + x^{-1}y^{-1}}{y - x^{-1}y^{-1}}$

7.7 FRACTIONAL EQUATIONS

To solve an equation containing fractions, it is usually easiest to first find an equivalent equation that is free of fractions. In Section 1.4, we used the multiplication property of equality to obtain such equivalent equations for simple fractional equations. In the same way, to solve the equation

$$\frac{x}{2} + 3 = \frac{1}{2},$$

we first multiply each member by 2 to obtain

$$2 \left(\frac{x}{2} + 3 \right) = 2 \left(\frac{1}{2} \right).$$

Applying the distributive law to the left member and simplifying yields

$$2 \left(\frac{x}{2} \right) + 2(3) = 2 \left(\frac{1}{2} \right),$$

$$x + 6 = 1,$$

or

$$x = -5.$$

In general, we multiply each member of an equation containing fractions by the lowest common denominator of the fractions. For example, to solve

$$\frac{x}{3} - 2 = \frac{4}{5},$$

we multiply each member by the L.C.D. 15 to obtain an equivalent equation that does not contain fractions as shown below.

$$15\left(\frac{x}{3} - 2\right) = 15\left(\frac{4}{5}\right),$$

$$(\overset{5}{\cancel{15}})\left(\frac{x}{\cancel{3}}\right) - 15(2) = \cancel{15}\left(\frac{4}{\cancel{5}}\right),$$

$$5x - 30 = 12,$$

$$5x = 42,$$

$$x = \frac{42}{5}.$$

Although we can apply the algebraic properties we have studied in any order, the following steps show the order most helpful in solving an equation when the solution is not obvious. Of course, not all the steps are always necessary.

To Solve an Equation:

1. "Clear fractions," if there are any, by multiplying each member of the equation by the L.C.D.
2. Use the distributive law to write any expression that contains parentheses as an expression without parentheses.
3. Combine any like terms in either member.
4. Obtain all terms containing the variable in one member and all terms not containing the variable in the other member.
5. Write the member containing the variable as the product of the variable and a coefficient.
6. Divide each member by the coefficient of the variable if it is different from 1.
7. Check the answer if each member of the equation has been multiplied by an expression containing a variable.

Thus to solve the equation

$$\frac{3}{4} = 8 - \frac{2x + 11}{x - 5}$$

we multiply each member by the L.C.D. $4(x - 5)$ to get

$$4(x - 5)\left(\frac{3}{4}\right) = 4(x - 5)(8) - 4(x - 5) \cdot \frac{(2x + 11)}{(x - 5)}$$

or

$$3(x - 5) = 32(x - 5) - 4(2x + 11).$$

Applying the distributive law to remove parentheses, we obtain

$$3x - 15 = 32x - 160 - 8x - 44,$$

$$3x - 15 = 24x - 204,$$

$$-21x = -189,$$

$$x = 9.$$

CHECKING SOLUTIONS

It is important to keep in mind that the multiplication property of equality allows us to multiply each member of an equation by a *nonzero* number in order to obtain an equivalent equation. In the previous example we multiplied each member by $4(x - 5)$. This operation is valid as long as $4(x - 5)$ does not equal zero; that is if x does not equal 5. Thus when we arrive at a solution we should check that it is actually a permissible value for x. For the foregoing example, we note that $4(x - 5)$ is *not* equal to zero for $x = 9$, so $x = 9$ is a valid solution for the equation.

However, consider the equation

$$6 + \frac{4}{x - 3} = \frac{x + 1}{x - 3}.$$

To solve this equation we multiply each member by $x - 3$, where x cannot equal 3, since $x - 3$ equals zero if $x = 3$. Applying the multiplication property of equality, we obtain

$$(x - 3)6 + (x - 3)\frac{4}{x - 3} = (x - 3)\frac{x + 1}{x - 3},$$

$$6x - 18 + 4 = x + 1,$$

$$5x = 15,$$

$$x = 3.$$

But 3 was not a permissible value for x, and in fact substituting 3 for x in the original equation yields

$$6 + \frac{4}{0} = \frac{3 + 1}{0}.$$

Since division by zero is undefined, it follows that 3 is *not* a solution. The equation does not have a solution.

17. $\dfrac{y + 12}{9} = \dfrac{y - 9}{2}$

18. $\dfrac{y + 1}{4} - \dfrac{3}{2} = \dfrac{2y - 9}{10}$

19. $\dfrac{x - 1}{10} + \dfrac{19}{15} = \dfrac{x}{3}$

20. $\dfrac{2x}{3} - \dfrac{2x + 5}{6} = \dfrac{1}{2}$

21. $\dfrac{x + 6}{2} - 1 = 5$

22. $\dfrac{2x - 2}{2} + 2 = \dfrac{1}{2}$

Sample Problem

$$\dfrac{2}{3}(x + 4) + \dfrac{1}{2}x = 2$$

Multiply each member by L.C.D. 6.

$$\overset{2}{\cancel{6}} \cdot \dfrac{2}{\cancel{3}}(x + 4) + \overset{3}{\cancel{6}} \cdot \dfrac{1}{\cancel{2}}x = 6 \cdot 2$$

$$4(x + 4) + 3x = 12$$

Complete the solution.

$$4x + 16 + 3x = 12$$

$$7x = -4$$

Ans. $x = \dfrac{-4}{7}$

23. $\dfrac{2}{3}(x - 1) + x = 6$

24. $2x + \dfrac{3}{4}(x + 4) = 14$

25. $\dfrac{2}{5}(3x + 2) - \dfrac{2}{3}(x - 1) = 4$

26. $\dfrac{3}{4}(3x - 2) - \dfrac{2}{3}(4x - 1) = 3$

27. $\dfrac{5}{6}(3x + 4) - \dfrac{3}{4}(2x + 3) = \dfrac{2}{3}(2x - 1)$

28. $\dfrac{2}{5}(2x - 3) + \dfrac{2}{3}(x - 3) = \dfrac{5}{6}(2x - 4)$

Sample Problem

$$\dfrac{1}{3} - \dfrac{1}{4} = \dfrac{10}{3x} - \dfrac{1}{18}$$

Multiply each member by L.C.D. $36x$.

$$(\cancel{36x})\overset{12}{\left(\dfrac{1}{\cancel{3}}\right)} - (\cancel{36x})\overset{9}{\left(\dfrac{1}{\cancel{4}}\right)} = (\cancel{36x})\overset{12}{\left(\dfrac{10}{\cancel{3x}}\right)} - (\cancel{36x})\overset{2}{\left(\dfrac{1}{\cancel{18}}\right)}$$

$$12x - 9x = 120 - 2x$$

Complete the solution.

$$5x = 120$$

Ans. $x = 24$ *Note.* This value does not make any denominator in the original equation zero.

29. $4 + \dfrac{4}{y} = \dfrac{12}{y}$

30. $1 + \dfrac{3}{y} = \dfrac{12}{y}$

31. $2 + \dfrac{5}{2x} = \dfrac{3}{x} + \dfrac{3}{2}$

32. $3 - \dfrac{1}{x} = \dfrac{7}{5x} - \dfrac{9}{5}$

33. $\dfrac{2}{t} + \dfrac{3}{5} = \dfrac{2}{3} + \dfrac{1}{2t}$

34. $\dfrac{2}{3t} - \dfrac{1}{2} + \dfrac{1}{6t} = \dfrac{1}{4}$

35. $\dfrac{x-2}{x} = \dfrac{14}{3x} - \dfrac{1}{3}$

36. $\dfrac{3}{2x} - \dfrac{x-3}{2x} = \dfrac{5}{2x} - 1$

37. $\dfrac{2-y}{5y} = \dfrac{4}{15y} - \dfrac{1}{6}$

38. $\dfrac{2z-5}{z} - \dfrac{3}{z} = -\dfrac{2}{3}$

39. $\dfrac{x-3}{2x} + \dfrac{3x-7}{2x} = \dfrac{1}{3}$

40. $\dfrac{4}{x} + \dfrac{5}{2} = \dfrac{4x+5}{2x} - \dfrac{2x-3}{5x}$

Sample Problem

$$\dfrac{2}{x+10} = \dfrac{1}{x+3}$$

Multiply each term in the equation by L.C.D. $(x + 10)(x + 3)$

$$(x+10)(x+3)\,\dfrac{2}{(x+10)} = (x+10)(x+3)\,\dfrac{1}{(x+3)}$$

$$2(x+3) = x + 10$$

Complete the solution.

$$2x + 6 = x + 10$$

Ans. $x = 4$ *Note.* This value does not make any denominator in the original equation zero.

41. $\dfrac{3}{5} = \dfrac{x}{x+2}$

42. $\dfrac{2}{x+4} = \dfrac{2}{3x}$

43. $\dfrac{3}{2y-1} = \dfrac{7}{3y+1}$

44. $\dfrac{7}{4-x} = \dfrac{4}{7+x}$

45. $\dfrac{2}{x-9} = \dfrac{9}{x+12}$

46. $\dfrac{4}{y-5} = \dfrac{-5}{y+4}$

47. $\dfrac{1}{x-2} + \dfrac{1}{x+2} = \dfrac{4}{x^2-4}$

48. $\dfrac{2}{x^2-3x} + \dfrac{4}{x} = \dfrac{6}{x-3}$

49. $\dfrac{4}{z-4} + \dfrac{3}{z+1} = \dfrac{20}{z^2-3z-4}$

50. $\dfrac{8}{z-3} - \dfrac{5}{z-2} = \dfrac{19}{z^2-5z+6}$

■ *Solve each formula for the symbol in color.*

Sample Problem

$$\dfrac{1}{T} = \dfrac{PR}{A-P}$$

Multiply each member by the L.C.D.
$T(A - P)$

$$\mathit{T}(A - P) \cdot \frac{1}{\mathit{T}} = T(A - P) \cdot \frac{PR}{A - P}$$

$$A - P = TPR$$

Add P to each member.

$$A = P + TPR$$

Factor out the desired variable.

$$A = P(1 + TR)$$

Divide each member by $1 + TR$.

Ans. $P = \dfrac{A}{1 + TR}$

51. $A = \dfrac{h}{2}(b + c)$ **52.** $S = \dfrac{n}{2}(a + l)$ **53.** $F = \dfrac{9}{5}C + 32$

54. $C = \dfrac{5}{9}(F - 32)$ **55.** $\dfrac{2S}{a + l} = n$ **56.** $\dfrac{2A}{b + c} = h$

57. $\dfrac{1}{R} = \dfrac{1}{A} + \dfrac{1}{B}$ **58.** $\dfrac{1}{x} + \dfrac{1}{y} = \dfrac{1}{z}$ **59.** $S = \dfrac{a(1 - r^3)}{1 - r}$

60. $S = \dfrac{a(1 - r^5)}{1 - r}$ **61.** $W = \dfrac{n + 1}{n}(Y - 1)$ **62.** $W = \dfrac{n + 1}{n}Y - 1$

63. $y = \dfrac{3x - 1}{2 - x}$ **64.** $y = \dfrac{2x + 3}{1 - x}$ **65.** $m = \dfrac{y - k}{x - h}$

66. $m = \dfrac{y - k}{x - h}$ **67.** $\dfrac{x}{a} + \dfrac{y}{b} = 1$ **68.** $\dfrac{x - h}{a} = \dfrac{y - k}{b}$

69. $r = \dfrac{de}{1 - ec}$ **70.** $r = \dfrac{de}{1 + es}$

7.8 APPLICATIONS

The word problems in the following exercises lead to equations involving fractions. At this time, you may want to review the steps suggested on page 84 to solve word problems and the steps suggested on page 241 to solve equations that contain fractions.

For example, consider the problem:

> If $\dfrac{2}{3}$ of a certain number is added to $\dfrac{1}{4}$ of the number, the result is 11. Find the number.

To solve this problem, we follow the six steps introduced in Chapter 3.

Steps 1–2   We first write what we want to find (the number) as a word phrase. Then, we represent the number in terms of a variable.

The number: x

Step 3   A sketch is not applicable.

Step 4   Now we can write an equation. Remember that "of" indicates multiplication.

$$\frac{2}{3} \cdot x + \frac{1}{4} \cdot x = 11.$$

Step 5   Solving the equation yields

$$(\cancel{12})\overset{4}{\frac{2}{3}} x + (\cancel{12})\overset{3}{\frac{1}{4}} x = (12)11$$
$$8x + 3x = 132$$
$$11x = 132$$
$$x = 12.$$

Step 6   **Ans.** The number is 12.

MOTION PROBLEMS

Equations for word problems concerned with motion sometimes include fractions. The basic idea of motion problems is that, at a constant speed, the distance traveled d equals the product of the rate of travel r and the time of travel t. Thus, $d = rt$. We can solve this formula for r or t to obtain:

$$r = \frac{d}{t} \quad \text{and} \quad t = \frac{d}{r}.$$

The following example illustrates these ideas.

> An express train travels 180 miles in the same time that a freight train travels 120 miles. If the express goes 20 miles per hour faster than the freight, find the rate of each.

Steps 1–2   We represent the two unknown quantities that we want to find as word phrases. Then, we represent the word phrases in terms of one variable.

Rate of freight train: r

Rate of express train: r + 20

Step 3   Next, we make a table showing the distances, rates, and times.

	Distance	Rate	Time: d/r
Freight train	120	r	$120/r$
Express train	180	$r + 20$	$180/(r + 20)$

Step 4 Because the times of both trains are the same, we can equate the expressions for time to obtain

$$\frac{120}{r} = \frac{180}{r + 20}.$$

Step 5 We can now solve for r by first multiplying each member by the L.C.D. $r(r + 20)$ and we get

$$r(r + 20)\,\frac{120}{r} = r(r + 20)\,\frac{180}{r + 20}$$

$$120(r + 20) = 180r$$

$$120r + 2400 = 180r$$

$$2400 = 60r$$

$$r = 40.$$

Step 6 **Ans.** The freight train's speed is 40 mph and the express train's speed is 40 + 20, or 60 mph.

EXERCISES 7.8

■ *Solve each of the following problems completely. Follow the six steps outlined on page 84.*

Sample Problem If two-thirds of a certain number is added to three-fourths of the number, the result is 17. Find the number.

Steps 1–2 The number: x

Step 3 A sketch is not applicable.

Step 4 $\dfrac{2}{3}x + \dfrac{3}{4}x = 17$

Step 5 $(12)\dfrac{2}{3}x + (12)\dfrac{3}{4}x = (12)(17)$

$$8x + 9x = 204$$

$$17x = 204$$

$$x = 12$$

Step 6 **Ans.** The number is 12.

1. If one-half of a certain number is added to three times the number, the result is $\frac{35}{2}$. Find the number.

2. If two-thirds of a certain number is subtracted from twice the number, the result is 20. Find the number.

3. Find two consecutive integers such that the sum of one-half the first and two-thirds of the next is 17.

4. Find two consecutive integers such that twice the second less one-half of the first is 14.

5. The denominator of a certain fraction is 6 more than the numerator, and the fraction is equivalent to $\frac{3}{4}$. Find the numerator.

6. The denominator of a certain fraction is eight more than the numerator, and the fraction is equivalent to $\frac{3}{5}$. Find the denominator.

7. A partner receives $\frac{2}{3}$ of the profits of the partnership. How much must the business make in profits if this partner is to receive $160 per week from the partnership?

8. A man owns a three-eighths interest in a lot that was purchased for $4000. What should the lot sell for if the man is to obtain a $600 profit on his investment?

9. Fahrenheit (F) and Celsius (C) temperatures are related by the equation $F = \frac{9}{5}C + 32$. What is the Celsius temperature in a room in which the Fahrenheit temperature is 68?

10. Using the equation in Exercise 9, find the Celsius temperature of boiling water at sea level (212°F).

11. A woman has 75 kg of ore of which $\frac{1}{15}$ is copper. How many more kilograms of the same ore does she need to have 15 kg of copper?

12. A man saves one-sixth of his weekly wages. How much more than $100 per week must he make if he is to save $25 per week?

13. A student received grades of 72, 78, 84, and 94 on the first four tests. What grade would she need on the next test to have an average of 80?

14. A boy got 17 hits on his first 60 times at bat. How many hits does he need in the next 20 times at bat to have an average of 0.300?

Sample Problem

A train travels 60 miles in the same time that a plane travels 360 miles. The plane goes 100 miles per hour faster than the train. Find the rate of each.

Steps 1–2   Rate of train: r

Rate of plane: $r + 100$

Step 3

	Distance	Rate	Time
Train	60	r	$\dfrac{60}{r}$
Plane	360	$r + 100$	$\dfrac{360}{r + 100}$

Step 4
$$\underbrace{\frac{60}{r}}_{\text{time of train}} = \underbrace{\frac{360}{r + 100}}_{\text{time of plane}}$$

Step 5 $\cancel{r}(r + 100)\,\dfrac{60}{\cancel{r}} = r\cancel{(r + 100)}\,\dfrac{360}{\cancel{r + 100}}$

$$60(r + 100) = 360r$$
$$60r + 6000 = 360r$$
$$6000 = 300r$$
$$r = 20$$

Step 6   **Ans.** The train's speed is 20 mph, and the plane's speed is $20 + 100$ or 120 mph.

15. A woman drives 120 miles in the same time that a man drives 80 miles. If the speed of the woman is 20 miles per hour greater than the speed of the man, find the speed of each.

16. An airplane travels 630 miles in the same time that an automobile covers 210 miles. If the speed of the airplane is 120 miles per hour greater than the speed of the automobile, find the speed of each.

17. A man rides 15 miles on his bicycle in the same time it takes him to walk 7 miles. If his rate riding is 2 miles per hour more than his rate walking, how fast does he walk?

18. A woman rides 10 miles in a car and then walks 4 miles on foot. If her rate driving is 20 times her rate walking, and if the whole trip takes her $2\frac{1}{4}$ hours, how fast does she walk?

19. Two men drive from town A to town B, a distance of 300 miles. If one man drives twice as fast as the other and arrives at town B 5 hours ahead of the other, how fast was each driving?

20. Two trains traveled from town A to town B, a distance of 400 miles. If one train traveled twice as fast as the other and arrived at town B 4 hours ahead of the other, how fast was each traveling?

21. A rocket travels 12,600 miles in the same time that a supersonic plane travels 3600 miles. If the rate of the rocket is 2700 miles per hour greater than the rate of the plane, find the rate of each.

22. Two rockets are propelled over a 5600-mile test range. One rocket travels twice as fast as the other. The faster rocket covers the entire distance in two hours less time than the slower. How fast in miles per hour is each rocket traveling?

7.9 RATIO AND PROPORTION

The quotient of two numbers, $a \div b$ or $\dfrac{a}{b}$, is sometimes referred to as a **ratio** and read "the ratio of a to b." This is a convenient way to compare two numbers. For example, the ratio of 8 to 12 can be expressed as $\frac{8}{12}$ or $\frac{2}{3}$, and the ratio of 3 in. to 5 in. can be expressed as $\frac{3}{5}$.

A statement that two ratios are equal, such as

$$\frac{2}{3} = \frac{4}{6} \qquad \text{or} \qquad \frac{a}{b} = \frac{c}{d},$$

is called a **proportion** and is read "2 is to 3 as 4 is to 6" and "a is to b as c is to d." The numbers a, b, c, and d are called the first, second, third, and fourth **terms** of the proportion, respectively. The first and fourth terms are called the **extremes** of the proportion, and the second and third terms are called the **means** of the proportion.

$$\text{means} \quad \boxed{\frac{a}{b} = \frac{c}{d}} \quad \text{extremes}$$

For example, the proportion "4 is to 5 as 12 is to 15" can be expressed as $\dfrac{4}{5} = \dfrac{12}{15}$, and "3 is to x as 9 is to 21" can be expressed as $\dfrac{3}{x} = \dfrac{9}{21}$.

If each ratio in the proportion

$$\frac{a}{b} = \frac{c}{d}$$

is multiplied by bd, we have

$$(bd)\,\frac{a}{b} = (bd)\,\frac{c}{d},$$

$$ad = bc.$$

Thus:

$$\text{If} \quad \frac{a}{b} = \frac{c}{d}, \quad \text{then} \quad ad = bc. \tag{1}$$

In any proportion, the product of the extremes is equal to the product of the means.

A proportion is a special type of fractional equation. The rule above for obtaining an equivalent equation without denominators is a special case of our general approach. For example, the equation

$$\frac{x-2}{3} = \frac{x}{4}$$

is a proportion and can be solved by using Property (1) to get

$$4(x-2) = 3x,$$
$$4x - 8 = 3x,$$
$$x = 8.$$

CONVERSIONS

We can use proportions to convert English units of measure into metric units and vice versa. The following basic relationships (in addition to the table of metric measures on page 534 in the Appendix) will be helpful in setting up appropriate proportions for conversions.

$$1 \text{ meter (m)} = 39.37 \text{ inches (in.)}$$
$$1 \text{ kilogram (kg)} = 2.2 \text{ pounds (lb)}$$
$$1 \text{ kilometer (km)} = 0.62 \text{ miles (mi)}$$
$$1 \text{ liter (1)} = 1.06 \text{ quarts (qt)}$$
$$1 \text{ pound (lb)} = 454 \text{ grams (g)}$$
$$1 \text{ inch (in.)} = 2.54 \text{ centimeters (cm)}$$

When converting units, it is easiest to follow the six steps for the solution of word problems suggested on page 84. For example, to change 8 inches to centimeters, we proceed as follows.

Steps 1–2   Represent what is to be found (centimeters) in terms of a word phrase and in terms of a variable

Centimeters: x

Step 3   Make a table showing the basic relationship between inches and centimeters. Use the data from page 253.

Inches	Centimeters
8	x
1	2.54

Step 4   Using the table from Step 3, write a proportion relating inches to centimeters.

$$\frac{8}{1} = \frac{x}{2.54}$$

Step 5   Solve for x by equating the product of the means to the product of the extremes.

$$8(2.54) = 1 \cdot x$$

$$20.32 = x$$

Step 6   **Ans.** Eight inches equals 20.32 centimeters.

EXERCISES 7.9

■ *Express each of the following as a ratio.*

Sample Problems

a. 1 in. to 8 in.

Ans. $\frac{1}{8}$

b. 10 cm to 40 cm

$\dfrac{\overset{1}{\cancel{10}}}{\underset{4}{\cancel{40}}}$

Ans. $\frac{1}{4}$

c. 12 to 30

$\dfrac{\overset{2}{\cancel{12}}}{\underset{5}{\cancel{30}}}$

Ans. $\frac{2}{5}$

1. 4 in. to 6 in.
4. 6 mm to 14 mm
7. 8 to 20
10. 12 to 30

2. 8 cm to 12 cm
5. 12 g to 30 g
8. 10 to 30
11. 16 to 10

3. 10 m to 18 m
6. 14 kg to 42 kg
9. 6 to 20
12. 20 to 6

■ *Express each of the following as a proportion.*

Sample Problems

a. 2 is to 5 as 4 is to 10.

Ans. $\frac{2}{5} = \frac{4}{10}$

b. 4 is to 9 as x is to 27.

Ans. $\frac{4}{9} = \frac{x}{27}$

13. 8 is to 3 as 24 is to 9.

14. 18 is to 4 as 9 is to 2.

15. 21 is to 24 as 7 is to 8.

16. 6 is to 12 as 4 is to 8.

17. 15 is to x as 10 is to 4.

18. 12 is to 6 as 4 is to x.

19. 6 is to 2 as x is to $x + 1$.

20. $x + 3$ is to x as 15 is to 3.

■ *Solve each proportion for x.*

Sample Problems

a. $\dfrac{4}{5} = \dfrac{x}{20}$

b. $\dfrac{14}{12} = \dfrac{x}{x - 1}$

Set the product of the extremes equal to the product of the means and solve.

$$(4)(20) = 5x$$

$$80 = 5x$$

$$14(x - 1) = 12x$$

$$14x - 14 = 12x$$

$$2x = 14$$

Ans. $x = 16$

Ans. $x = 7$

21. $\dfrac{3}{x} = \dfrac{1}{5}$

22. $\dfrac{2}{7} = \dfrac{x}{28}$

23. $\dfrac{x}{21} = \dfrac{5}{7}$

24. $\dfrac{6}{11} = \dfrac{x}{22}$

25. $\dfrac{3}{14} = \dfrac{x}{7}$

26. $\dfrac{12}{5} = \dfrac{6}{x}$

27. $\dfrac{x}{x + 2} = \dfrac{2}{3}$

28. $\dfrac{x}{x - 2} = \dfrac{14}{10}$

29. $\dfrac{1}{3} = \dfrac{x + 3}{x + 5}$

30. $\dfrac{1}{2} = \dfrac{x}{6 - x}$

31. $\dfrac{3}{4} = \dfrac{x + 2}{12 - x}$

32. $\dfrac{x - 7}{14 + x} = \dfrac{-3}{4}$

■ *Make each of the following conversions using a proportion. (Round off answers to the nearest hundredth when appropriate.) Follow the six steps on page 84.*

Sample Problem

18.48 pounds to kilograms.

Steps 1–2   Number of kilograms: x

Step 3

Pounds	Kilograms
18.48	x
2.2	1

Step 4 $\dfrac{18.48}{2.2} = \dfrac{x}{1}$

Step 5 $18.48(1) = 2.2x$

$$\frac{18.48}{2.2} = x$$

$$x = 8.4$$

Step 6   **Ans.** 18.48 pounds equals 8.4 kilograms.

33. 22 inches to centimeters.

34. 3.6 pounds to grams.

35. 24.6 kilometers to miles.

36. 12.4 liters to quarts.

37. 36 pounds to kilograms.

38. 40 miles to kilometers.

39. 32 quarts to liters.

40. 100 inches to meters.

■ *Solve each problem using a proportion.*

Sample Problem   It takes 2 hours to address 70 envelopes. At the same rate, how many envelopes can be addressed in 5 hours?

Steps 1–2   Number of envelopes addressed in 5 hours: x

Step 3

Time	Envelopes
2	70
5	x

Step 4 $\dfrac{2}{5} = \dfrac{70}{x}$

Step 5 $2x = 5(70)$

$2x = 350$

$x = 175$

Step 6   **Ans.** 175 envelopes can be addressed in 5 hours.

41. How many pounds of coffee are needed to make 3000 cups of coffee if 3 pounds make 225 cups?

42. Five kilograms of sugar cost $3.10. How many kilograms can be purchased for $21.70?

43. A car uses 32 liters of gas to travel 184 kilometers. How many liters would be required to drive 460 kilometers?

44. A woman earns $4200 in 30 weeks. At the same rate of earnings, how much could she anticipate earning in one year (52 weeks)?

45. A family uses 3 liters of milk every 2 days. How many liters will be used in 3 months (90 days)?

46. If $\frac{3}{4}$ centimeters on a map represents 10 kilometers, how many kilometers does 6 centimeters represent?

47. If 660 bricks are required for 740 centimeters of a wall, how many bricks will be required for 925 centimeters?

48. The sum of two numbers is 48 and their ratio is $\frac{5}{19}$. What are the numbers?

49. If 20 pounds of apples cost $1.60, how much would 28 pounds of apples cost?

50. A typist takes 1 hour and 20 minutes to type 12 pages of manuscript. At the same rate, how long would it take him (in hours) to type 50 pages?

CHAPTER SUMMARY

[7.1–7.2]   The following properties are used to rewrite products and quotients of fractions.

$$\frac{a}{b} \cdot \frac{c}{d} = \frac{ac}{bd} \quad \text{and} \quad a\left(\frac{b}{c}\right) = \frac{ab}{c};$$

$$\frac{a}{b} \div \frac{c}{d} = \frac{a}{b} \cdot \frac{d}{c}$$

A power of a quotient can be rewritten as a quotient of powers according to the rule

$$\left(\frac{a}{b}\right)^n = \frac{a^n}{b^n}, \quad b \neq 0.$$

[7.3–7.5]   The smallest natural number that is a multiple of each of the denominators of a set of fractions is called the **lowest common denominator (L.C.D.)** of the fractions. The following properties are used to rewrite sums and differences of fractions.

$$\frac{a}{c} + \frac{b}{c} = \frac{a+b}{c} \quad \text{and} \quad \frac{a}{c} - \frac{b}{c} = \frac{a}{c} + \frac{-b}{c} = \frac{a-b}{c};$$

$$\frac{a}{b} + \frac{d}{c} = \frac{(c)a}{(c)b} + \frac{d(b)}{c(b)} = \frac{ca+db}{bc}$$

[7.6]   A fraction that contains one or more fractions in either numerator or denominator or both is called a **complex fraction.** We can simplify a complex fraction by multiplying the numerator and denominator by the L.C.D. of all fractions in the numerator and denominator.

[7.7–7.8]   We can solve an equation containing fractions by obtaining an equivalent equation in which the solution is evident by inspection. Generally it is best to obtain an equivalent equation that is free of fractions by multiplying each member of the equation by the L.C.D. of the fractions.

[7.9]   The quotient of two numbers is called a **ratio;** a statement that two ratios are equal is called a **proportion.** In the proportion

$$\frac{a}{b} = \frac{c}{d},$$

a and d are called the **extremes** of the proportion and b and c are called the **means.** In any proportion of this form,

$$ad = bc.$$

CHAPTER REVIEW

■ *Simplify.*

1. a. $\dfrac{2xy^2}{3} \cdot \dfrac{x}{4y^2}$ **b.** $\dfrac{x^2 - 2x}{5} \cdot \dfrac{25}{x^2}$ **c.** $\dfrac{x^2 - 7x + 6}{x^2 - 1} \cdot \dfrac{x + 1}{x - 6}$

2. a. $\dfrac{2r}{3s} \div \dfrac{2r^2}{21s^2}$ **b.** $\dfrac{a^2 - b^2}{4} \div \dfrac{a^2 + ab}{4a - 4}$ **c.** $\dfrac{2x^2 - 5x - 3}{2x^2 + x} \div \dfrac{x - 3}{x^4}$

3. a. $\dfrac{2}{5} - \dfrac{1}{5} + \dfrac{3}{5}$ **b.** $\dfrac{x + 3}{y} - \dfrac{3}{y}$ **c.** $\dfrac{a - 2}{3} - \dfrac{a + 3}{3}$

4. a. $\dfrac{1}{3} + \dfrac{5}{6}$ **b.** $\dfrac{7}{8} + \dfrac{1}{4}$ **c.** $\dfrac{5}{2y} - \dfrac{1}{y}$

5. a. $\dfrac{3}{x} - \dfrac{2}{3x}$ **b.** $\dfrac{3}{r} + \dfrac{5}{2s}$ **c.** $\dfrac{2}{ab^2} - \dfrac{3}{a^2b}$

6. a. $\dfrac{3}{a - b} + \dfrac{1}{a + b}$ **7. a.** $\dfrac{3a - 6b}{8b - 4a}$

b. $\dfrac{a}{a^2 - 1} - \dfrac{1}{a^2 + a}$ **b.** $\dfrac{x}{4 - x^2} - \dfrac{1}{2x^2 - 4x}$

c. $\dfrac{1}{x^2 - 25} + \dfrac{5}{x^2 - 4x - 5}$ **c.** $\dfrac{3}{9 - x^2} + \dfrac{2}{x^2 - 2x - 3}$

8. a. $\dfrac{\dfrac{3}{6}}{\dfrac{2}{9}}$ **b.** $\dfrac{\dfrac{2}{3} + \dfrac{1}{6}}{\dfrac{1}{3} + \dfrac{5}{6}}$ **c.** $\dfrac{1 + \dfrac{1}{2}}{3 - \dfrac{1}{4}}$

9. a. $\dfrac{1 - \dfrac{a}{b}}{1 + \dfrac{2}{b}}$ **b.** $\dfrac{x - \dfrac{x}{y}}{y - \dfrac{y}{x}}$ **c.** $\dfrac{\dfrac{1}{y} + 3}{2 - \dfrac{3}{y}}$

10. a. $\dfrac{x^{-1} - 2}{x^{-2} + 3}$ b. $\dfrac{xy^{-1} - y}{yx^{-1} - x}$ c. $\dfrac{y^{-2} + x^{-1}}{x^{-2} + y^{-1}}$

■ *Solve.*

11. a. $\dfrac{x}{2} = -1 + \dfrac{2x}{3}$

  b. $\dfrac{x}{3} + \dfrac{7}{9} = \dfrac{1}{3}$

  c. $\dfrac{x + 1}{2} = \dfrac{2x - 9}{5} + 3$

12. a. $\dfrac{2}{3}(x - 1) - 4x = 2$

  b. $\dfrac{1}{5}(2x + 3) - \dfrac{3}{4}(x + 1) = 3$

  c. $\dfrac{2}{3}x + \dfrac{3}{5}(6 - x) = 5$

13. a. $\dfrac{y - 2}{2y} = \dfrac{5}{2}$

  b. $1 - \dfrac{3 + y}{2y} = \dfrac{3 - y}{y}$

  c. $\dfrac{2}{3y} - \dfrac{1}{4} = \dfrac{5}{6y}$

14. a. $\dfrac{6}{x} = \dfrac{16}{x + 5}$

  b. $\dfrac{10}{x + 4} - \dfrac{6}{x} = \dfrac{-4}{x}$

  c. $\dfrac{4}{x - 5} - \dfrac{3}{x + 2} = \dfrac{2}{x^2 - 3x - 10}$

15. Write without parentheses.

  a. $\left(\dfrac{6a^2}{bc^3}\right)^3$ b. $\left(\dfrac{-2x^3y}{5z^5}\right)^3$ c. $\left(\dfrac{24x^8y^2z^5}{16x^3y^7z}\right)^4$

16. Solve for p: $\dfrac{3}{p} = \dfrac{2}{q} - \dfrac{3}{r}$.

17. If three times a certain number is divided by 10 more than that number, the result is $\frac{1}{2}$. What is the number?

18. Convert 12 inches to centimeters by using a proportion.

19. A sample of 92 parts in a manufacturing plant proved to contain 3 defective parts. If the sample was a valid sample, how many defective parts would you expect to find in a run of 276 parts?

20. One car travels 90 miles in the same time that another car travels 60 miles. If the slower car is traveling 10 mph slower than the other car, find the rate of each.

CUMULATIVE REVIEW

1. The relationship between Celsius and Fahrenheit temperatures is given by $F = \frac{9}{5}C + 32$. Solve for C in terms of F.

2. Show by direct substitution that 4 is a solution of the equation

$$\dfrac{2x}{5} + \dfrac{2(x - 3)}{5} = 2.$$

3. The reciprocal of $\frac{13}{42}$ is __?__ .

4. Simplify: $\dfrac{x^2 - 10x + 24}{x^2 - 5x - 6} \div \dfrac{x^2 - 2x - 8}{x^2 - 2x - 3}$.

5. Find the L.C.D. for the fractions

$$\frac{1}{3x^2 - 3}, \qquad \frac{2}{x^2 - x}, \qquad \text{and} \qquad \frac{1}{x^2 + 7x + 6}.$$

6. Simplify: $(3x^2 - x + 6) - (x^2 - x + 2) - (3x + 5)$.

7. Simplify: $\dfrac{x - 3}{x^2 - 5x + 4} + \dfrac{4}{x - 1}$.

8. Write $4x^3y^2 - 2xy^3$ in factored form.

9. Reduce $\dfrac{x^2 - 9}{x^2 + 4x + 3}$.

■ *In Exercises 10–15, solve for x.*

10. $1 - \dfrac{x}{3} = \dfrac{2}{5}$.

11. $\dfrac{x + 1}{2x + 4} = \dfrac{3}{7}$.

12. $\dfrac{x}{3} + \dfrac{x}{2} = x - 1$.

13. $\dfrac{x}{b} = \dfrac{a}{c}$.

14. $2 - 3x > -6 + 5x$

15. $\dfrac{2x + 5}{3} \le 1$

16. Write as a basic numeral: $(3y)^0 \cdot 4^{-2}$.

17. Write in decimal form: 3.25×10^{-4}.

18. Express as a polynomial: $\dfrac{2x^3 + 4x^2 + 6x}{2x}$.

19. If $\frac{1}{4}$ centimeter on a map represents 8 kilometers, how many kilometers does 5 centimeters represent?

20. A man gave $\frac{1}{3}$ of his money to his son and $\frac{1}{4}$ to his daughter. He has $10 left. How much did he have to start with?

■ *True or false.*

21. $x^4 \cdot x^3$ is equivalent to x^{12}.

22. $-\dfrac{3 - x}{2}$ is equivalent to $\dfrac{-3 + x}{2}$.

23. $3x^{-1}$ is equivalent to $\dfrac{1}{3x}$.

24. $(x + 3)^2$ is equivalent to $x^2 + 9$.

25. $\dfrac{4 + y}{8 + y}$ is equivalent to $\dfrac{1}{2}$.

8 GRAPHING LINEAR EQUATIONS AND INEQUALITIES

The language of mathematics is particularly effective in representing relationships between two or more variables. As an example, let us consider the distance traveled in a certain length of time by a car moving at a constant speed of 40 miles per hour. We can represent this relationship by:

1. A word sentence: The distance traveled in miles is equal to forty times the number of hours traveled.
2. An equation: $d = 40t$.
3. A tabulation of values.
4. A graph showing the relationship between time and distance.

We have already used word sentences and equations to describe such relationships; in this chapter we will deal with tabular and graphical representations.

8.1 SOLVING EQUATIONS IN TWO VARIABLES

ORDERED PAIRS

The equation $d = 40t$ pairs a distance d for each time t. For example,

$$\text{if } t = 1, \quad \text{then} \quad d = 40,$$

$$\text{if } t = 2, \quad \text{then} \quad d = 80,$$

$$\text{if } t = 3, \quad \text{then} \quad d = 120,$$

and so on.

The pair of numbers 1 and 40, considered together, is called a **solution** of the equation $d = 40t$ because when we substitute 1 for t and 40 for d in the equation, we get a true statement. If we agree to refer to the paired numbers in a specified order, where the first number refers to time and the second number refers to distance, we can abbreviate the solutions as (1, 40), (2, 80), (3, 120), and so on. We call such pairs of numbers **ordered pairs,** and we refer to the first and second numbers in the pairs as **components.** With this agreement, solutions of the equation $d = 40t$ are ordered pairs (t, d) whose components satisfy the equation. Some ordered pairs for t equal to 0, 1, 2, 3, 4, and 5 are

$$(0, 0), \quad (1, 40), \quad (2, 80), \quad (3, 120), \quad (4, 160), \quad \text{and} \quad (5, 200).$$

Such pairings are sometimes shown in one of the following tabular forms.

t	0	1	2	3	4	5
d	0	40	80	120	160	200

t	d
0	0
1	40
2	80
3	120
4	160
5	200

In any particular equation involving two variables, when we assign a value to one of the variables, the value for the other variable is determined and therefore dependent on the first. It is convenient to speak of the variable associated with the *first component* of an ordered pair as the **independent variable** and the variable associated with the *second component* of an ordered pair as the **dependent variable.** If the variables x and y are used in an equation, it is understood that replacements for x are first components, and hence x is the independent variable; replacements for y are second components, and hence y is the dependent variable.

For example, we can obtain pairings for equation

independent variable — dependent variable

$$2x + y = 4 \tag{1}$$

by substituting a particular value of one variable into Equation (1) and solving for the other variable. For example:

$$\text{if} \quad x = 0, \quad \text{then} \quad 2(0) + y = 4,$$
$$y = 4;$$

$$\text{if} \quad x = 1, \quad \text{then} \quad 2(1) + y = 4,$$
$$y = 2;$$

$$\text{if} \quad x = 2, \quad \text{then} \quad 2(2) + y = 4,$$
$$y = 0.$$

The three pairings can now be displayed as the three ordered pairs

$$(0, 4), \quad (1, 2), \quad \text{and} \quad (2, 0),$$

or in the tabular forms

x	0	1	2
y	4	2	0

or

x	y
0	4
1	2
2	0

Equation (1) is called *first-degree* because the variables each have an exponent of 1.

EXPRESSING A VARIABLE EXPLICITLY

We can add $-2x$ to both members of Equation (1) to get

$$-2x + 2x + y = -2x + 4,$$
$$y = -2x + 4. \tag{2}$$

In Equation (2), where y is by itself, we say that y is expressed *explicitly* in terms of x. It is often easier to obtain solutions if equations are first expressed in such form because the dependent variable is expressed explicitly in terms of the independent variable.

For example, in Equation (2) above,

$$\text{if} \quad x = 0, \quad \text{then} \quad y = -2(0) + 4 = 4$$
$$\text{if} \quad x = 1, \quad \text{then} \quad y = -2(1) + 4 = 2$$
$$\text{if} \quad x = 2, \quad \text{then} \quad y = -2(2) + 4 = 0$$

We get the same pairings that we obtained using Equation (1)

$$(0, 4), \quad (1, 2), \quad \text{and} \quad (2, 0).$$

We obtained Equation (2) by adding the same quantity, $-2x$, to each member, in that way getting y by itself. In general, we can write equivalent equations in two variables by using the properties we introduced in Chapter 1, where we solved first-degree equations in one variable.

Equations are equivalent if:

1. The same quantity is added to or subtracted from both sides of the equation.
2. Both sides of the equation are multiplied or divided by the same nonzero quantity.

For example, we can solve

$$2y - 3x = 4$$

for y explicitly in terms of x by first adding $3x$ to each member to obtain

$$2y - 3x + 3x = 4 + 3x,$$
$$2y = 4 + 3x,$$

and then dividing each member by 2 to obtain

$$\frac{2y}{2} = \frac{4 + 3x}{2},$$

$$y = \frac{4 + 3x}{2}.$$

In this form we obtain values of y for given values of x as follows:

$$\text{if} \quad x = 0, \quad y = \frac{4 + 3(0)}{2} = \frac{4 + 0}{2} = 2;$$

$$\text{if} \quad x = 1, \quad y = \frac{4 + 3(1)}{2} = \frac{4 + 3}{2} = \frac{7}{2};$$

$$\text{if} \quad x = 2, \quad y = \frac{4 + 3(2)}{2} = \frac{4 + 6}{2} = 5.$$

In this case, three solutions are (0, 2), (1, 7/2), and (2, 5).

EXERCISES 8.1

1. The equation $d = 4t$ relates the distance d traveled by a person walking 4 miles per hour to the length of time t he walks.

a. Which symbols are variables?

b. What is the effect on d of increasing t?

c. Which symbol is the independent variable if a solution is given by (t, d)?

d. Which symbol is the dependent variable?

e. If t is assigned the value 3, what is the value of d?

2. The equation $F = 32m$ relates the force F of an object on earth to the mass m of the object.

a. Which symbols are variables?

b. What is the effect on F of increasing m?

c. Which symbol is the independent variable if a solution is given by (m, F)?

d. Which symbol is the dependent variable?

e. If m is assigned the value 4, what is the value of F?

■ *Find the missing component so that the ordered pair is a solution to the given equation.*

Sample Problem

$-2x + 3y = 6$

a. $(3, ?)$ b. $(-2, ?)$

$\qquad$ Replace x with 3. $\qquad$ Replace x with -2.

$-2(3) + 3y = 6$ $\qquad\qquad$ $-2(-2) + 3y = 6$

$3y = 12$ $\qquad\qquad\qquad$ $3y = 2$

$y = 4$ $\qquad\qquad\qquad\qquad$ $y = \dfrac{2}{3}$

Ans. $(3, 4)$ $\qquad\qquad\qquad$ *Ans.* $(-2, 2/3)$

3. $x + y = 3$ $\qquad\qquad\qquad$ **4.** $3x + y = 10$

$\quad$ a. $(2, ?)$ $\quad$ b. $(-1, ?)$ $\qquad\quad$ a. $(-3, ?)$ $\quad$ b. $(3, ?)$

5. $4x + 2y = -2$ $\qquad\qquad\quad$ **6.** $x + 2y = -6$

$\quad$ a. $(2, ?)$ $\quad$ b. $(-2, ?)$ $\qquad\quad$ a. $(2, ?)$ $\quad$ b. $(4, ?)$

7. $3x + 2y = 7$ $\qquad\qquad\quad$ **8.** $2x - 2y = -4$

$\quad$ a. $(-3, ?)$ $\quad$ b. $(3, ?)$ $\qquad\quad$ a. $(-2, ?)$ $\quad$ b. $(2, ?)$

9. $2y - x = 1$ $\qquad\qquad\qquad$ **10.** $3y - x = -3$

$\quad$ a. $(-4, ?)$ $\quad$ b. $(4, ?)$ $\qquad\quad$ a. $(-5, ?)$ $\quad$ b. $(5, ?)$

Sample Problem

$y = 2x + 3$

a. $(-4, ?)$ $\qquad\qquad\qquad\qquad$ b. $(0, ?)$

$\qquad$ Replace x with -4. $\qquad\qquad$ Replace x with 0.

$y = 2(-4) + 3$ $\qquad\qquad\qquad$ $y = 2(0) + 3$

$y = -5$ $\qquad\qquad\qquad\qquad$ $y = 3$

Ans. $(-4, -5)$ $\qquad\qquad\qquad$ *Ans.* $(0, 3)$

11. $y = 2x + 3$ $\qquad\qquad\qquad$ **12.** $y = 3x - 2$

$\quad$ a. $(2, ?)$ $\quad$ b. $(-3, ?)$ $\qquad\quad$ a. $(4, ?)$ $\quad$ b. $(-1, ?)$

13. $y = 4x - 1$ $\qquad\qquad\qquad$ **14.** $y = 4 - 2x$

$\quad$ a. $(3, ?)$ $\quad$ b. $(1, ?)$ $\qquad\quad$ a. $(-1, ?)$ $\quad$ b. $(-2, ?)$

15. $y = 3x$ $\qquad\qquad\qquad\qquad$ **16.** $y = -x$

$\quad$ a. $(0, ?)$ $\quad$ b. $(-3, ?)$ $\qquad\quad$ a. $(0, ?)$ $\quad$ b. $(2, ?)$

■ *Express y explicitly in terms of x.*

Sample
Problem

$$3x + 4y = 8$$

Solve for y in terms of x: subtract $3x$ from each member.

$$3x + 4y - 3x = 8 - 3x$$
$$4y = 8 - 3x$$

Divide each member by 4.

$$\frac{4y}{4} = \frac{-3x + 8}{4}$$

Ans. $y = \dfrac{-3x + 8}{4}$

17. $3x + y = 10$ **18.** $7 - 2x + y = 0$
19. $x + 2y = -6$ **20.** $4x + 2y = -2$
21. $2x - 3y = 7$ **22.** $5x - 4y = 2$
23. $3y - x + 3 = 0$ **24.** $2y - x - 1 = 0$

■ *Express y explicitly in terms of x. Then find the missing component so that the ordered pair is a solution to the given equation.*

Sample
Problem

$$y - 2x = 5; \quad (2, ?)$$

Solve $y - 2x = 5$ for y in terms of x: add $2x$ to each member.

$$y - 2x + 2x = 5 + 2x$$
$$y = 2x + 5$$

Replace x with 2.

$$y = 2(2) + 5$$
$$y = 9$$

Ans. $(2, 9)$

25. $y - 2x = 5$ **26.** $y + 3x - 5 = 0$
    a. $(4, ?)$ b. $(1, ?)$ a. $(2, ?)$ b. $(1, ?)$
27. $y + 2x + 4 = 0$ **28.** $y - 5x + 6 = 0$
    a. $(5, ?)$ b. $(3, ?)$ a. $(6, ?)$ b. $(4, ?)$
29. $y - 4x + 2 = 0$ **30.** $y - 2x - 4 = 0$
    a. $(-2, ?)$ b. $(0, ?)$ a. $(-1, ?)$ b. $(0, ?)$

Sample Problem

$$2x - 3y = 6; \quad (2, ?)$$

Solve $2x - 3y = 6$ for y in terms of x: subtract $2x$ from each member.

$$2x - 3y - 2x = 6 - 2x$$

$$-3y = 6 - 2x$$

Divide each member by -3.

$$\frac{-3y}{-3} = \frac{-2x + 6}{-3}$$

$$y = \frac{2x - 6}{3}$$

Replace x with 2.

$$y = \frac{2(2) - 6}{3}$$

$$y = \frac{-2}{3}$$

Ans. $\left(2, \dfrac{-2}{3}\right)$

31. $3x + 4y = 8$
   a. $(-2, ?)$ b. $(5, ?)$

32. $4x - 3y = 9$
   a. $(1, ?)$ b. $(-1, ?)$

33. $2x - 3y = 6$
   a. $(10, ?)$ b. $(5, ?)$

34. $3x + 2y = 4$
   a. $(2, ?)$ b. $(1, ?)$

35. $4x + 2y = 7$
   a. $(7, ?)$ b. $(5, ?)$

36. $3x - 4y = 5$
   a. $(-3, ?)$ b. $(5, ?)$

37. $4x - 5y = 3$
   a. $(-3, ?)$ b. $(0, ?)$

38. $3x + 5y = -2$
   a. $(0, ?)$ b. $(-2, ?)$

39. $\dfrac{y}{2} + 3x = -2$
   a. $(-2, ?)$ b. $(1, ?)$

40. $\dfrac{y}{3} - 2x = 4$
   a. $(4, ?)$ b. $(-1, ?)$

8.2 GRAPHS OF ORDERED PAIRS

In Section 2.1, we saw that every number corresponds to a point in a line. Similarly, every ordered pair of numbers (x, y) corresponds to a point in a plane. To graph an ordered pair of numbers, we begin by constructing a pair of perpendicular number lines, called **axes.** The horizontal axis is called the **x-axis,** the vertical axis is called the **y-axis,** and their point of intersection is called the **origin.** These axes divide the plane into four **quadrants,** as shown in Figure 8.1.

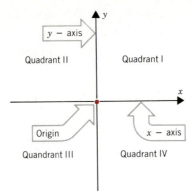

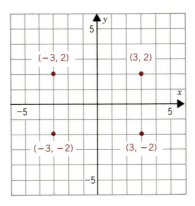

Figure 8.1

Now we can assign an ordered pair of numbers to a point in the plane by referring to the perpendicular distance of the point from each of the axes. If the first component is positive, the point lies to the right of the vertical axis; if negative, it lies to the left. If the second component is positive, the point lies above the horizontal axis; if negative, it lies below. The point in the plane is called the **graph** of the ordered pair.

Figure 8.2

For example, as shown in Figure 8.2:

the graph of (3, 2) lies 3 units *to the right* of the y-axis and 2 units *above* the x-axis;

the graph of (−3, 2) lies 3 units *to the left* of the y-axis and 2 units *above* the x-axis;

the graph of (−3, −2) lies 3 units *to the left* of the y-axis and 2 units *below* the x-axis;

the graph of (3, −2) lies 3 units *to the right* of the y-axis and 2 units *below* the x-axis.

The distance y from the point to the x-axis is called the **ordinate** of the point, and the distance x from the point to the y-axis is called the **abscissa** of the point.

The abscissa and ordinate together are called the **rectangular** or **Cartesian coordinates** of the point (see Figure 8.3).

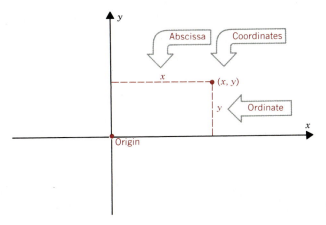

Figure 8.3

EXERCISES 8.2

■ *In exercises 1–8, graph each set of ordered pairs on a rectangular coordinate system.*

Sample Problem

a. (2, 3)
b. (−2, 1)
c. (0, −2)
d. (−4, 4)
e. (3, −1)
f. (−3, −3)

Ans.

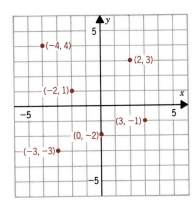

1. a. (1, 2)   b. (−2, 3)   c. (3, −1)
    d. (−4, 5)   e. (4, 4)   f. (0, 5)

2. a. (3, 4)   b. (−2, 0)   c. (2, −1)
    d. (0, 5)   e. (4, −1)   f. (−5, 1)

3. a. (0, 0)   b. (0, 2)   c. (0, 5)
    d. (0, −2)   e. (−5, 0)   f. (5, 0)

■ *In exercises 4 and 5, let the distance between successive marks on your axes represent 5 units.*

4. a. $(10, 5)$ b. $(-10, 5)$ c. $(25, -5)$

     d. $(0, 20)$ e. $(-20, -20)$ f. $(25, 15)$

5. a. $(0, 30)$ b. $(-25, 0)$ c. $(0, -20)$

     d. $(30, -25)$ e. $(-5, 0)$ f. $(5, -30)$

■ *In exercises 6, 7, and 8, let the distance between successive marks on the x-axis represent 1 unit and on the y-axis 5 units.*

6. a. $(3, 25)$ b. $(-1, 20)$ c. $(5, 0)$

     d. $(-2, -20)$ e. $(0, -10)$ f. $(-5, -30)$

7. a. $(2, -10)$ b. $(1, -5)$ c. $(0, 0)$

     d. $(-1, 5)$ e. $(-2, 10)$ f. $(-3, 15)$

8. a. $(3, -24)$ b. $(2, -16)$ c. $(1, -8)$

     d. $(0, 0)$ e. $(-1, 8)$ f. $(-2, 16)$

9. Connect the points plotted in exercise 7. What do you observe?

10. Connect the points plotted in exercise 8. What do you observe?

11. In the rectangular coordinate system, name the (perpendicular) distance to a given point from: a. The *x*-axis.   b. The *y*-axis.

12. Which component of the ordered pair (x, y) represents the (perpendicular) distance of a point from the *x*-axis? The *y*-axis?

13. Name the point corresponding to $(0, 0)$.

14. Describe the location of the graphs of all ordered pairs of the form $(0, y)$ and $(x, 0)$.

15. Describe the location of the graphs of all ordered pairs (a, a), that is, all points whose first and second components are equal.

16. Graph $(1, 2)$ and $(3, 6)$ and draw a line connecting the points.

     a. Does the graph of $(2, 4)$ lie on the line?

     b. If you extended the graph in both directions, would the graphs of $(4, 8)$ and $(-1, -2)$ lie on the line?

     b. If you extended the graph in both directions, would the graphs of $(4, 8)$ and $(-1, -2)$ lie on the line?

     c. Is the graph of $(0, 0)$ on this line?

17. Locate the graphs of $(2, 3)$ and $(2, 5)$ and draw a line connecting them.

     a. Does the graph of $(2, 4)$ lie on this line?

     b. Does the graph of $(1, 3)$ lie on this line?

    c. If you extended the graph in both directions, would the graphs of (2, 6) and (2, −1) lie on the line?

    d. Is the graph of (0, 0) on the line?

18. How many points lie on any line in the plane?

19. What would be the least number of points necessary to determine a line?

20. Which axis, the horizontal or the vertical, is usually used to represent:
    a. The independent variable?   b. The dependent variable?

8.3 GRAPHING FIRST-DEGREE EQUATIONS

In Section 8.1, we saw that a solution of an equation in two variables is an ordered pair. The **graph** of an equation in two variables is the graph of all the ordered pairs that are solutions of the equation. For example, we can find some solutions to the first-degree equation

$$y = x + 2$$

by letting x equal 0, −3, −2, and 3. Then,

$$\text{for } x = 0, \qquad y = 0 + 2 = 2;$$
$$\text{for } x = -3, \qquad y = -3 + 2 = -1;$$
$$\text{for } x = -2, \qquad y = -2 + 2 = 0;$$
$$\text{for } x = 3, \qquad y = 3 + 2 = 5;$$

and we obtain the solutions

$$(0, 2), \qquad (-3, -1), \qquad (-2, 0), \qquad \text{and} \qquad (3, 5),$$

which can be displayed in a tabular form as shown below.

x	y
0	2
−3	−1
−2	0
3	5

x	0	−3	−2	3
y	2	−1	0	5

    If we graph the points determined by these ordered pairs, we see that they lie in a straight line. In fact, all the solutions to the equation $y = x + 2$ will lie on this same straight line. So if we draw the line determined by the points we plotted, we obtain the graph of $y = x + 2$, as shown in Figure 8.4. *The graphs of first-degree equations in two variables are always straight lines;* therefore such equations are also called **linear equations.**

    In the foregoing example, the values we used for x were chosen at random; we could have used any values of x to find solutions to the equation. Each

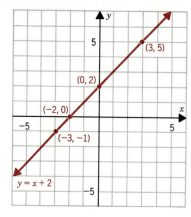

Figure 8.4

linear equation in two variables has an infinite number of solutions whose graphs lie on a line. However, we only need to find two solutions because only two points are necessary to determine a straight line. A third point can be obtained as a check.

To graph a first-degree equation:

1. Construct a set of rectangular axes showing the scale and the variable represented by each axis.
2. Find two ordered pairs that are solutions of the equation to be graphed by assigning any convenient value to one variable and determining the corresponding value of the other variable.
3. Graph these ordered pairs.
4. Draw a straight line through the points.
5. Check by graphing a third ordered pair that is a solution of the equation and verify that it lies on the line.

For example, to graph the equation

$$y = 2x - 6,$$

we first select *any two values* of x to find the associated values of y. We will use 1 and 4 for x.

$$\text{If } x = 1, \quad y = 2(1) - 6 = -4;$$

$$\text{if } x = 4, \quad y = 2(4) - 6 = 2.$$

Thus, two solutions of the equation are $(1, -4)$ and $(4, 2)$.

Next we graph these ordered pairs and draw a straight line through the points as shown in Figure 8.5. We use arrowheads to show that the line extends infinitely far in both directions.

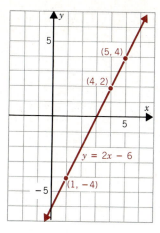

Figure 8.5

Any third ordered pair that satisfies the equation can be used as a check:

$$\text{if } x = 5, \qquad y = 2(5) - 6 = 4.$$

We then note that the graph of (5, 4) also lies on the line.

As noted earlier, to find solutions to an equation it is often easiest to first solve for y in terms of x. For example, to graph

$$x + 2y = 4$$

we first solve for y in terms of x to get

$$2y = -x + 4$$

$$y = \frac{-x + 4}{2}$$

We now select *any two* values of x to find the associated values of y. We will use 2 and 0 for x.

$$\text{For } \quad x = 2, \quad y = \frac{-(2) + 4}{2} = 1$$

$$\text{for } \quad x = 0, \quad y = \frac{-(0) + 4}{2} = 2$$

Thus, two solutions of the equation are (2, 1) and (0, 2).

Next, we graph these ordered pairs and pass a straight line through the points, as shown in Figure 8.6 on page 274.

Any third ordered pair that satisfies the equation can be used as a check:

$$\text{if } \quad x = -2, \qquad y = \frac{-(-2) + 4}{2} = 3$$

We then note that the graph of (−2, 3) also lies on the line.

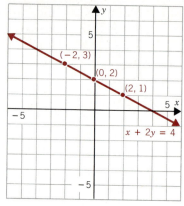

Figure 8.6

SPECIAL CASES OF LINEAR EQUATIONS

The equation $y = 2$ can be written as

$$0x + y = 2 \qquad\qquad (1)$$

and can be considered a linear equation in two variables where the coefficient of x is 0. Some solutions of $0x + y = 2$ are

$$(1, 2), \qquad (-1, 2), \qquad \text{and} \qquad (4, 2)$$

In fact, any ordered pair of the form $(x, 2)$ is a solution of (1). Graphing the solutions yields a horizontal line as shown in Figure 8.7. Similarly, an equation

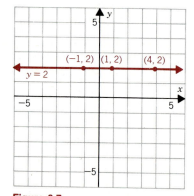

Figure 8.7

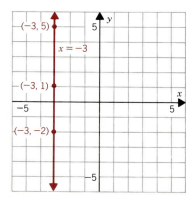

Figure 8.8

such as $x = -3$ can be written as

$$x + 0y = -3 \qquad\qquad (2)$$

and can be considered a linear equation in two variables where the coefficient of y is 0.

Some solutions of $x + 0y = -3$ are $(-3, 5)$, $(-3, 1)$, and $(-3, -2)$. In fact, any ordered pair of the form $(-3, y)$ is a solution of (2). Graphing the solutions yields a vertical line as shown in Figure 8.8.

EXERCISES 8.3

1. Given $d = 4t$ (see Section 8.1, exercise 1), find the value of d corresponding to each value of t and express your answer in the form of an ordered pair (t, d). Then graph each of the ordered pairs and connect them with a straight line.

   a. $(0, ?)$ b. $(2, ?)$ c. $(4, ?)$

   d. Where are all points located whose coordinates satisfy $d = 4t$?

   e. Check by obtaining additional solutions $(1, ?)$ and $(3, ?)$ and graphing these ordered pairs.

   f. What is the least number of points necessary to determine the line representing $d = 4t$?

2. Given $d = 2t$, find the value of d corresponding to each value of t and express your answer in the form of an ordered pair (t, d). Then graph each of the ordered pairs and connect them with a straight line.

   a. $(0, ?)$ b. $(2, ?)$ c. $(4, ?)$

   d. Where are all points located whose coordinates satisfy $d = 2t$?

   e. Check by obtaining additional solutions $(1, ?)$ and $(3, ?)$ and graphing these ordered pairs.

   f. What is the least number of points necessary to determine the line representing $d = 2t$?

■ *In exercises 3–14:*

1. Find any two ordered pairs that are solutions of the given equation.

2. Graph these ordered pairs and draw a straight line through them.

3. Check your result by finding a third solution of the equation and verifying that its graph is a point on the line. Graph each equation on a separate set of axes.

Sample Problem $2x + 3y = 6$

Solve for y in terms of x.

$$y = \frac{6 - 2x}{3}$$

Take *any* two numbers for the first components, say 0 and 6. Find the second components of (0, ?) and (6, ?) so that the ordered pairs are solutions of the equation:

$$y = \frac{6 - 2(0)}{3} = 2,$$

and

$$y = \frac{6 - 2(6)}{3} = -2.$$

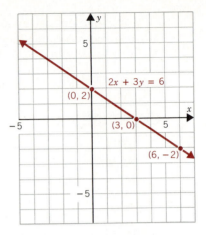

x	0	6
y	2	-2

Graph the ordered pairs (0, 2) and (6, −2) and draw a straight line through them. Check by noting that the graph of a third pair that satisfies the equation, say with first component 3 and second component

$$y = \frac{6 - 2(3)}{3} = 0$$

also lies on the line.

3. $y = x + 2$ **4.** $y = x - 2$ **5.** $y = 2x + 1$

6. $y = 3x - 1$ **7.** $y = 2x - 1$ **8.** $y + x = 4$

9. $y - 3x = 0$ **10.** $2y = 3x + 4$ **11.** $3y = 4 - x$

12. $3y + 2x = 12$ **13.** $2y + x - 6 = 0$ **14.** $y - 2x - 6 = 0$

15. Consider the two ordered pairs (5, 3) and (2, 3).

     a. Graph these ordered pairs and draw a straight line through their graphs.

     b. Are the graphs of (1, 3), (2, 3), (5, 3), (−2, 3), and (−1, 3) on the line?

     c. Would the graph of $(x, 3)$ lie on the line for any (all) x?

     d. Does the value of x have anything to do with the fact that a point lies on this line?

     e. Is $y = 0x + 3$ an equation for the line?

     f. Does $y = 3$ give a complete description of the line?

16. Consider the two ordered pairs (3, 5) and (3, 2).

     a. Graph these ordered pairs and draw a straight line through their graphs.

     b. Are the graphs of (3, −1), (3, 4), (3, 6) on the line?

     c. Would the graphs of $(3, y)$ lie on the line for any (all) y?

     d. Does the value of y have anything to do with the fact that a point lies on this line?

e. Is $x = 0y + 3$ an equation for the line?

f. Does $x = 3$ give a complete description of the line?

■ *Graph each of the following equations.*

Sample Problem

$y = 3$

See exercise 15. The equation $y = 3$ is equivalent to $y = 0x + 3$. Thus,

if $x = 1$, $y = 0(1) + 3 = 3$;

if $x = 2$, $y = 0(2) + 3 = 3$;

if $x = 5$, $y = 0(5) + 3 = 3$;

and so on.

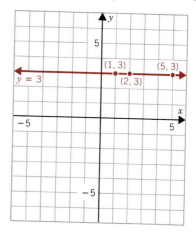

x	y
1	3
2	3
5	3

Thus, the graph is a line parallel to the x-axis at a distance of 3 units above this axis.

17. $x + 0y = 4$ **18.** $0x + y = -1$ **19.** $0x - 2y = 8$

20. $3x + 0y = 6$ **21.** $x = -2$ **22.** $y = -5$

23. $x = 0$ **24.** $y = 0$ **25.** $3y = 12$

26. $2x = 6$ **27.** $4x = -20$ **28.** $4y = -12$

8.4 INTERCEPT METHOD OF GRAPHING

In Section 8.3, we assigned values to x in equations in two variables to find the corresponding values of y. The solutions of an equation in two variables that are generally easiest to find are those in which either the first or second component is 0. For example, if we substitute 0 for x in the equation

$$3x + 4y = 12 \tag{1}$$

we have

$$3(0) + 4y = 12,$$

$$y = 3.$$

Thus, a solution of Equation (1) is (0, 3). We can also find ordered pairs that are solutions of equations in two variables by assigning values to y and determining the corresponding values of x. In particular, if we substitute 0 for y in Equation (1), we get

$$3x + 4(0) = 12,$$

$$x = 4$$

and a second solution of the equation is (4, 0). We can now use the ordered pairs (0, 3) and (4, 0) to graph Equation (1). The graph is shown in Figure 8.9. Notice that the line crosses the x-axis at 4 and the y-axis at 3. For this reason, the number 4 is called the **x-intercept** of the graph, and the number 3 is called the **y-intercept.**

This method of drawing the graph of a linear equation is called the **intercept method of graphing.** Note that when we use this method of graphing a linear equation, there is no advantage in first expressing y explicitly in terms of x.

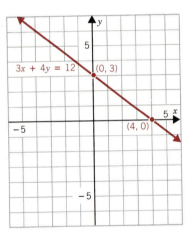

Figure 8.9

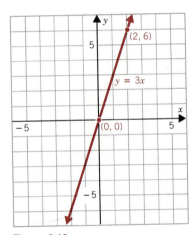

Figure 8.10

If the graph intersects the axes at or near the origin, the intercept method *is not* satisfactory. We must then graph an ordered pair that is a solution of the equation and whose graph is not the origin or is not too close to the origin.

For example, to graph the equation $y = 3x,$ we can substitute 0 for x and find

$$y = 3(0) = 0.$$

Similarly, substituting 0 for y, we get

$$0 = 3x, \qquad x = 0.$$

Thus, 0 is both the x-intercept and the y-intercept. Since one point is not sufficient to graph $y = 3x,$ we resort to the methods outlined in Section 8.3.

Choosing any other value for x, say 2, we get

$$y = 3(2) = 6.$$

Thus, (0, 0) and (2, 6) are solutions to the equation. The graph of $y = 3x$ is shown in Figure 8.10.

EXERCISES 8.4

■ *Graph each equation by the intercept method if possible.*

Sample Problem

$2x - y = 6$ (1)

Substitute 0 for x and solve for y.

$2(0) - y = 6$

$\quad -y = 6$

$\quad\quad y = -6$

The ordered pair (0, −6) is a solution of Equation (1).
Substitute 0 for y in Equation (1) and solve for x.

$2x - (0) = 6$

$\quad 2x = 6$

$\quad\quad x = 3$

The ordered pair (3, 0) is a solution of Equation (1).
Graph (0, −6) and (3, 0) and complete the graph of Equation (1).

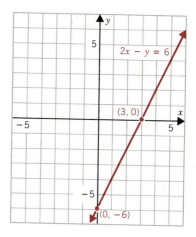

1. $x + y = 5$
2. $x - y = 4$
3. $2x + y = 8$
4. $x + 2y = 6$
5. $3x - y = 6$
6. $x - 2y = 4$
7. $2x + 3y = 12$
8. $3x + 5y = 15$
9. $3x - 4y = 12$
10. $4x - 5y = 20$
11. $y = x + 6$
12. $y = x + 4$
13. $y = 2x - 4$
14. $y = 3x + 9$
15. $y = 2x + 5$
16. $y = 3x - 7$
17. $x = 4 + y$
18. $x = 6 - 2y$

19. $x = 3y - 10$ **20.** $x = 5y + 5$ **21.** $2x + 3y = 1$

22. $4x - 3y = 1$ **23.** $4x + 5y = 1$ **24.** $5x - 4y = 1$

25. $2x - y = 0$ **26.** $x - 3y = 0$ **27.** $x - 2y = 0$

28. $x - 4y = 0$ **29.** $2x - 3y = 0$ **30.** $3x - 2y = 0$

8.5   SLOPE OF A LINE

SLOPE FORMULA

In this section, we will study an important property of a line. We will assign a number to a line, which we call *slope,* that will give us a measure of the "steepness" or "direction" of the line.

It is often convenient to use a special notation to distinguish between the rectangular coordinates of two different points. We can designate one pair of coordinates by (x_1, y_1) (read "*x* sub one, *y* sub one"), associated with a point P_1, and a second pair of coordinates by (x_2, y_2), associated with a second point P_2, as shown in Figure 8.11. Note that when going from P_1 to P_2, the vertical change (or vertical distance) between the two points is $y_2 - y_1$, and the horizontal change (or horizontal distance) is $x_2 - x_1$.

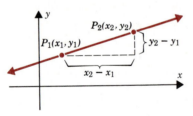

Figure 8.11

The ratio of the vertical change to the horizontal change is called the **slope** of the line containing the points P_1 and P_2. This ratio is usually designated by m. Thus,

$$m = \frac{\text{vertical change}}{\text{horizontal change}} = \frac{y_2 - y_1}{x_2 - x_1} \quad (x_2 \neq x_1). \quad \text{(1)}$$

For example, the slope of the line containing the two points with coordinates $(-4, 2)$ and $(3, 5)$, as shown in Figure 8.12, is given by

$$m = \frac{5 - 2}{3 - (-4)} = \frac{3}{7},$$

where we have substituted 3 and 5 for the coordinates x_2 and y_2, and -4 and 2 for the coordinates x_1 and y_1 in Equation (1). Note that we get the same result if we substitute -4 and 2 for x_2 and y_2, and 3 and 5 for x_1 and y_1.

$$m = \frac{2 - 5}{-4 - 3} = \frac{-3}{-7} = \frac{3}{7}.$$

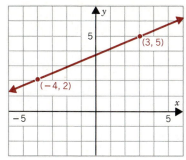

Figure 8.12

Lines with various slopes are shown in Figure 8.13. Slopes of lines that go up to the right are positive (Figure 8.13*a*), and the slopes of lines that go down to the right are negative (Figure 8.13*b*). Note (Figure 8.13*c*) that because all points on a horizontal line have the same *y* value, $y_2 - y_1$ equals zero for any two points, and the slope of a horizontal line is simply

$$m = \frac{y_2 - y_1}{x_2 - x_1} = \frac{0}{x_2 - x_1} = 0.$$

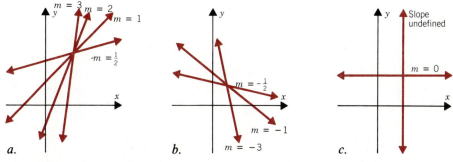

a. *b.* *c.*

Figure 8.13

Also note (Figure 8.13*c*) that since all points on a vertical line have the same *x* value, $x_2 - x_1$ equals zero for any two points. However,

$$m = \frac{y_2 - y_1}{x_2 - x_1} = \frac{y_2 - y_1}{0}$$

is undefined, so that a *vertical line does not have a slope.*

PARALLEL AND PERPENDICULAR LINES

Consider the lines shown in Figure 8.14. Line l_1 has slope $m_1 = 3$, and line l_2 has slope $m_2 = 3$. In this case,

$$m_1 = m_2 = 3.$$

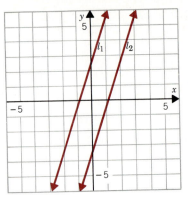

Figure 8.14

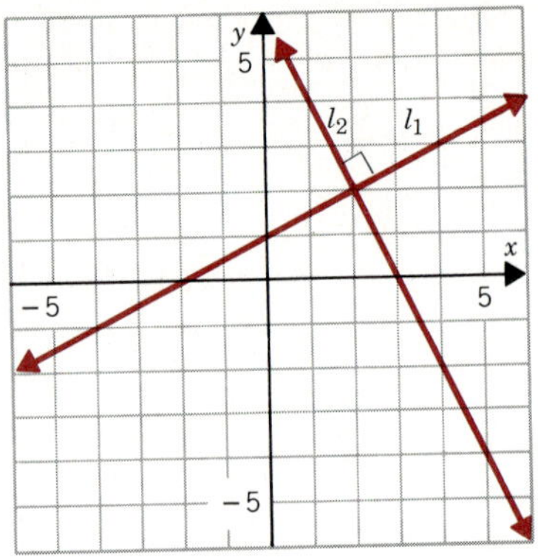

Figure 8.15

These lines will never intersect and are called **parallel lines.** Now consider the lines shown in Figure 8.15. Line l_1 has slope $m_1 = 1/2$ and line l_2 has slope $m_2 = -2$. In this case

$$m_1 \cdot m_2 = \frac{1}{2} \cdot (-2) = -1.$$

These lines meet to form a right angle and are called **perpendicular lines.**

In general:

> If two lines have slopes m_1 and m_2:
>
> **a.** The lines are parallel if they have the same slope, that is, if $m_1 = m_2$.
>
> **b.** The lines are perpendicular if the product of their slope is -1, that is, if $m_1 \cdot m_2 = -1$.

EXERCISES 8.5

■ *Find the slope of the line segment joining the given pairs of points.*

Sample Problem

$(2, -5), (4, 3)$

Consider $(2, -5)$ as P_1 and $(4, 3)$ as P_2 and use $m = \dfrac{y_2 - y_1}{x_2 - x_1}$.

$$m = \frac{3 - (-5)}{4 - 2}$$

$$= \frac{3 + 5}{2} = 4$$

Ans. Slope: 4

1. $(5, 2), (8, 7)$ 2. $(4, 1), (6, 3)$ 3. $(2, 4), (6, 1)$
4. $(4, 1), (2, 5)$ 5. $(3, -2), (0, 1)$ 6. $(-4, 0), (3, 5)$
7. $(-3, -4), (-7, 1)$ 8. $(6, -2), (-3, -3)$ 9. $(2, 5), (2, -3)$
10. $(3, -1), (3, 5)$ 11. $(3, 4), (-2, 4)$ 12. $(-2, 1), (3, 1)$

13. a. Graph the points $A(-2, 5)$, and $B(1, -1)$, and $C(3, -5)$.
    b. Find the slope of the line passing through points A and B.
    c. Find the slope of the line passing through points B and C.
    d. Do points A, B, and C lie on the same line?

14. a. Graph the points $A(-1, 5)$, $B(1, -1)$ and $C(2, -4)$.
    b. Find the slope of the line passing through points A and B.
    c. Find the slope of the line passing through points B and C.
    d. Do points A, B, and C lie on the same line?

15. a. Graph the points $A(2, 2)$, $B(3, 5)$ and $C(4, 7)$.
    b. Find the slope of the line passing through points A and B.
    c. Find the slope of the line passing through points B and C.
    d. Do points A, B, and C lie on the same line?

16. a. Graph the points $A(-1, 3)$, $B(0, 5)$ and $C(2, 6)$.
    b. Find the slope of the line passing through points A and B.
    c. Find the slope of the line passing through points B and C.
    d. Do points A, B, and C lie on the same line?

17. a. Find the slope of the line passing through the points $(5, 4)$ and $(3, 0)$.
    b. Find the slope of the line passing through the points $(-1, 8)$ and $(-4, 2)$.
    c. Graph the two lines.
    d. Are the lines parallel?

18. a. Find the slope of the line passing through the points $(-4, 2)$ and $(2, -2)$.
    b. Find the slope of the line passing through the points $(3, 0)$ and $(-3, 4)$.
    c. Graph the two lines.
    d. Are the lines parallel?

19. a. Find the slope of the line passing through the points $(-1, 2)$ and $(3, 4)$.
    b. Find the slope of the line passing through the points $(-2, 3)$ and $(6, 8)$.
    c. Graph the two lines.
    d. Are the lines parallel?

20. a. Find the slope of the line passing through the points $(-2, -3)$ and $(3, 4)$.
    b. Find the slope of the line passing through the points $(-1, -2)$ and $(5, 5)$.
    c. Graph the two lines.
    d. Are the lines parallel?

21. a. Find the slope of the line passing through the points $(0, -7)$ and $(8, -5)$.
   b. Find the slope of the line passing through the points $(5, 7)$ and $(8, -5)$.
   c. Graph the two lines.
   d. Are the lines perpendicular (does the product of their slopes equal -1)?

22. a. Find the slope of the line passing through the points $(8, 0)$ and $(6, 6)$.
   b. Find the slope of the line passing through the points $(-3, 3)$ and $(6, 6)$.
   c. Graph the two lines.
   d. Are the lines perpendicular (does the product of their slopes equal -1)?

23. a. Find the slope of the line passing through the points $(-1, -2)$ and $(3, 3)$.
   b. Find the slope of the line passing through the points $(0, 4)$ and $(-6, -8)$.
   c. Graph the two lines.
   d. Are the lines perpendicular (does the product of their slopes equal -1)?

24. a. Find the slope of the line passing through the points $(-4, -6)$ and $(2, 1)$.
   b. Find the slope of the line passing through the points $(-3, 5)$ and $(1, -2)$.
   c. Graph the two lines.
   d. Are the lines perpendicular (does the product of their slopes equal -1)?

25. If one line has a slope of -2, what must the slope of a second line be if the two lines are perpendicular?

26. If one line has a slope of 3, what must the slope of a second line be if the two lines are perpendicular?

27. If one line has a slope of $2/3$, what must the slope of a second line be if the two lines are perpendicular?

28. If one line has a slope of $-3/5$, what must the slope of a second line be if the two lines are perpendicular?

29. The slope of a horizontal line is _?_ .

30. The slope of a vertical line is _?_ .

8.6 EQUATIONS OF STRAIGHT LINES

POINT-SLOPE FORM

In Section 8.5, we found the slope of a straight line by using the formula

$$m = \frac{y_2 - y_1}{x_2 - x_1} \qquad (x_2 \neq x_1).$$

Suppose we know that a line goes through the point $(2, 3)$ and has a slope of 2. If we denote *any* other point on the line as $P(x, y)$ (see Figure 8.16a), by the slope formula

$$m = 2 = \frac{y - 3}{x - 2},$$

from which

$$2(x - 2) = y - 3,$$
$$2x - 4 = y - 3,$$
$$2x - 1 = y,$$

or

$$y = 2x - 1. \tag{1}$$

Thus, Equation (1) is the equation of the line that goes through the point $(2, 3)$ and has a slope of 2.

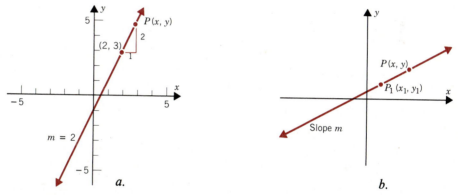

Figure 8.16

In general, suppose we know that a line passes through a point $P_1(x_1, y_1)$ and has slope m. If we denote *any* other point on the line as $P(x, y)$ (see Figure 8.16b), by the slope formula

$$m = \frac{y - y_1}{x - x_1},$$

from which

$$m(x - x_1) = y - y_1,$$

or

$$y - y_1 = m(x - x_1). \tag{2}$$

Equation (2) is called the **point-slope form** for a linear equation. In Equation (2), m, x_1, and y_1 are known and x and y are variables that represent the coordinates of *any* point on the line. Thus, whenever we know the slope of a line and a point on the line, we can find the equation of the line by using Equation (2).

For example, if a line has slope -2 and passes through point $(2, 4)$, we can find the equation of the line by substituting -2 for m and $(2, 4)$ for (x_1, y_1) in Equation (2):

y_1 value ⟶   slope   ⟶ x_1 value

$$y - 4 = -2(x - 2),$$

$$y - 4 = -2x + 4,$$

$$y = -2x + 8.$$

Thus, a line with slope -2 that passes through the point $(2, 4)$ has the equation $y = -2x + 8$. We could also write the equation in equivalent forms

$$y + 2x = 8, \qquad 2x + y = 8 \qquad \text{or} \qquad 2x + y - 8 = 0.$$

SLOPE-INTERCEPT FORM

Now consider the equation of a line with slope m and y-intercept b as shown in Figure 8.17. Substituting 0 for x_1 and b for y_1 in the point-slope form of a linear equation, we have

$$y - b = m(x - 0),$$

$$y - b = mx,$$

or

$$y = mx + b. \qquad (3)$$

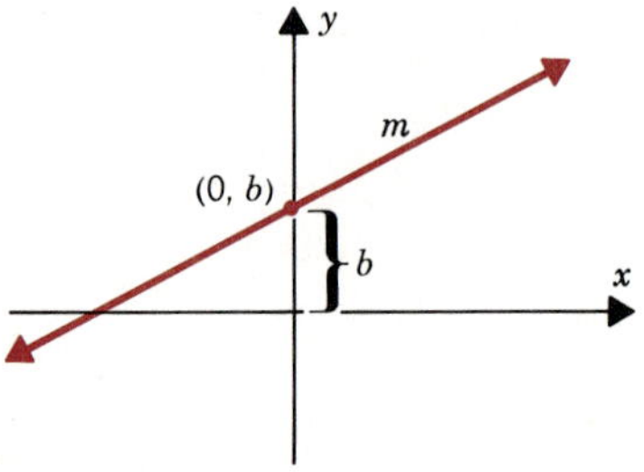

Figure 8.17

Equation (3) is called the **slope-intercept form** for a linear equation. The slope and y-intercept can be obtained directly from an equation in this form. For example, if a line has the equation

$$y = -2x + 8,$$

slope ⟶   ⟵ y-intercept

then the slope of the line must be -2 and the y-intercept must be 8. Similarly, the graph of

$$y = -3x + 4$$

has a slope -3 and a y-intercept 4; and the graph of

$$y = \frac{1}{4}x - 2$$

has a slope $1/4$ and a y-intercept -2.

To find the slope of the graph of an equation such as $2x - 3y = 6$, we solve for y in terms of x by first adding $-2x$ to each member.

$$2x - 3y - 2x = 6 - 2x,$$

$$-3y = 6 - 2x.$$

Now dividing each member by -3, we have

$$\frac{-3y}{-3} = \frac{-2x}{-3} + \frac{6}{-3},$$

$$y = \frac{2}{3}x - 2.$$

Comparing this equation with the form $y = mx + b$, we note that the slope m (the coefficient of x) is 2/3, and the y-intercept is -2.

EXERCISES 8.6

■ *Find an equation of the line with the given slope and passing through the given point. Write the equation in the form $y = mx + b$.*

Sample Problem

$m = 3;\quad (-2, 4)$

Substitute -2 for x_1, 4 for y_1, and 3 for m in the point-slope form $y - y_1 = m(x - x_1)$.

$y - 4 = 3[x - (-2)]$

Remove parentheses and simplify.

$y - 4 = 3(x + 2)$

$y - 4 = 3x + 6$

Ans. $y = 3x + 10$

1. $m = 3;\quad (-1, 1)$ **2.** $m = -2;\quad (1, -1)$ **3.** $m = -2;\quad (-3, 4)$

4. $m = -3;\quad (-2, 4)$ **5.** $m = 0;\quad (-2, 1)$ **6.** $m = 0;\quad (-3, 2)$

7. $m = \dfrac{1}{2};\quad (4, -3)$ **8.** $m = \dfrac{-1}{2};\quad (2, -6)$ **9.** $m = -4;\quad (-2, 3)$

10. $m = 5;\quad (2, -5)$ **11.** $m = \dfrac{2}{3};\quad (-3, 1)$ **12.** $m = \dfrac{-2}{3};\quad (-6, 2)$

■ *Find an equation of the line passing through the two given points. Write the equation in the form $y = mx + b$.*

Sample Problem

$(-1, 3), (2, 6)$

Using the slope formula, first find the slope.

$$m = \frac{y_2 - y_1}{x_2 - x_1} = \frac{6 - 3}{2 - (-1)}$$

$$= \frac{3}{3} = 1$$

Use the point-slope form for a linear equation with $m = 1$ and (x_1, y_1) equal to $(-1, 3)$ or $(2, 6)$. We use $(-1, 3)$.

$$y - 3 = 1[x - (-1)]$$

Remove parentheses and simplify.

$$y - 3 = x + 1$$

Ans. $y = x + 4$

13. $(-2, 4), (1, 7)$	**14.** $(2, 5), (4, 9)$	**15.** $(-1, 3), (0, 7)$
16. $(1, 2), (0, -6)$	**17.** $(0, 0), (3, 5)$	**18.** $(0, 0), (-2, -7)$
19. $(2, 4), (-3, 4)$	**20.** $(1, 3), (-2, 3)$	**21.** $(6, 4), (-2, 5)$
22. $(5, -2), (-1, 3)$		

■ *Find the slope and y-intercept of the graph of the equation.*

Sample Problem

$$2x + 3y = 6$$

Solve for y in terms of x. Add $-2x$ to each member.

$$2x + 3y + (-2x) = 6 + (-2x)$$

$$3y = -2x + 6$$

Divide each member by 3.

$$\frac{3y}{3} = \frac{-2x}{3} + \frac{6}{3}$$

$$y = \frac{-2}{3}x + 2$$

Compare with slope-intercept form $y = mx + b$.

Ans. Slope: $m = -2/3$; y-intercept: 2

23. $y = 5x - 2$	**24.** $y = 3x + 4$	**25.** $y = -4x + 3$
26. $y = -6x - 3$	**27.** $3x + y = 4$	**28.** $4x + y = -5$
29. $6x + 3y = 5$	**30.** $4x + 2y = 7$	**31.** $5x + 4y = -3$
32. $4x + 3y = -2$	**33.** $2x - 3y = 6$	**34.** $x - 2y = -7$

35. a. What is the slope of the line $y = -3x + 2$?

  b. What is the slope of a line parallel to $y = -3x + 2$?

  c. A line is parallel to the line $y = -3x + 2$ and passes through the point $(1, 3)$. What is its equation?

36. a. What is the slope of the line $y = 2x + 1$?

b. What is the slope of a line parallel to $y = 2x + 1$?

c. A line is parallel to the line $y = 2x + 1$ and passes through the point $(-2, 1)$. What is its equation?

37. a. What is the slope of the line $y = -2x + 4$?

b. What is the slope of a line perpendicular to $y = -2x + 4$?

c. A line is perpendicular to the line $y = -2x + 4$ and passes through the point $(4, 2)$. What is its equation?

38. a. What is the slope of the line $y = -3x - 2$?

b. What is the slope of a line perpendicular to $y = -3x - 2$?

c. A line is perpendicular to the line $y = -3x - 2$ and passes through the point $(6, -2)$. What is its equation?

8.7 INEQUALITIES IN TWO VARIABLES

In Sections 8.3 and 8.4, we graphed equations in two variables. In this section we will graph inequalities in two variables. For example, consider the inequality

$$y \leq -x + 6.$$

The solutions are ordered pairs of numbers that "satisfy" the inequality. For example, the ordered pair $(1, 1)$ is a solution because, when 1 is substituted for x and 1 is substituted for y, we get

$$(1) \leq -(1) + 6, \quad \text{or} \quad 1 \leq 5,$$

which is a true statement. On the other hand, $(2, 5)$ is *not* a solution because when 2 is substituted for x and 5 is substituted for y, we obtain

$$(5) \leq -(2) + 6, \quad \text{or} \quad 5 \leq 4,$$

which is a false statement.

To graph the inequality $y \leq -x + 6$, we first graph $y = -x + 6$ as indicated in Figure 8.18. Notice that $(3, 3)$, $(3, 2)$, $(3, 1)$, $(3, 0)$, and so on, associated with the points that are on or below the line, are all solutions of the inequality $y \leq -x + 6$, whereas $(3, 4)$, $(3, 5)$, and $(3, 6)$, associated with points above the line, are not solutions of the inequality. In fact, all ordered pairs associated with points on or below the line are solutions of $y \leq -x + 6$. Thus, *every* point on or below the line is in the graph of the inequality. We represent this by shading the region below the line (see Figure 8.19).

In general, to graph a first-degree inequality in two variables of the form $Ax + By \leq C$ or $Ax + By \geq C$, we first graph the equation $Ax + By = C$ and then determine which half-plane (a region above or below the line) contains the solutions. We then shade this half-plane. We can always determine which half-plane to shade by selecting a point that is not on the graph of $Ax + By = C$

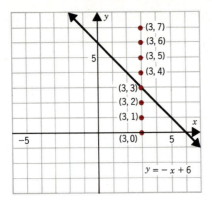

Figure 8.18

Figure 8.19

and testing to see if the ordered pair associated with the point is a solution of the given inequality. If so, we shade the half-plane containing the test point; otherwise, we shade the other half-plane. Often, $(0, 0)$ is a convenient test point as long as the line does not contain the origin.

Thus, to graph $2x + 3y \geq 6$, we first graph the equation $2x + 3y = 6$ (see Figure 8.20). Using the origin as a test point, we determine whether $(0, 0)$ is a solution of $2x + 3y \geq 6$. Since the statement

$$2(0) + 3(0) \geq 6$$

is false, $(0, 0)$ is not a solution and we shade the half-plane that does not contain the origin (see Figure 8.21).

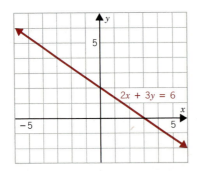

Figure 8.20

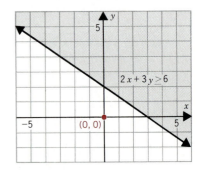

Figure 8.21

As another example, to graph $y \geq 2x$, we first graph $y = 2x$ (see Figure 8.22). Since the line passes through the origin, we must choose another point not on the line as our test point. We will use $(0, 1)$. Since the statement

$$(1) \geq 2(0)$$

is true, $(0, 1)$ is a solution and we shade the half-plane that contains $(0, 1)$ (see Figure 8.23).

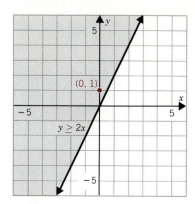

Figure 8.22   **Figure 8.23**

If the inequality symbol is $<$ or $>$, rather than $\leq$ or $\geq$, the points on the graph of $Ax + By = C$ are *not* solutions of the inequality. We then use a dashed line for the graph of $Ax + By = C$. An example is shown in the exercises.

EXERCISES 8.7

■ *Determine if the given point is a solution of the inequality.*

Sample Problems

a. $y < 4x$; $(1, -3)$

Substitute 1 for x and -3 for y.

$(-3) < 4(1)$ is true.

Ans. $(1, -3)$ is a solution.

b. $2x + y \geq 4$; $(2, -2)$

Substitute 2 for x and -2 for y.

$2(2) + (-2) \geq 4$ is false.

Ans. $(2, -2)$ is not a solution.

1. $y < x$; $(0, 1)$
2. $y > x$; $(1, 3)$
3. $y < x + 3$; $(-1, 2)$
4. $y \geq x - 5$; $(3, 2)$
5. $x + y < -2$; $(-2, -1)$
6. $x + y > -1$; $(-1, 3)$
7. $2x + y \leq 1$; $(-1, -1)$
8. $2x - y > 3$; $(-2, -4)$
9. $x < 4y + 1$; $(0, -3)$
10. $x < 3y + 5$; $(2, 0)$
11. $0 \leq x + 4y$; $(-1, 1)$
12. $2 > x + 2y$; $(0, -2)$

■ *Graph each inequality.*

Sample Problem

$-2x + y < 4$

Graph the equation $-2x + y = 4$.

Test $(0, 0)$: $-2(0) + (0) < 4$ is true. Shade the half-plane containing $(0, 0)$. Since the inequality symbol is $<$, use a dashed line.

Ans.

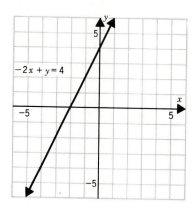

13. $y - 3x \leq 5$ **14.** $y + 3x \leq 6$ **15.** $3x - 4y \geq 12$

16. $2x + 5y \geq 10$ **17.** $y + 3x < 6$ **18.** $2x - y < 4$

19. $y > 3x + 2$ **20.** $y > -2x + 1$ **21.** $y < x - 3$

22. $y < -x - 3$ **23.** $6x + 4y \leq 12$ **24.** $3x + 2y \leq 12$

25. $y \leq 2$ **26.** $y \leq -1$ **27.** $x > -3$

28. $x > 2$ **29.** $y \leq 2x$ **30.** $y \geq -2x$

                             (*Hint.* $(0, 0)$ is not a   (*Hint.* $(0, 0)$ is not a
                             good test point.)   good test point.)

31. $y < 3x$ **32.** $y > 3x$ **33.** $y > 4x$ **34.** $y < -4x$

8.8 DIRECT VARIATION

A special case of a first-degree equation in two variables where the y-intercept is zero is given by

$$y = kx \qquad (k \text{ is a constant}).$$

Such a relationship is called a **direct variation.** We say that the variable y *varies directly as x.* For example, in the formula

$$d = 40t,$$

which relates the distance traveled at a constant rate of 40 miles per hour to the time traveled, d varies directly as t. As the time increases, the distance increases.

In a direct variation, if we know a set of conditions on the two variables, and if we further know another value for one of the variables, we can find the value of the second variable for this new set of conditions. For example, the pressure P in a liquid varies directly as the depth d below the surface of the liquid. We can state this relationship in symbols as

$$P = kd.$$

Solving for the constant k we obtain

$$k = \frac{P}{d}.$$

Since the ratio $\frac{P}{d}$ is constant for each set of conditions, we can use a proportion to solve problems involving direct variation. For example, if $P = 40$ when $d = 10$, we can find P when $d = 15$ as follows. Substituting the values for P and d, we obtain the proportion

$$k = \frac{40}{10} = \frac{P}{15},$$

from which

$$40 \cdot 15 = 10 \cdot P,$$

$$\frac{40 \cdot 15}{10} = P,$$

$$60 = P.$$

Thus, $P = 60$ when $d = 15$.

EXERCISES 8.8

■ *Write an equation expressing the relationship between the variables using k as the constant of variation.*

Sample Problem

The distance d that a car travels at a constant speed varies directly with the time t.

Ans. $d = kt$

1. y varies directly with x.　　　　**2.** P varies directly with R.

3. R varies directly with L.　　　　**4.** I varies directly with d.

5. The cost c in dollars of a certain item varies directly with the weight w in kilograms.

6. The current I in an electric circuit varies directly with the voltage E.

7. The tension T on a spring varies directly with the distance s it is stretched.

8. The pressure p exerted on a liquid at a given point varies directly with the depth d of the point beneath the surface of the liquid.

■ *Solve.*

Sample Problem   If y varies directly as x, and $y = 22$ when $x = 33$, find y if $x = 40$.

Write the relationship between y and x.

$$y = kx$$

Solve for k. Divide each member by x.

$$\frac{y}{x} = k$$

Since the ratio y/x is a constant, ratios are equal for different sets of conditions. Substitute appropriate values in each of two ratios to obtain a proportion.

$$\frac{22}{33} = \frac{y}{40}$$

Solve for y.

$$(22)(40) = 33y$$

$$\frac{(22)(40)}{33} = y$$

Ans. $y = \dfrac{80}{3}$

9. If y varies directly with x, and $y = 6$ when $x = 4$, find y when $x = 14$.

10. If z varies directly with x, and $z = 9$ when $x = 12$, find z when $x = 3$.

11. If y varies directly with t, and $y = \frac{1}{5}$ when $t = \frac{3}{10}$, find y when $t = \frac{3}{4}$.

12. If s varies directly with t, and $s = \frac{15}{2}$ when $t = 6$, find s when $t = \frac{4}{3}$.

13. If P varies directly with T, and $P = 13$ when $T = 20$, find P when $T = 26$.

14. If W varies directly with d, and $W = 37\frac{1}{2}$ when $d = \frac{3}{4}$, find W when $d = \frac{5}{3}$.

15. If R varies directly with L, and $R = 42$ when $L = 30$, find R when $L = 45$.

16. If I varies directly with d, and $I = 640$ when $d = 120$, find I when $d = 600$.

Sample Problem   The force F exerted by a spring varies directly with the distance d the spring is stretched. If the force exerted by the spring is 48 pounds when the distance is 8 feet, find the force when the distance is 10 feet.

Write an equation for the variation.

$$F = kd$$

Solve for k.

$$\frac{F}{d} = k$$

The ratio F/d is constant. Substitute given values.

$$\frac{48}{8} = \frac{F}{10}$$

Solve for F.

$$8F = (48)(10)$$

$$F = \frac{(48)(10)}{8}$$

Ans. $F = 60$ pounds.

17. The amount of work W (measured in foot-pounds) done in moving an object varies directly with the distance the object is moved (in feet). If 1200 foot-pounds are required to move the object 15 feet, how much work is done in moving the object 8 feet?

18. If 72 foot-pounds are required to move an object 3 feet, how much work is done in moving the object $5\frac{1}{2}$ feet?

19. If the force exerted by a spring is 24 pounds when the spring is stretched 20 inches, find the force exerted when the spring is stretched 1 foot.

20. A spring exerts $17\frac{1}{2}$ pounds of force when it is stretched 10 inches. How far must it be stretched to exert 14 pounds of force?

21. The weight W of a steel beam varies directly with its length l. If a 20-foot beam weighs 1400 pounds, how much does a 26-foot beam weigh?

22. What would be the length of a beam in exercise 21 if it weighed 900 pounds?

23. The distance d in miles that a car travels varies directly with the time t in hours. If $d = 96$ when $t = 2$, find the distance traveled in $3\frac{1}{2}$ hours.

24. The cost C in dollars of a certain vitamin varies directly with the weight w in kilograms. If $C = 60$ when $w = 24$, find the cost of 54 kilograms.

25. The current I in an electric circuit varies directly with the voltage E. If $I = 4.2$ when $E = 60$, find E when $I = 6.6$.

26. The tension T on a spring varies directly as the distance s it is stretched. If $T = 54$ when $s = 12$, find s if $T = 20$.

27. The pressure p exerted by a liquid at a given point varies directly with the depth d of the point beneath the surface of the liquid. If a certain liquid exerts a pressure of 20 pounds per square foot at a depth of 6 feet, what would be the pressure at 9 feet?

28. What would be the pressure of the liquid in exercise 27 at a depth of 15 feet?

29. The circumference C of a circle varies directly with the diameter D. What is the effect on the circumference if the diameter is doubled in size?

30. What is the effect on the diameter of a circle if the circumference is reduced to one-fourth of its original length? (See exercise 29.)

CHAPTER SUMMARY

[8.1]   A **solution** of an equation in two variables is an **ordered pair** of numbers. In the ordered pair (x, y), x is called the **first component** and y is called the **second component.** For an equation in two variables, the variable associated with the first component of a solution is called the **independent variable** and the variable associated with the second component is called the **dependent variable.**

[8.2]   The intersection of the two perpendicular axes in a coordinate system is called the **origin** of the system, and each of the four regions into which the plane is divided is called a **quadrant.** The components of an ordered pair (x, y) associated with a point in the plane are called the **coordinates** of the point; x is called the **abscissa** of the point and y is called the **ordinate** of the point.

[8.3-8.4]   The graph of a first-degree equation in two variables is a straight line. That is, every ordered pair that is a solution of the equation has a graph that lies in a line, and every point in the line is associated with an ordered pair that is a solution of the equation.

    The graphs of *any two solutions* of an equation in two variables can be used to obtain the graph of the equation. However, the two solutions of an equation in two variables that are generally easiest to find are those in which either the first or second component is 0. The *x*-coordinate of the point where a line crosses the *x*-axis is called the **x-intercept** of the line, and the *y*-coordinate of the point where a line crosses the *y*-axis is called the **y-intercept** of the line. Using the intercepts to graph an equation is called the **intercept method of graphing.**

[8.5]   The **slope** of a line containing the points $P_1(x_1, y_1)$ and $P_2(x_2, y_2)$ is given by

$$m = \frac{y_2 - y_1}{x_2 - x_1} \quad (x_2 \neq x_1).$$

Two lines are parallel if they have the same slope $(m_1 = m_2)$.
Two lines are perpendicular if the product of their slopes is -1 $(m_1 \cdot m_2 = -1)$.

[8.6]   The **point-slope form** of a line with slope m and passing through the point (x_1, y_1) is

$$y - y_1 = m(x - x_1).$$

The **slope-intercept form** of a line with slope m and y-intercept b is

$$y = mx + b.$$

[8.7]   A **solution** of an inequality in two variables is an ordered pair of numbers that, when substituted into the inequality, makes the inequality a true statement. The graph of a linear inequality in two variables is a half-plane.

[8.8]   A relationship determined by an equation of the form

$$y = kx \qquad (k \text{ a constant})$$

is called a **direct variation.**

■ *The symbols introduced in this chapter appear on the inside front cover.*

CHAPTER REVIEW

1. Given $I = 0.05P$. The yearly interest (I) on an investment equals the rate (0.05) times the principal (P).
   a. Which of the symbols are constants?
   b. Which symbols are variables?
   c. If the principal increases, what happens to the interest?
2. Solve $2x - y = 4$ for y in terms of x.
3. Solve $2y - 3x = 6$ for y in terms of x.
4. If $y = 2x + 1$, find the solutions with specified first components.
   a. (3, ?)   b. (−2, ?)   c. (0, ?)   d. $(-\frac{1}{2}, ?)$
5. If $2x - 3y = 8$, find the missing component so that the ordered pair is a solution to the equation.
   a. (6, ?)   b. (?, −6)
6. Graph the following ordered pairs on a set of rectangular axes.
   a. (3, 4)   b. (−2, 3)   c. (3, −2)   d. (0, 4)

■ *In exercises 7–10, graph each equation.*

7. $x + y = 3$ 8. $2y - x = 4$
9. $x = 3$ 10. $3x + 2y = 6$

11. Where does the graph of $x - y = 8$ cross the x-axis? The y-axis?
12. Find the slope of the line containing the points (2, 3) and (−4, 5).
13. Is the line passing through the points (2, 4) and (−1, 3) parallel to the line passing through (5, −1) and (−3, 2)?
14. What is the equation of a line with slope −2 and passing through the point (−2, −5)? Write your answer in $y = mx + b$ form.
15. What is the equation of a line passing through the points (−2, 3) and (1, 5)? Write your answer in $y = mx + b$ form.

16. What is the slope and y-intercept of the graph of $2y - 5x = 0$?

17. If y varies directly with x, and $y = 20$ when $x = 6$, find x when $y = 44$.

■ *In exercises 18–20, graph each inequality.*

18. $2x + 3y \geq 6$ **19.** $x \leq -2$ **20.** $y > -3x$

CUMULATIVE REVIEW

1. Graph all integers between $-\frac{3}{2}$ and $\frac{15}{4}$ on a number line.

2. If $a = 0$, $b = 1$, $c = -1$, $d = 2$, find the value of $\dfrac{bc + d}{ab + d}$.

3. If $x = 1$, $y = -1$, and $z = -2$, find the value of $-xy^2 + (xz)^2 - yz^2$.

4. Given $s = \frac{1}{2} gt^2 + c$. Find s, if $c = 2000$, $g = 32$, and $t = 2$.

■ *Simplify.*

5. $\dfrac{3x + 6}{2x + 4}$

6. $\dfrac{a^2 - b^2}{a^2 - 2ab + b^2}$

7. $\dfrac{6a - 5}{8} + \dfrac{3a + 5}{12}$

8. $\dfrac{2}{x - 2} - \dfrac{3}{x + 1}$

9. $\dfrac{7a - 14}{5a - 10} \cdot \dfrac{3a - 3}{7a - 7}$

10. $\dfrac{3a + 3}{2a - 6} \div \dfrac{6a + 6}{3a - 9}$

11. $\dfrac{x - \dfrac{x}{y}}{1 - \dfrac{1}{y}}$

12. $(x^2 - 6x + 8) \div (x - 2)$

■ *Solve each equation for x.*

13. $\dfrac{3x}{4} - 9 = 0$

14. $x - \dfrac{3x - 2}{2} = \dfrac{1}{2}$

15. $\dfrac{3}{5} = \dfrac{x + 1}{x + 3}$

16. $\dfrac{1}{2} + \dfrac{1}{x} = \dfrac{1}{3}$

17. Graph: $y = 2x - 6$.

18. Find the slope and y-intercept for the graph of $2x - 3y = 9$.

19. Find the slope of a line passing through the points $(-5, 1)$ and $(3, 7)$.

20. Graph: $2x + 4y < 10$.

■ *True or false.*

21. $\dfrac{x}{2} + \dfrac{y}{2}$ is equivalent to $\dfrac{x + y}{4}$.

22. $\dfrac{3 + 2x}{3}$ is equivalent to $2x$.

23. $\dfrac{4 + 8x}{4}$ is equivalent to $1 + 2x$.

24. $(x + 2)^2$ is equivalent to $x^2 + 4$.

25. $\dfrac{x^2 + x - 6}{x^2 + x - 12}$ is equivalent to $\dfrac{1}{2}$.

9 SYSTEMS OF LINEAR EQUATIONS

9.1 GRAPHICAL SOLUTIONS

Often we want to find a single ordered pair that is a solution to two different linear equations. One way to obtain such an ordered pair is by graphing the two equations on the same set of axes and determining the coordinates of the point where they intersect. For example, consider the equations

$$x + y = 5 \qquad \text{and} \qquad x - y = 1.$$

Using the intercept method of graphing, we find that two ordered pairs that are solutions of $x + y = 5$ are

$$(0, 5) \qquad \text{and} \qquad (5, 0).$$

And two ordered pairs that are solutions of $x - y = 1$ are

$$(0, -1) \qquad \text{and} \qquad (1, 0).$$

The graphs of the equations are shown in Figure 9.1. The point of intersection is (3, 2). Thus, (3, 2) should satisfy each equation. In fact,

$$3 + 2 = 5 \qquad \text{and} \qquad 3 - 2 = 1.$$

In general, graphical solutions are only approximate. We will develop methods for exact solutions in later sections.

Linear equations considered together in this fashion are said to form a **system of equations.** In the example above, the solution of the system of linear equations is a single ordered pair. The components of this ordered pair satisfy *each* of the two equations.

Some systems have no solutions, while others have an infinite number of solutions. If the graphs of the equations in a system do not intersect—that is, if the lines are parallel (see Figure 9.2a)—the equations are said to be **inconsistent,** and there is no ordered pair that will satisfy both equations. If the graphs

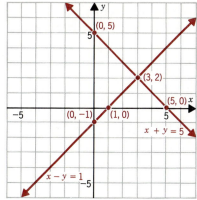

Figure 9.1

of the equations are the same line (see Figure 9.2b), the equations are said to be **dependent,** and each ordered pair which satisfies one equation will satisfy both equations. Notice that when a system is inconsistent, the slopes of the lines are the same but the y-intercepts are different. When a system is dependent, the slopes and y-intercepts are the same.

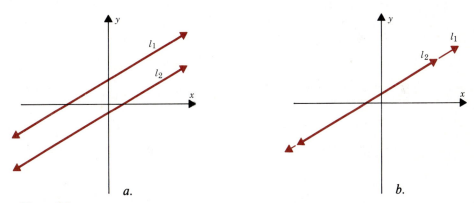

a.   b.

Figure 9.2

In our work we will be primarily interested in systems that have one and only one solution. Such systems are said to be **consistent and independent.** The graph of such a system is shown in Figure 9.1.

EXERCISES 9.1

■ *State whether or not the given point is a solution of the system of equations.*

Sample Problem

$$\begin{aligned} 3x - 4y &= -19 \\ x + 2y &= 5 \end{aligned} \quad ; \quad (-1, 4)$$

$3(-1) - 4(4) = -3 - 16 = -19$

> The ordered pair is a solution of the first equation.

$-1 + 2(4) = -1 + 8 = 7 \neq 5$

> The ordered pair is not a solution of the second equation.

Ans. The ordered pair is *not* a solution of the system.

1. $x + y = 8$
$\quad x - y = 2$; $(-2, 10)$

2. $x + 2y = -8$
$\quad 2x - y = 4$; $(4, -2)$

3. $3x - 2y = -5$
$\quad x + 3y = -9$; $(-3, -2)$

4. $x - 5y = 13$
$\quad 2x + 3y = 0$; $(3, -2)$

5. $4x - y = -6$
$\quad 2x - 3y = 8$; $(1, -2)$

6. $8x + 3y = 21$
$\quad 5x - y = -16$; $(-3, 1)$

■ *Find the solution of each system by graphical methods. If the system is inconsistent or dependent, so state.*

Sample Problem

$x + y = 8$

$5x - 2y = 5$

Using the intercept method of graphing, we find that two solutions of $x + y = 8$ are

$(0, 8)$ and $(8, 0)$.

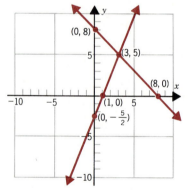

Two solutions of $5x - 2y = 5$ are $\left(0, \dfrac{-5}{2}\right)$ and $(1, 0)$. The graphs of the two equations intersect at the point corresponding to $(3, 5)$.

Ans. $(3, 5)$

7. $x - y = 2$
$\quad x + y = 8$

8. $x - y = -2$
$\quad 3x - y = 8$

9. $2x - y = 0$
$\quad 2x - 3y = -4$

10. $y + 3x = 0$
$\quad 4x - 3y = 13$

11. $x - 3y = 0$
$\quad 2x - y = -5$

12. $2y + x = 0$
$\quad y - x = 6$

13. $x + y = 6$
$\quad x - y = 2$

14. $5y - x = 0$
$\quad -y + x = 4$

15. $y = x$
$\quad y = 4 - x$

16. $y = -x$
$\quad y = 3 - 2x$

17. $y - x = 1$
$\quad y + x = -5$

18. $3x - 3y = 0$
$\quad x + y = 2$

19. $2x - y + 5 = 0$
$x - 3y = -20$

20. $2x + 7y - 12 = 0$
$x + 5y = 0$

21. $5x - 2y + 8 = 0$
$3x + y + 7 = 0$

22. $7y - 2x + 6 = 0$
$8y - 5x - 4 = 0$

23. $x + y = 4$
$2x + 2y = 8$

24. $x - 2y = 3$
$2x - 4y = 6$

25. $x + 3y = 5$
$2x + 6y = 5$

26. $x - 3y = -4$
$-2x + 6y = 5$

27. $3x - y = 1$
$6x = 2y + 2$

28. $-x + 2y = 3$
$3x = 6y - 9$

29. $x - 2y = 4$
$-4y = 5 - 2x$

30. $x + y = 7$
$-2y = 5 + 2x$

9.2 SOLVING SYSTEMS BY ADDITION I

We can solve systems of equations algebraically. What's more, the solutions we obtain by algebraic methods are exact.

For example, consider the system we solved graphically in Section 9.1 on page 300.

$$x + y = 5 \quad (1)$$
$$\underline{x - y = 1} \quad (2)$$
$$2x \quad = 6.$$

Obtain an equation in one variable by adding Equations (1) and (2).

Solve the resulting equation for x.

$$x = 3.$$

$$(3) + y = 5,$$
$$y = 2.$$

Substitute 3 for x in either Equation (1) or Equation (2) to obtain the corresponding value of y. In this case we have selected Equation (1).

Thus the solution is $x = 3$, $y = 2$; or (3, 2).

Notice that we are simply applying the addition property of equality so we can obtain an equation containing a single variable. The equation in one variable, together with either of the original equations, then forms an equivalent system whose solution is easily obtained.

In the example above, we were able to obtain an equation in one variable by adding Equations (1) and (2) because the terms $+y$ and $-y$ are the negatives of each other. Sometimes it is necessary to multiply each member of one of the equations by -1 so that terms in the same variable will have opposite signs. For example, in the system

$$2a + b = 4 \qquad (3)$$

$$a + b = 3, \qquad (4)$$

by multiplying *each member* of Equation (4) by -1, we obtain

$$2a + b = 4 \qquad (3)$$

$$-a - b = -3, \qquad (4')$$

where $+b$ and $-b$ are negatives of each other.

The symbol ′, called "prime," indicates an equivalent equation; that is, an equation that has the same solutions as the original equation. Thus, Equation (4′) is equivalent to Equation (4). Now, adding Equations (3) and (4′), we get

$$
\begin{array}{rl}
2a + b = & 4 \qquad\qquad (3) \\
-a - b = & -3 \qquad\quad (4') \\
\hline
a \quad\quad = & 1.
\end{array}
$$

Substituting 1 for a in Equation (3) or Equation (4) [say, Equation (4)], we obtain

$$(1) + b = 3,$$

$$b = 2,$$

and our solution is $a = 1$, $b = 2$ or (1, 2). Where the variables are a and b, the ordered pair is given in the form (a, b).

EXERCISES 9.2

■ *Solve each system.*

Sample Problem

$$
\begin{array}{rl}
2x - y = 0 & (1) \\
2x + y = 4 & (2) \\
\hline
4x \quad\;\; = 4 &
\end{array}
$$

Obtain an equation in one variable by adding Equations (1) and (2).

Solve the resulting equation for x.

$$x = 1$$

Substitute 1 for x in either Equation (1) or Equation (2) and solve for y. Here we select Equation (1).

$$2(1) - y = 0$$

$$2 = y$$

Ans. $x = 1$, $y = 2$; or (1, 2)

1. $x + y = 5$
$\quad\; x - y = 1$

2. $x + y = 10$
$\quad\; x - y = 4$

3. $\quad x + y = 3$
$\quad -x + y = 5$

4. $-x + y = 7$
$\quad\;\; x + y = 5$

5. $x + 2y = 10$
$\quad\; x - 2y = 2$

6. $\quad 2x + y = 3$
$\quad -2x + y = -1$

7. $\quad 3x + 2y = 5$
$\quad -2x - 2y = -4$

8. $5x - 3y = -1$
$\quad 3x + 3y = 9$

9. $3x + 3y = 15$
$\quad 3x - 3y = 27$

10. $6x - y = 4$
$\quad\;\; 2x + y = 4$

11. $7x - 3y = -10$
$\quad\;\; x + 3y = 2$

12. $3x - 2y = 11$
$\quad 3x + 2y = 19$

Sample Problem

$$
\begin{array}{rl}
2a + b = -2 & (1) \\
a + b = \;\; 1 & (2)
\end{array}
$$

Multiply each member of Equation (2) by -1 to obtain (2′). Add Equations (1) and (2′).

$$-a - b = -1 \quad (2')$$
$$\underline{2a + b = -2} \quad (1)$$
$$a \quad\quad = -3$$

Substitute -3 for a in either Equation (1) or Equation (2). We select Equation (2).

$$(-3) + b = 1$$
$$b = 4$$

Ans. $a = -3, \quad b = 4; \quad$ or $(-3, 4)$

13. $2a + b = 3$
$\quad\quad a + b = 2$

14. $\quad a + b = 6$
$\quad\quad 3a + b = 10$

15. $a + \quad b = 7$
$\quad\quad a + 3b = 11$

16. $\quad a + 2b = -1$
$\quad\quad 3a + 2b = 1$

17. $3a + 2b = 1$
$\quad\quad 2a + 2b = 0$

18. $\quad 6a + 5b = 6$
$\quad\quad -4a + 5b = -4$

19. $4a + 2b = 4$
$\quad\quad 3a + 2b = 8$

20. $4a + 2b = 4$
$\quad\quad 6a + 2b = 8$

21. $\quad a + 3b = 2$
$\quad\quad 2a + 3b = 7$

22. $\quad 3a - 4b = 19$
$\quad\quad -2a - 4b = -6$

23. $3a - 4b = 0$
$\quad\quad 3a + 2b = 18$

24. $\quad 3a - 2b = 5$
$\quad\quad -a - 2b = -15$

25. $\quad x + y = -1$
$\quad\quad 2x + y = -5$

26. $2x + 3y = -5$
$\quad\quad 2x - 4y = 16$

27. $3x + 4y = -10$
$\quad\quad x + 4y = -6$

28. $4x - 4y = -4$
$\quad\quad 4x - 3y = -3$

29. $7x + 6y = 12$
$\quad\quad 9x + 6y = 12$

30. $3a + 2b = -12$
$\quad\quad a + 2b = -4$

Sample Problem

$$\frac{x}{6} + \frac{y}{4} = \frac{3}{2} \quad\quad (1)$$

$$\frac{2x}{3} - \frac{y}{2} = 0 \quad\quad (2)$$

Remove fractions by multiplying each equation by L.C.D. of fractions in that equation.

$$(\overset{2}{\cancel{12}})\frac{x}{\cancel{6}} + (\overset{3}{\cancel{12}})\frac{y}{\cancel{4}} = (\overset{6}{\cancel{12}})\frac{3}{\cancel{2}} \quad (1)$$

$$(\overset{2}{\cancel{6}})\frac{2x}{\cancel{3}} - (\overset{3}{\cancel{6}})\frac{y}{\cancel{2}} = (6)0 \quad (2)$$

Simplify Equations (1) and (2).

$$2x + 3y = 18 \quad\quad (1')$$
$$\underline{4x - 3y = 0} \quad\quad (2')$$
$$6x \quad\quad = 18$$

Solve for x.

$$x = 3$$

Substitute 3 for x in Equation $(2')$.

$$4(3) - 3y = 0$$

$$-3y = -12$$

$$y = 4$$

Ans. $x = 3$, $y = 4$; or (3, 4)

31. $y + \dfrac{1}{2}x = 0$

$\dfrac{1}{3}y - \dfrac{1}{3}x = 2$

32. $\dfrac{a}{5} + \dfrac{2b}{5} = 2$

$\dfrac{a}{2} - b = 1$

33. $\dfrac{2x}{3} + y = 3$

$\dfrac{x}{2} - \dfrac{y}{4} = \dfrac{5}{4}$

34. $x - \dfrac{2}{3}y = -4$

$\dfrac{x}{4} + \dfrac{y}{2} = -1$

35. $\dfrac{a}{4} - \dfrac{b}{3} = 0$

$\dfrac{a}{2} + \dfrac{b}{3} = 3$

36. $a + \dfrac{b+1}{3} = 0$

$\dfrac{5a}{2} - \dfrac{b+8}{4} = -5$

9.3 SOLVING SYSTEMS BY ADDITION II

As we saw in Section 9.2, solving a system of equations by addition depends on the same variable in both equations having coefficients that are the negatives of each other. If this is not the case, we can find equivalent equations that do have variables with such coefficients. Consider the system

$$-5x + 3y = -11 \tag{1}$$

$$-7x - 2y = -3. \tag{2}$$

If we multiply each member of Equation (1) by 2 and each member of Equation (2) by 3, we obtain the equivalent system

$$(2)(-5x) + (2)(3y) = (2)(-11)$$

$$(3)(-7x) - (3)(2y) = (3)(-3),$$

or

$$-10x + 6y = -22, \tag{1'}$$

$$-21x - 6y = -9. \tag{2'}$$

Now, adding Equations (1') and (2'), we get

$$-31x = -31,$$

$$x = 1.$$

Substituting 1 for x in Equation (1) yields

$$-5(1) + 3y = -11,$$
$$3y = -6,$$
$$y = -2.$$

The solution is $x = 1$, $y = -2$ or $(1, -2)$.

Note that in Equations (1) and (2), the terms involving variables are in the left-hand member and the constant term is in the right-hand member. We will refer to such an arrangement as **standard form** for systems. It is convenient to arrange systems in standard form before proceeding with their solution. For example, if we want to solve the system

$$3y = 5x - 11 \tag{3}$$
$$-7x = 2y - 3, \tag{4}$$

we would first write the system in standard form by adding $-5x$ to each member of Equation (3) and by adding $-2y$ to each member of Equation (4). Thus, we get

$$-5x + 3y = -11$$
$$-7x - 2y = -3,$$

and we can now proceed as shown in the example above.

EXERCISES 9.3

■ *Solve.*

Sample Problem

$$3x + 2y = 11 \quad (1)$$
$$5x - 4y = 11 \quad (2)$$

Multiply each member of Equation (1) by 2. Add Equations (1′) and (2).

$$6x + 4y = 22 \quad (1')$$
$$\underline{5x - 4y = 11} \quad (2)$$
$$11x \qquad = 33$$

Solve for x.

$$x = 3$$

Substitute 3 for x in Equation (1) and solve for y.

$$3(3) + 2y = 11$$
$$2y = 2$$
$$y = 1$$

Ans. $x = 3$, $y = 1$; or $(3, 1)$

1. $3x + 2y = 7$
$x + y = 3$

2. $2x - 3y = 8$
$x + y = -1$

3. $2x - y = 2$
$3x + 2y = 10$

4. $a - 4b = 9$
$3a + 2b = 13$

5. $3a - b = -5$
$2a + 3b = -7$

6. $-x + 3y = -1$
$-6x + y = -6$

7. $3a - 3b = -3$
$-6a + 2b = 14$

8. $3x - 6y = 6$
$x - 2y = 3$

9. $5x + 3y = 19$
$2x - y = 12$

10. $5x - 3y = 32$
$2x + 6y = -16$

11. $3x - 5y = -1$
$x + 2y = 18$

12. $3x + 3y = 0$
$6x + 9y = -6$

Sample Problem

$2y = 11 - 3x$　(1)

$5x = 11 + 4y$　(2)

Add $3x$ to each member of Equation (1) and add $-4y$ to each member of Equation (2) to arrange system in standard form.

$3x + 2y = 11$

$5x - 4y = 11$

Solve as described in the previous example.

13. $3x + 2y = 7$
$x = y - 1$

14. $3x + 9 = -2y$
$x = -3y + 25$

15. $a + 5b = 0$
$-3a + 10 = 10b$

16. $x = 8 - 2y$
$2x = 6 + y$

17. $8b - 3a = 5$
$a - b = 0$

18. $6x = 22 + 2y$
$8x = 33 - y$

19. $8a = 4b + 4$
$3a = 2b + 3$

20. $x + 3y = 35$
$0 = 2x - y$

21. $5b = 3a + 8$
$2b = -a + 1$

22. $x - y = 15$
$2x = -3y$

23. $3x - 2y = 3$
$2x = y + 2$

24. $2x = 2y + 2$
$4x = 5 + 4y$

25. $2x + 3y = -1$
$3x + 5y = -2$

26. $3x - 2y = 13$
$7x + 3y = 15$

27. $2x = 3y - 1$
$3x + 4y = 24$

28. $5x - 2y = 0$
$2x - 3y = -11$

29. $4b - a = 8$
$2b + a = 10$

30. $3b - 4a = 3$
$2b = 4a - 2$

31. $2a + 3b = 0$
$5a - 2b = -19$

32. $8x - 7y = 0$
$7x = 8y + 15$

Sample Problem

$$\frac{x + 1}{2} - \frac{2y - 1}{6} = \frac{7}{6} \qquad (1)$$

$$\frac{2x - 1}{4} - \frac{y - 3}{4} = 1 \qquad (2)$$

Express each equation without fractions by multiplying by L.C.D. of fractions in each equation.

$$(6)\frac{x + 1}{2} - (6)\frac{2y - 1}{6} = (6)\frac{7}{6} \qquad (1)$$

$$(4)\frac{2x-1}{4} - (4)\frac{y-3}{4} = (4)\ 1 \quad (2)$$

$$3(x+1) - (2y-1) = 7$$

$$(2x-1) - (y-3) = 4$$

Write system in standard form.

$$3x - 2y = 3 \qquad (1')$$

$$2x - y = 2 \qquad (2')$$

Multiply Equation (2') by -2, and solve for x.

$$3x - 2y = 3 \qquad (1')$$

$$\underline{-4x + 2y = -4} \qquad (2'')$$

$$-x \qquad\quad = -1$$

$$x = 1$$

Substitute 1 for x in Equation (1') and solve for y.

$$3(1) - 2y = 3$$

$$-2y = 0$$

$$y = 0$$

Ans. $x = 1$, $y = 0$; or $(1, 0)$

33. $\dfrac{5a}{4} + b = \dfrac{11}{2}$

$a + \dfrac{b}{3} = 3$

34. $2a - \dfrac{5b}{2} = 13$

$\dfrac{a}{3} + \dfrac{b}{5} = \dfrac{14}{15}$

35. $\dfrac{x}{4} + \dfrac{y}{5} = 1$

$\dfrac{2x}{9} - \dfrac{y}{9} = -2$

36. $\dfrac{5x}{8} + y = \dfrac{1}{4}$

$\dfrac{5x}{4} - \dfrac{3y}{2} = 4$

37. $\dfrac{x}{3} - \dfrac{y}{2} = 1$

$\dfrac{2x+3}{5} - \dfrac{y}{2} = 2$

38. $\dfrac{2x+1}{7} + \dfrac{3y+2}{5} = \dfrac{1}{5}$

$\dfrac{3x-2}{2} + \dfrac{y+4}{4} = 4$

9.4   SOLVING SYSTEMS BY SUBSTITUTION

In Sections 9.2 and 9.3, we solved systems of first-degree equations in two variables by the addition method. Another method, called the **substitution**

method, can also be used to solve such systems. Consider the system

$$-2x + y = 1 \tag{1}$$

$$x + 2y = 17. \tag{2}$$

Solving Equation (1) for y in terms of x, we obtain

$$y = 2x + 1. \tag{1'}$$

We can now substitute $2x + 1$ for y in Equation (2) to obtain

$$x + 2(2x + 1) = 17,$$

$$x + 4x + 2 = 17,$$

$$5x = 15,$$

$$x = 3.$$

Substituting 3 for x in Equation (1'), we have

$$y = 2(3) + 1 = 7.$$

Thus, the solution of the system is $x = 3$, $y = 7$; or (3, 7).

In the example above, it was easy to express y explicitly in terms of x using Equation (1). But we could also have used Equation (2) to write x explicitly in terms of y:

$$x = -2y + 17. \tag{2'}$$

Now substituting $-2y + 17$ for x in Equation (1), we get

$$-2(-2y + 17) + y = 1,$$

$$4y - 34 + y = 1,$$

$$5y = 35,$$

$$y = 7.$$

Substituting 7 for y in Equation (2'), we have

$$x = -2(7) + 17 = 3.$$

The solution of the system is again (3, 7).

Note that the substitution method is useful if we can easily express one variable in terms of the other variable.

EXERCISES 9.4

■ *Solve each system using the substitution method.*

Sample Problem

$2y + 3x = 7$ (1)

$y = x + 1$ (2)

              Substitute $x + 1$ for y in Equation (1).

$$2(x + 1) + 3x = 7$$

Solve for x.

$$2x + 2 + 3x = 7$$

$$5x = 5$$

$$x = 1$$

Substitute 1 for x in Equation (2).

$$y = (1) + 1 = 2$$

Ans. $x = 1$, $y = 2$; or $(1, 2)$

1. $y = 2x$
   $3x + y = 10$

2. $y = 3x$
   $x + 2y = 14$

3. $y = x - 2$
   $2x + y = 7$

4. $y = x - 4$
   $x + 4y = 9$

5. $2x + 3y = 4$
   $y = 2x + 4$

6. $x + 2y = -1$
   $y = x - 5$

7. $x - y = 6$
   $y = x - 6$

8. $x - 2y = 1$
   $y = 2x - 5$

9. $3x - 2y = 10$
   $y = -x$

10. $2x + y = -3$

   $y = -x$

11. $\dfrac{x}{2} - 2y = \dfrac{17}{2}$

   $y = x - 5$

12. $\dfrac{x}{3} + \dfrac{y}{2} = \dfrac{3}{2}$

   $y = 2x - 5$

Sample Problem

$$x - 10 = 3y \quad (1)$$

$$2x + 3y = -7 \quad (2)$$

Since x has a coefficient 1 in Equation (1), solve for x in terms of y.

$$x = 3y + 10 \quad (1')$$

Substitute $3y + 10$ for x in Equation (2).

$$2(3y + 10) + 3y = -7$$

Solve for y.

$$6y + 20 + 3y = -7$$

$$9y = -27$$

$$y = -3$$

Substitute -3 for y in Equation (1').

$$x = 3(-3) + 10 = 1$$

Ans. $x = 1$, $y = -3$; or $(1, -3)$

13. $x = y + 2$
   $2x - y = 7$

14. $x = y - 4$
   $3x - 7y = -8$

15. $x + 1 = 3y$
   $2x - 3y = 4$

16. $x - 3 = 2y$
   $5x + 5y = 0$

17. $x + 5 = 2y$
   $2x = 1 - 7y$

18. $x - 3 = 5y$
   $3x = 2y - 4$

19. $2y + x = -8$
$4x + 7 = -3y$

20. $3y + 11 = x$
$8x - 3y = 4$

21. $\dfrac{x}{4} + y = \dfrac{5}{2}$
$x = y$

22. $\dfrac{x}{2} + \dfrac{y}{3} = \dfrac{1}{2}$
$x = -y$

23. $\dfrac{x}{4} + y = \dfrac{1}{4}$
$2x + 3y = -3$

24. $\dfrac{x}{2} - \dfrac{y}{4} = \dfrac{5}{2}$
$4x - 7y = 10$

25. $\dfrac{x}{2} - \dfrac{y}{3} = \dfrac{1}{2}$
$\dfrac{5y}{2} - 2x = \dfrac{-7}{2}$

26. $\dfrac{5x}{4} + \dfrac{3y}{4} = 1$
$x + \dfrac{y}{6} = -\dfrac{1}{2}$

27. $\dfrac{x}{3} - 1 = \dfrac{4y}{9}$
$\dfrac{2x}{7} - 1 = y$

28. $\dfrac{y}{4} + x = -\dfrac{3}{4}$
$\dfrac{1}{3} + \dfrac{x}{3} = \dfrac{-y}{5}$

29. $y + \dfrac{3x}{4} = -1$
$\dfrac{x}{2} = \dfrac{4}{3} - \dfrac{y}{6}$

30. $\dfrac{3x}{4} = 2 + \dfrac{y}{2}$
$y = \dfrac{32}{3} - \dfrac{x}{3}$

9.5 APPLICATIONS USING TWO VARIABLES

If two variables are related by a single first-degree equation, there are infinitely many ordered pairs that are solutions of the equation. But if the two variables are related by two independent first-degree equations, there can be only one ordered pair that is a solution of both equations. Therefore, *to solve problems using two variables, we must represent two independent relationships using two equations.* We can often solve problems more easily by using a system of equations than by using a single equation involving one variable. We will follow the six steps outlined on page 84, with minor modifications as shown in the following example.

⌈ The sum of two numbers is 26. The larger number is 2 more than three times the ⌉
⌊ smaller number. Find the numbers. ⌋

Steps 1–2   We represent what we want to find as *two* word phrases. Then, we represent the word phrases in terms of *two* variables.

Smaller number: x

Larger number: y

Step 3   A sketch is not applicable.

Step 4   Now we must write *two* equations representing the conditions stated.

The sum of two numbers is 26.

$$x + y = 26$$

The larger number is two more than three times the smaller number.

$$y = 2 + 3 \cdot x$$

Step 5   To find the numbers, we solve the system

$$x + y = 26 \tag{1}$$

$$y = 2 + 3x \tag{2}$$

Since Equation (2) shows y explicitly in terms of x, we will solve the system by the substitution method. Substituting $2 + 3x$ for y in Equation (1), we get

$$x + (2 + 3x) = 26$$

$$4x = 24$$

$$x = 6$$

Substituting 6 for x in Equation (2), we get

$$y = 2 + 3(6) = 20$$

Step 6   **Ans.** The smaller number is 6 and the larger number is 20.

EXERCISES 9.5

■ *In each of the following exercises, represent two independent conditions of the problem by a system of equations using two variables. Then solve the system. Follow the six steps outlined on page 84.*

Sample Problem   A 12-foot board is cut into two parts so that one part is 2 feet longer than the other. How long is each part?

Steps 1–2   The longer part: x

The shorter part: y

Step 3   A sketch is helpful.

Step 4

$$x + y = 12 \tag{1}$$

$$x = y + 2 \tag{2}$$

Step 5   We use the substitution method. Substituting $y + 2$ for x in Equation (1), we get

$$y + 2 + y = 12$$
$$2y = 10$$
$$y = 5$$
$$x = 5 + 2 = 7$$

Step 6   **Ans.** The longer part is 7 feet; the shorter part is 5 feet.

1. The sum of two numbers is 25 and their difference is 9. What are the numbers?

2. The sum of two numbers is 21 and their difference is 13. What are the numbers?

3. A 20-meter board is cut into two pieces, one of which is 2 meters longer than the other. How long is each piece?

4. A 30-meter board is cut into two pieces, one of which is 6 meters shorter than the other. How long is each piece?

5. Two packages weighed together total 28 kilograms. One of the packages weighs 8 kilograms less than twice the other. How much does each weigh?

6. Two packages weighed together total 45 kilograms. One of the packages weighs 11 kilograms more than the other. How much does each weigh?

7. A car and trailer together sold for $12,000. The car was valued at $5000 more than the trailer. What was the value of each separately?

8. A guitar and amplifier together cost $356. The amplifier cost $20 more than two times the guitar. What was the price of each?

9. There were 7672 votes cast in a recent election. The winning candidate received 12 votes more than her opponent. How many votes did each candidate receive?

10. At a recent election, the winning candidate received 122 votes more than his opponent. If there were a total of 10,764 votes cast, how many votes did each candidate receive?

11. A certain fishing spot is located 45 kilometers from town. Part of the distance can be driven in a car, but part of it must be traveled on foot. If it is possible to drive 19 more kilometers than must be walked, how far must be walked?

12. Two trains left towns *A* and *B*, which are 240 kilometers apart, at the same time and proceeded toward each other on parallel tracks. At the time they met, the train from town *A* had traveled 10 kilometers farther than the train from town *B*. How many kilometers from town *A* were the trains when they met?

13. The sum of two numbers is 24. One half of one number is 3 more than the other number. What are the numbers?

14. The difference of two numbers is 13. If the smaller number is 2 more than one fourth of the larger, what are the numbers?

Sample Problem

A collection of nickels and dimes has a value of $2.55. How many nickels and dimes are in the collection if there are 3 more dimes than nickels?

Steps 1–2   Number of dimes: x

Number of nickels: y

Step 3   A table is helpful.

Denomination	Value of One Coin	Number of Coins	Value of Coins
Dimes	0.10	x	$0.10x$
Nickels	0.05	y	$0.05y$

Step 4   We write one equation relating the *number* of each kind of coin, and a second equation relating their *values*.

$$x - y = 3$$
$$0.10x + 0.05y = 2.55$$

Step 5   We use the addition method.

$$\begin{aligned} 5x - 5y &= 15 \\ 10x + 5y &= 255 \\ \hline 15x &= 270 \end{aligned}$$

$$x = 18$$
$$18 - y = 3$$
$$y = 15$$

Step 6   **Ans.** There are 18 dimes and 15 nickels in the collection.

15. A collection of 34 coins consists of dimes and quarters. How many coins of each kind are in the collection if the total value is $5.50?

16. A man has $355 in ten-dollar and five-dollar bills. There are ten more ten-dollar bills than five-dollar bills. How many of each kind does he have?

17. Admission fees at a football game were $1.25 for adults and $.55 for children. The receipts were $530.40 for 454 paid admissions. How many adults and children attended the game?

18. Three pounds of bacon and 4 pounds of coffee cost $4.32. One pound of bacon and 5 pounds of coffee cost $3.97. How much are the costs of a pound of bacon and a pound of coffee?

Sample Problem

A sum of $2000 is invested, part at 7% and the remainder at 8%. Find the amount invested at each rate if the yearly income from the two investments is $151.

Steps 1–2   Amount invested at 7%: x

Amount invested at 8%: y

Step 3   A table is helpful.

Amount Invested	Interest Rate	Amount of Interest
x	0.07	$0.07x$
y	0.08	$0.08y$

Step 4   We write one equation representing the total amount *invested*, and a second equation relating the *interest* from each investment and the total interest.

$$x + y = 2000$$

$$0.07x + 0.08y = 151$$

Step 5   We use the addition method.

$$\begin{array}{r} 8x + 8y = 16{,}000 \\ -7x - 8y = -15{,}100 \\ \hline x = 900 \end{array}$$

$$900 + y = 2000$$

$$y = 1100$$

Step 6   **Ans.** $900 is invested at 7% and $1100 at 8%.

19. A sum of $3600 is invested, part at 8% and the remainder at 12%. Find the amount of each investment if the interest on each investment is the same.

20. The total income from two investments is $750. One investment yields 9% and the second investment yields 10%. How much is invested at each rate if the total investment was $8000?

21. A man has twice as much money invested at 8% as he has invested at 7%. If his yearly income is $345, how much does he have invested at each rate?

Now we factor the left-hand side and set each factor equal to zero.

$$(2x + 3)(x - 4) = 0,$$

$$2x + 3 = 0 \quad \text{or} \quad x - 4 = 0.$$

Solving these equations gives us the solutions.

$$x = -\frac{3}{2} \quad \text{and} \quad x = 4.$$

EXERCISES 10.2

■ *Solve by factoring.*

Sample Problem

$$x^2 = 5x$$

Write in standard form.

$$x^2 - 5x = 0$$

Factor left-hand member.

$$x(x - 5) = 0$$

Determine solutions by inspection or set each factor equal to 0 and solve the resulting equations.

$$x = 0 \qquad x - 5 = 0$$

Ans. $x = 0, \quad x = 5$

1. $x^2 + 3x = 0$ **2.** $x^2 - 2x = 0$ **3.** $2y^2 - 5y = 0$

4. $3y^2 - 7y = 0$ **5.** $2y^2 = 9y$ **6.** $4y^2 = 3y$

7. $4x^2 = 16x$ **8.** $5x^2 = 10x$

Sample Problem

$$x^2 = 25$$

Write in standard form.

$$x^2 - 25 = 0$$

Factor left-hand member.

$$(x - 5)(x + 5) = 0$$

Determine solutions by inspection or set each factor equal to 0 and solve the resulting equations.

$$x - 5 = 0 \qquad x + 5 = 0$$

Ans. $x = 5, \quad x = -5$

9. $x^2 - 1 = 0$ **10.** $x^2 - 36 = 0$ **11.** $x^2 - 4 = 0$

12. $x^2 - 9 = 0$ **13.** $x^2 - 16 = 0$ **14.** $x^2 - 100 = 0$

15. $x^2 = 64$ **16.** $x^2 = 49$ **17.** $3x^2 = 27$

18. $2x^2 = 32$ **19.** $5x^2 = 45$ **20.** $7x^2 = 63$

Sample Problem

$$x^2 - 4x - 5 = 0$$

Factor left-hand member.

$$(x - 5)(x + 1) = 0$$

Determine solutions by inspection or set each factor equal to 0 and solve the resulting equations.

$$x - 5 = 0 \qquad x + 1 = 0$$

Ans. $x = 5, \quad x = -1$

21. $x^2 - 3x + 2 = 0$ **22.** $x^2 + 3x + 2 = 0$ **23.** $y^2 + 4y + 4 = 0$

24. $y^2 - 8y + 12 = 0$ **25.** $y^2 - 3y - 4 = 0$ **26.** $y^2 - 3y - 10 = 0$

27. $x^2 + 4x - 21 = 0$ **28.** $x^2 + x - 42 = 0$ **29.** $x^2 + 5x - 14 = 0$

30. $x^2 + 8x + 15 = 0$ **31.** $x^2 + 12x + 36 = 0$ **32.** $x^2 + 14x + 49 = 0$

Sample Problem

$$2x^2 - 6x = 8$$

Write equation in standard form.

$$2x^2 - 6x - 8 = 0$$

Factor left-hand member completely.

$$2(x^2 - 3x - 4) = 0$$
$$2(x - 4)(x + 1) = 0$$

Determine solutions by inspection or set each factor containing a variable equal to 0 and solve the resulting equations. The constant 2 has no effect on the solution.

$$x - 4 = 0 \qquad x + 1 = 0$$

Ans. $x = 4, \quad x = -1$

33. $2x^2 - 10x = 12$ **34.** $3x^2 - 6x = -3$ **35.** $4y^2 - 12y = 16$

36. $4y^2 - 24y = 28$ **37.** $2y^2 + 2y = 60$ **38.** $3y^2 + 6y = 45$

39. $3x^2 - x = 4$ **40.** $4x^2 + 4x = 3$ **41.** $6x^2 = 11x - 3$

42. $4x^2 = 4x + 3$ **43.** $4x^2 = 4x - 1$ **44.** $12x^2 = 8x + 15$

45. $12x - 9 = 4x^2$ **46.** $x + 15 = 2x^2$

Sample Problem

$$x(x + 2) = 8$$

Remove parentheses.

$$x^2 + 2x = 8$$

Write in standard form.

$$x^2 + 2x - 8 = 0$$

Factor left-hand member.

$$(x + 4)(x - 2) = 0$$

Determine solutions by inspection or set each factor equal to 0 and solve the resulting equations.

$$x + 4 = 0 \qquad x - 2 = 0$$

Ans. $x = -4, \quad x = 2$

47. $y(2y - 3) = -1$ **48.** $x(x + 2) = 3$
49. $2(x^2 - 1) = 3x$ **50.** $y(y - 2) = 6 - y$
51. $x(x + 2) - 3x - 2 = 0$ **52.** $2y(y - 2) = y + 3$

Sample Problem

$$(x - 4)(x + 3) = -10$$

Multiply the factors in the left-hand member.

$$x^2 - x - 12 = -10$$

Write in standard form.

$$x^2 - x - 2 = 0$$

Factor left-hand member.

$$(x - 2)(x + 1) = 0$$

Determine solutions by inspection or set each factor equal to 0 and solve the resulting equations.

$$x - 2 = 0 \qquad x + 1 = 0$$

Ans. $x = 2, \quad x = -1$

53. $(x - 2)(x + 1) = 4$ **54.** $(x - 5)(x + 1) = -8$
55. $(x - 2)(x - 1) = 1 - x$ **56.** $(x + 3)^2 = 2x + 14$
57. $(2x + 5)(x - 4) = -18$ **58.** $(2x - 1)(x - 2) = -1$
59. $(6x + 1)(x + 1) = 4$ **60.** $(x - 2)(x + 1) = x(2 - x)$

10.3 SOLVING QUADRATIC EQUATIONS BY FACTORING II

In Section 7.7, we cleared a first-degree equation of fractions by multiplying each member by the L.C.D. of the fractions. We can use the same procedure for second-degree equations. Thus, to solve the equation

$$\frac{x^2}{2} + \frac{5x}{4} = 3,$$

we first multiply each member by 4 and then rewrite the equation in standard form:

$$(\overset{2}{\cancel{4}})\frac{x^2}{\cancel{2}} + (\cancel{4})\frac{5x}{\cancel{4}} = (\cancel{4})3,$$

$$2x^2 + 5x = 12,$$

$$2x^2 + 5x - 12 = 0.$$

Then, factoring the left-hand member, we have

$$(2x - 3)(x + 4) = 0,$$

from which

$$2x - 3 = 0 \qquad \text{or} \qquad x + 4 = 0,$$

$$2x = 3, \qquad\qquad x = -4.$$

$$x = \frac{3}{2},$$

Thus, the solutions are $x = 3/2$ and $x = -4$.

Remember that if we multiply each member of an equation by an expression containing a variable we must check all solutions to the new equation in the original equation to make sure no denominator is zero. For example, in the equation

$$\frac{7}{x - 3} - \frac{3}{x - 4} = \frac{1}{2},$$

x cannot be 3 or 4, since either of these values would make a denominator zero. To solve, we multiply each member by the L.C.D. $2(x - 3)(x - 4)$.

$$2(x - 3)(x - 4)\,\frac{7}{(x - 3)} - 2(x - 3)(x - 4)\,\frac{3}{(x - 4)} = 2(x - 3)(x - 4)\,\frac{1}{2},$$

$$14(x - 4) - 6(x - 3) = (x - 3)(x - 4).$$

Now we remove parentheses to obtain

$$14x - 56 - 6x + 18 = x^2 - 7x + 12,$$

$$8x - 38 = x^2 - 7x + 12.$$

Writing the equation in standard form yields

$$x^2 - 15x + 50 = 0,$$

and factoring the left-hand member, we get

$$(x - 10)(x - 5) = 0.$$

Finally, we can determine solutions by inspection or set each factor equal to 0 and solve

$$x - 10 = 0 \qquad \text{or} \qquad x - 5 = 0$$

$$x = 10 \qquad\qquad x = 5.$$

Since no denominator of the original equation equals zero for either of these values, both are solutions.

EXERCISES 10.3

■ *Solve.*

Sample Problem

$$\frac{15}{2} x^2 - \frac{10}{3} = 0$$

Multiply each term by the L.C.D. 6.

$$(\overset{3}{\cancel{6}}) \frac{15}{2} x^2 - (\overset{2}{\cancel{6}}) \frac{10}{\cancel{3}} = (6)0$$

$$45x^2 - 20 = 0$$

Factor left-hand member completely.

$$5(9x^2 - 4) = 0$$

$$5(3x - 2)(3x + 2) = 0$$

Determine solutions by inspection or set each factor containing a variable equal to 0 and solve the resulting equations. The constant 5 has no effect on the solution.

$$3x - 2 = 0 \qquad 3x + 2 = 0$$

$$3x = 2 \qquad\qquad 3x = -2$$

Ans. $x = \dfrac{2}{3}, \quad x = \dfrac{-2}{3}$

1. $x^2 - \dfrac{1}{9} = 0$ **2.** $\dfrac{1}{3} x^2 - \dfrac{4}{3} = 0$ **3.** $\dfrac{2}{3} x^2 - \dfrac{3}{2} = 0$

4. $\dfrac{5}{2} y^2 - 10 = 0$ **5.** $\dfrac{x^2}{2} = 8$ **6.** $3x^2 = \dfrac{75}{4}$

7. $\dfrac{x^2}{2} + x = 0$ **8.** $\dfrac{x^2}{3} - 2x = 0$ **9.** $\dfrac{x^2}{4} + \dfrac{x}{2} = 0$

10. $\dfrac{x^2}{18} + \dfrac{x}{3} = 0$ **11.** $\dfrac{x^2}{5} = x$ **12.** $\dfrac{x^2}{6} = \dfrac{x}{2}$

Sample Problem

$$y^2 = \frac{13}{6} y - 1$$

Multiply each member by L.C.D. 6.

$$(6)y^2 = (\cancel{6}) \frac{13}{\cancel{6}} y - (6)1$$

$$6y^2 = 13y - 6$$

Write in standard form; factor left-hand member.

$$6y^2 - 13y + 6 = 0$$

$$(2y - 3)(3y - 2) = 0$$

Determine solutions by inspection or set each factor equal to 0 and solve the resulting equations.

$$2y - 3 = 0 \qquad 3y - 2 = 0$$

$$2y = 3 \qquad 3y = 2$$

Ans. $y = \dfrac{3}{2}, \quad y = \dfrac{2}{3}$

13. $\dfrac{2}{3}x^2 + \dfrac{1}{3}x - 2 = 0$

14. $\dfrac{3}{4}x^2 + \dfrac{5}{2}x - 2 = 0$

15. $x^2 + 3x + \dfrac{9}{4} = 0$

16. $\dfrac{3}{2}x^2 - \dfrac{1}{4}x - \dfrac{1}{2} = 0$

17. $4x^2 + 13x + \dfrac{15}{2} = 0$

18. $\dfrac{1}{3}x^2 - \dfrac{5}{2}x + 3 = 0$

19. $\dfrac{x^2}{2} + x = \dfrac{15}{2}$

20. $x - 1 = \dfrac{x^2}{4}$

21. $\dfrac{x^2}{3} + x = \dfrac{-2}{3}$

22. $\dfrac{21}{2} + 2y = \dfrac{y^2}{2}$

23. $\dfrac{x^2}{6} + \dfrac{x}{3} = \dfrac{1}{2}$

24. $\dfrac{x^2}{15} = \dfrac{x}{5} + \dfrac{2}{3}$

Sample Problem

$$\frac{1}{8x^2} - \frac{13}{24x} = -\frac{1}{2}$$

Multiply each term by L.C.D. $24x^2$.

$$(24x^2)\overset{3}{\frac{1}{8x^2}} - (24x^2)\overset{x}{\frac{13}{24x}} = -(24x^2)\overset{12}{\frac{1}{2}}$$

$$3 - 13x = -12x^2$$

Write in standard form.

$$12x^2 - 13x + 3 = 0$$

Factor left-hand member.

$$(4x - 3)(3x - 1) = 0$$

Determine solutions by inspection or set each factor equal to 0 and solve the resulting equations.

$$4x - 3 = 0 \qquad 3x - 1 = 0$$

$$4x = 3 \qquad 3x = 1$$

Ans. $x = \dfrac{3}{4}, \quad x = \dfrac{1}{3}$

Note: These values for x do not make any denominator in the original equation zero.

25. $x + \dfrac{1}{x} = 2$ **26.** $\dfrac{x}{4} - \dfrac{3}{4} = \dfrac{1}{x}$ **27.** $1 - \dfrac{2}{x} = \dfrac{15}{x^2}$

28. $\dfrac{1}{2} + \dfrac{1}{2x} = \dfrac{1}{x^2}$ **29.** $1 + \dfrac{1}{x(x-1)} = \dfrac{3}{x}$ **30.** $\dfrac{4}{x} - 3 = \dfrac{5}{2x+3}$

Sample Problem $\dfrac{5}{2x+1} - \dfrac{4}{x-4} = 3$

Multiply each term by L.C.D. $(2x + 1)(x - 4)$.

$(2x+1)(x-4)\,\dfrac{5}{2x+1} - (2x+1)(x-4)\,\dfrac{4}{x-4} = (2x+1)(x-4)\cdot 3$

$5(x-4) - 4(2x+1) = 3(2x+1)(x-4)$

Remove parentheses and simplify.

$5x - 20 - 8x - 4 = 3(2x^2 - 7x - 4)$

$-3x - 24 = 6x^2 - 21x - 12$

Write in standard form.

$6x^2 - 18x + 12 = 0$

Factor the left-hand member.

$6(x-2)(x-1) = 0$

Set each factor equal to 0 and solve the resulting equations.

$x - 2 = 0 \qquad x - 1 = 0$

Ans. $x = 2, \quad x = 1$

Note: These values for x do not make any denominator in the original equation zero.

31. $\dfrac{14}{x-6} - \dfrac{6}{x-8} = \dfrac{1}{2}$ **32.** $\dfrac{12}{x-3} + \dfrac{12}{x+4} = 1$

33. $\dfrac{2}{x-3} - \dfrac{6}{x-8} = -1$ **34.** $\dfrac{4}{x-2} - \dfrac{7}{x-3} = \dfrac{2}{15}$

35. $\dfrac{4}{x-1} - \dfrac{4}{x+2} = \dfrac{3}{7}$ **36.** $\dfrac{3}{x+6} - \dfrac{2}{x-5} = \dfrac{5}{4}$

37. $\dfrac{4}{x-2} - \dfrac{8}{x^2+x-6} = 1$ **38.** $\dfrac{12}{x+4} - \dfrac{10}{x^2+3x-4} = 1$

Sample Problem

$$\frac{1}{x^2 - 3x} - \frac{2}{x^2 - 9} = \frac{1}{x + 3}$$

Multiply each term by L.C.D. $x(x + 3)(x - 3)$.

$$x(x + 3)(x - 3)\frac{1}{x(x - 3)} - x(x + 3)(x - 3)\frac{2}{(x + 3)(x - 3)}$$

$$= x(x + 3)(x - 3)\frac{1}{x + 3}$$

$$(x + 3) - 2x = x^2 - 3x$$

Write in standard form.

$$x^2 - 2x - 3 = 0$$

Factor left-hand member.

$$(x - 3)(x + 1) = 0$$

Determine solutions.

$$x - 3 = 0 \qquad x + 1 = 0$$

$$x = 3 \qquad\quad x = -1$$

Note: The left-hand member of the original equation is undefined for $x = 3$, since both denominators equal 0 when $x = 3$. Thus $x = 3$ is *not* a solution of the equation.

Ans. $x = -1$

39. $\dfrac{9}{x^2 + x - 2} + \dfrac{1}{x^2 - x} = \dfrac{4}{x - 1}$

40. $\dfrac{2}{x^2 - 2x} + \dfrac{1}{2x} = \dfrac{-1}{x^2 + 2x}$

41. $\dfrac{4}{x^2 - 5x + 6} - \dfrac{5x}{x^2 - 2x - 3} = \dfrac{-6x}{x^2 - x - 2}$

42. $\dfrac{3x}{x^2 - 2x - 8} - \dfrac{1}{x^2 + 5x + 6} = \dfrac{2x}{x^2 - x - 12}$

43. $\dfrac{3x}{6x^2 + 19x + 15} + \dfrac{1}{2x^2 - x - 6} = \dfrac{x}{3x^2 - x - 10}$

44. $\dfrac{24}{4x^2 - 9} - \dfrac{x}{2x^2 - 9x + 9} = \dfrac{-3x}{2x^2 - 3x - 9}$

45. $\dfrac{x + 2}{x - 3} - \dfrac{8x + 11}{2x^2 - 5x - 3} = \dfrac{x + 1}{2x + 1}$

46. $\dfrac{2x + 1}{2x - 3} + \dfrac{x^2 + 3x - 7}{2x^2 - 7x + 6} = \dfrac{x + 1}{x - 2}$

47. $\dfrac{x - 7}{x^2 + x - 6} + \dfrac{x + 11}{x^2 + 2x - 3} = \dfrac{x - 7}{x^2 - 3x + 2}$

48. $\dfrac{x-1}{x^2-2x-8} + \dfrac{x-6}{x^2-4} = \dfrac{x-3}{x^2-6x+8}$

49. $\dfrac{x-1}{x^2-x-6} - \dfrac{x+1}{2x^2-5x-3} = \dfrac{3}{2x^2+5x+2}$

50. $\dfrac{x+2}{2x^2-x-3} - \dfrac{x+2}{3x^2+5x+2} = \dfrac{1}{6x^2-5x-6}$

10.4 APPLICATIONS

A variety of word problems lead to quadratic equations. Again, we will follow the six steps outlined on page 84 when solving word problems. It is important to check solutions in the original problem to make sure they fulfill the physical conditions stated in the problem. Consider the following example.

> An object is thrown off a building 48 feet high. The object's height h above the ground, at a particular time t, is given by
>
> $$h = 48 + 32t - 16t^2 \tag{1}$$
>
> Determine the number of seconds t it will take the object to strike the ground.

Steps 1–2 Time to strike ground: t

Step 3 A sketch is helpful to visualize the problem.

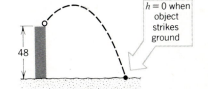

Step 4 When the object strikes the ground, $h = 0$. Substituting $h = 0$ in Equation (1) yields

$$0 = 48 + 32t - 16t^2.$$

Step 5 Solving for t, we get

$$16t^2 - 32t - 48 = 0.$$

Dividing each member by 16 produces

$$t^2 - 2t - 3 = 0.$$

Factoring, we have

$$(t - 3)(t + 1) = 0.$$

Then

$$t - 3 = 0 \quad \text{or} \quad t + 1 = 0$$
$$t = 3 \qquad\qquad t = -1.$$

Step 6   **Ans.** In this case, -1 does not meet the physical requirements of the problem, since time t must be positive. Thus, the object would strike the ground 3 seconds after it was thrown.

Here is another example.

> The length of a rectangle is 3 centimeters greater than the width, and the area is 54 square centimeters. Find the dimensions of the rectangle.

Steps 1–2   Width: x

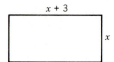

Length: $x + 3$

Step 3   A sketch is helpful.

Step 4   The area of a rectangle is equal to the product of its length and its width, or

$$x(x + 3) = 54.$$

Step 5   This is a quadratic equation. We put it in standard form and factor to get

$$x^2 + 3x = 54$$
$$x^2 + 3x - 54 = 0$$
$$(x + 9)(x - 6) = 0$$
$$x + 9 = 0 \qquad x - 6 = 0$$
$$x = -9 \qquad x = 6$$

Step 6   **Ans.** Since the width must be positive, -9 does not meet the conditions of the problem, Thus, the width is 6 centimeters, and the length is $6 + 3$ or 9 centimeters.

EXERCISES 10.4

■ *Solve each word problem. Follow the six steps outlined on page 84.*

Sample Problem   The square of an integer is 7 less than eight times the integer. Find the integer.

Steps 1–2   The integer: x

Step 3   A sketch is not applicable.

Step 4 $x^2 = 8x - 7$

Step 5 $x^2 - 8x + 7 = 0$

$(x - 7)(x - 1) = 0$

$x - 7 = 0$ $x - 1 = 0$

$x = 7$ $x = 1$

Step 6   **Ans.** Since 1 and 7 *both* meet the conditions of the problem, *both* are valid solutions. The integer is either 1 or 7.

1. The square of an integer is equal to five times the integer. Find the integer.

2. If three times the square of a certain integer is increased by the integer itself, the sum is 10. What is the integer?

3. Find two consecutive positive integers whose product is 72.

4. Find two consecutive positive integers whose product is 132.

5. The square of a positive integer increased by twice the square of the next consecutive integer gives 66. Find the integer.

6. The square of a positive integer is 79 less than twice the square of the next consecutive integer. Find the integers.

7. The distance h above the ground of a certain projectile launched upward from the top of a 160-foot building time at time $t = 0$ is given by the equation $h = 160 + 48t - 16t^2$, where t is in seconds. Find the time at which the projectile will strike the ground.

8. In exercise 7, find the time at which the projectile is again at 160 feet above the ground.

9. The cost C of producing a certain radio set is related to the number of hours t it takes to manufacture the set by $C = 8t^2 - 32t - 16$. How many hours would be devoted to producing a set at a cost of $80?

10. In exercise 9, how many hours would be devoted to producing a set at a cost of $24?

Sample Problem

The sum of a certain integer and twice its reciprocal is 19/3. Find the integer.

Steps 1–2   The integer: x

The reciprocal of the integer: $\dfrac{1}{x}$

Step 3   A sketch is not applicable.

Step 4 $x + 2 \left(\dfrac{1}{x}\right) = \dfrac{19}{3}$

Step 5 $(3x)x + (3x)2 \left(\dfrac{1}{x}\right) = (3x)\dfrac{19}{3}$

$$3x^2 + 6 = 19x$$

$$3x^2 - 19x + 6 = 0$$

$$(3x - 1)(x - 6) = 0$$

$$3x - 1 = 0 \qquad x - 6 = 0$$

$$x = \dfrac{1}{3} \qquad x = 6$$

Step 6.   **Ans.** Since 1/3 is not an integer, the only solution that meets the conditions of the original problem is 6. The integer is 6.

11. The sum of a certain number and its reciprocal is $\frac{17}{4}$. What is the number?

12. The sum of a certain number and twice its reciprocal is $\frac{9}{2}$. What is the number?

13. The sum of the reciprocals of two consecutive odd integers is $\frac{8}{15}$. What are the integers?

14. The sum of the reciprocals of two consecutive even integers is $\frac{5}{12}$. Find the integers.

15. The sum of the reciprocal of a positive integer and twice the reciprocal of the next consecutive integer is $\frac{5}{8}$. What are the integers?

16. Twice the reciprocal of a positive integer is subtracted from three times the reciprocal of the next successive integer and the difference is $\frac{2}{21}$. What are the integers?

Sample Problem   A boat travels 18 miles downstream and back in $4\frac{1}{2}$ hours. If the speed of the current is 3 miles per hour, what is the speed of the boat in still water?

Steps 1–2   Speed of boat in still water: x

Step 3

	d	r	$t = d/r$
Downstream	18	$x + 3$	$\dfrac{18}{x + 3}$
Upstream	18	$x - 3$	$\dfrac{18}{x - 3}$

Step 4 $\dfrac{18}{x + 3} + \dfrac{18}{x - 3} = \dfrac{9}{2}$

Step 5 $2(x + 3)(x - 3) \dfrac{18}{x + 3} + 2(x + 3)(x - 3) \dfrac{18}{x - 3} = 2(x + 3)(x - 3) \dfrac{9}{2}$

$$36(x - 3) + 36(x + 3) = 9(x + 3)(x - 3)$$

$$36x - 108 + 36x + 108 = 9x^2 - 81$$

$$9x^2 - 72x - 81 = 0$$

$$9(x^2 - 8x - 9) = 0$$

$$9(x - 9)(x + 1) = 0$$

$$x = 9 \qquad x = -1$$

Step 6   **Ans.** Since -1 does not meet the conditions of the problem, only the positive solution is used. The boat travels 9 miles per hour in still water.

17. A motor boat travels 24 miles downstream on a river and returns to its starting place. If the speed of the current is 2 miles per hour, and the round trip takes 5 hours, what is the speed of the boat in still water?

18. A crew rows a boat 6 miles downstream and then rows back to its starting place. If the speed of the current is 2 miles per hour, and the total trip takes 4 hours, how fast would the crew row in still water?

19. A man drove 180 miles from town A to town B and returned to town A. If he drove 15 miles per hour faster on the return trip than he did on the initial trip, and the initial trip took one hour longer, what was the man's speed on each trip?

20. A plane flew 480 miles at a certain speed, then increased its speed by 20 miles per hour and continued on the same course. After having flown a distance of 840 miles in a total of 5 hours, the plane landed. What was its original speed?

Sample Problem   The pages of a book measure 25 centimeters by 18 centimeters and have margins of equal width on all four sides. If the area of the printed region is 330 square centimeters, how wide is the margin?

Steps 1–2   Width of margin: x

Step 3

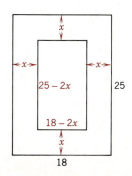

Step 4 $(25 - 2x)(18 - 2x) = 330$

Step 5 $450 - 86x + 4x^2 = 330$

$4x^2 - 86x + 120 = 0$

$2(2x - 3)(x - 20) = 0$

$x = \frac{3}{2} \qquad x = 20$

Step 6   **Ans.** The margin cannot be 20 centimeters wide since the page is only 18 centimeters wide. Hence the margin is $\frac{3}{2}$ or 1.5 centimeters wide.

21. The area of a rectangle is 60 square centimeters. The length is 2 centimeters longer than twice the width. Find the dimensions of the rectangle.

22. Find the dimensions of a rectangle if the area is 240 square centimeters and the width is $\frac{3}{5}$ of the length.

23. A rectangular lawn whose dimensions are 40 and 60 feet is to have its area increased 498 square feet by a border of uniform width along both 60-foot sides and one end. Find the width of the border.

24. Debra's living room is 18 feet long by 12 feet wide, but she can only afford 135 square feet of carpet. If she carpets the center of the room, leaving a border of uniform width on the edges, how wide will the border be?

25. The base of a triangle is 11 inches longer than its altitude. Find the base and altitude if the area of the triangle is 40 square inches.

26. The base of a triangle is 4 feet shorter than twice its altitude. If the area of the triangle is 48 square feet, find its dimensions.

CHAPTER SUMMARY

[10.1–10.3] A quadratic equation written in the form

$$ax^2 + bx + c = 0$$

is said to be in **standard form.**

The factoring method of solving a quadratic equation is based on the following principle:

If the product of two factors is 0, at least one of the factors is zero.

To solve a quadratic equation by factoring:

1. Write the equation in standard form.
2. Factor the left-hand member.

3. Set each factor equal to zero.

4. Solve each of the resulting equations.

[10.4] When using quadratic equations to solve word problems, check both solutions against the original word problem to make sure they fulfill the conditions in the problem.

CHAPTER REVIEW

■ *In exercises 1–12, solve for x or y.*

1. a. $(x - 2)(x + 5) = 0$ b. $y(y + 3) = 0$

2. a. $x^2 - 2x = 0$ b. $3x^2 = 6x$

3. a. $x^2 - 49 = 0$ b. $3y^2 = 27$

4. a. $y^2 - 4y - 5 = 0$ b. $y^2 - 7y - 18 = 0$

5. a. $x^2 - 2x = 3$ b. $x^2 = 4x - 4$

6. a. $2x^2 + 5x = 12$ b. $2x^2 = 3 - 5x$

7. a. $3x^2 + 5x = 2$ b. $3x^2 = 14x - 8$

8. a. $y(y - 6) = 16$ b. $x(x + 2) = 8$

9. a. $(x + 5)(x - 8) = -36$ b. $(x + 1)(x - 2) = 4$

10. a. $\dfrac{x^2}{4} - 9 = 0$ b. $3x^2 = \dfrac{27}{25}$

11. a. $\dfrac{y^2}{6} + \dfrac{y}{6} = 2$ b. $\dfrac{y}{2} - 1 = \dfrac{y^2}{16}$

12. a. $\dfrac{15}{y^2} + \dfrac{2}{y} = 1$ b. $\dfrac{2}{x} - \dfrac{1}{6} = \dfrac{2}{x + 2}$

13. a. $\dfrac{2x}{x^2 - 25} + \dfrac{2}{x^2 + 5x} = \dfrac{1}{x - 5}$

    b. $\dfrac{9x}{x^2 - 5x + 6} - \dfrac{14}{x^2 + x - 12} = \dfrac{2x}{x^2 + 2x - 8}$

14. a. $\dfrac{6}{x - 4} + \dfrac{5}{x^2 - 3x - 4} = \dfrac{-1}{x^2 + 3x + 2}$

    b. $\dfrac{4}{x^2 - 2x - 8} + \dfrac{1}{2x + 1} = \dfrac{2}{2x^2 + 5x + 2}$

15. The difference of two positive numbers is 7 and their product is 60. Find the numbers.

16. The sum of two positive numbers is 11 and their product is 30. Find the numbers.

17. The sum of the reciprocals of two consecutive integers is $\frac{11}{30}$. Find the integers.

18. The area of a rectangle is 45 square centimeters. The length is 6 centimeters longer than three times the width. Find the dimensions of the rectangle.

19. Tri rode his bicycle for 6 miles and then walked an additional 4 miles. His walking rate was 2 miles per hour less than his riding rate, and the entire trip took 6 hours. What was Tri's rate for each part of the trip?

20. Laura made a trip of 360 miles. If her average speed had been 5 miles per hour faster, the trip would have taken 1 hour less. What was her average speed?

CUMULATIVE REVIEW

1. Simplify: $\dfrac{4^2 - 2^2}{3} - \dfrac{4^2 + 2^2}{4}$.

2. If $a = 0$, $b = -1$, and $c = 2$, find the value of $a(b^2 + c^2)$.

3. Write $-3, \dfrac{3}{8}, \dfrac{3}{7}, 3, 0, \dfrac{-5}{2}$, and $\dfrac{-5}{3}$ in order from smallest to largest.

4. Factor: $x^3 - 3x^2 + 2x$.

5. Multiply: $(2x - 3)(x + 4)$.

6. Write $-\dfrac{x + 2}{-3}$ in standard form.

7. Simplify: $(x - 3)(x + 2) - (x^2 - 3x)$.

8. Represent $\dfrac{3}{x^2 + x} - \dfrac{2}{x + 1}$ as a single fraction.

9. Express $\dfrac{3}{x - 1}$ as a fraction with a denominator of $x^2 + 2x - 3$.

10. Simplify: $\dfrac{3x^2 - x - 2}{2x^2 + x - 3}$.

11. Solve the system $2x - 3 = y$ by algebraic methods.
$$3x + 2y = 15$$

12. Graph the equation: $x + 2y \le 4$.

13. Solve: $\dfrac{7}{8} = \dfrac{42}{y + 4}$.

14. Find the equation of a line passing through the points $(1, 1)$ and $(3, 5)$.

15. The numerator of a certain fraction is five more than the denominator, and the fraction is equivalent to $\frac{4}{3}$. Find the fraction.

16. Two packages together weigh 140 kilograms. If one package weighs 30 kilograms more than the other, what is the weight of each?

■ *In exercises 17–19, solve each equation.*

17. $3y^2 - 2y = y$ **18.** $25x^2 - 4 = 5$ **19.** $x(x + 3) = 5x + 3$

20. If S varies directly with T, and $S = 4.2$ when $T = 12.3$, find T when $S = 10.6$.

■ *True or false.*

21. $\dfrac{x^2 + 2x}{x^2 + x}$ is equivalent to 2.

22. $-\dfrac{2x - 1}{3}$ is equivalent to $\dfrac{-2x - 1}{3}$.

23. $-a^2$ is equivalent to $(-a)^2$.

24. $\dfrac{(x - 2)^2}{x^2 - 2^2}$ is equivalent to 1.

25. $\dfrac{3 + 2x}{2x}$ is equivalent to 3.

11 RADICAL EXPRESSIONS

11.1 RADICALS

In this chapter we study a new kind of number—numbers that will enable us to solve additional quadratic equations.

SQUARE ROOTS

A **square root** of a nonnegative number a is a number whose square is a. For example:

$$\text{a square root of 9 is 3 because } 3^2 = 9;$$

$$\text{a square root of 4 is 2 because } 2^2 = 4;$$

and

$$\text{a square root of 16 is } -4 \text{ because } (-4)^2 = 16.$$

Notice that

$$(-2)^2 = 4 \quad \text{and} \quad (2)^2 = 4.$$

Thus, -2 and 2 are both square roots of 4. In fact, since $(a)(a) = a^2$ and $(-a)(-a) = a^2$, every positive number has two square roots: one positive and one negative.

RADICAL SIGN

We use a special symbol, $\sqrt{}$, called a **radical sign,** to denote the positive or **principal square root** of a positive number. That is, for all positive numbers a, $\sqrt{a}$ (read "the square root of a") is the positive number whose square is a. The number a is called the **radicand.** In symbols:

$$\sqrt{a} \cdot \sqrt{a} = (\sqrt{a})^2 = a$$

where $\sqrt{a}$ and a are positive. For example,

$$\sqrt{4} = 2 \quad \text{because} \quad 2^2 = 4;$$
$$\sqrt{9} = 3 \quad \text{because} \quad 3^2 = 9;$$

and

$$\sqrt{16} = 4 \quad \text{because} \quad 4^2 = 16.$$

In the special case of zero,

$$\sqrt{0} = 0 \quad \text{because} \quad 0^2 = 0.$$

NEGATIVE SQUARE ROOTS

To represent the negative square root of a, we use the symbol $-\sqrt{a}$. For example,

$$-\sqrt{4} = -2, \quad -\sqrt{9} = -3, \quad \text{and} \quad -\sqrt{16} = -4.$$

Using radical notation, we represent the two square roots of a positive number a by $\sqrt{a}$ and $-\sqrt{a}$. The symbol $\pm\sqrt{a}$ is sometimes used to denote both the positive and negative square roots of a. For example,

$$\pm\sqrt{16} = \pm 4, \quad \pm\sqrt{81} = \pm 9, \quad \text{and} \quad \pm\sqrt{36} = \pm 6.$$

THE RADICAND

Since the square of any number is positive or zero, a symbol such as $\sqrt{-4}$ is not meaningful, because there is no number among the numbers we are studying whose square is -4. In general, $\sqrt{a}$ represents one of the numbers we are studying only if $a \geq 0$. In this book, *we will assume that all variables and expressions under radicals of the form $\sqrt{}$ represent nonnegative numbers*. For example:

for $\sqrt{x}$, we assume that $x \geq 0$;

for $\sqrt{x - 2}$, we assume that $x - 2 \geq 0$, or $x \geq 2$;

for $\sqrt{x + 2}$, we assume that $x + 2 \geq 0$, or $x \geq -2$.

With this agreement that radicands are positive or zero, we can rewrite some radical expressions that contain variables as equivalent expressions without radical notation. For example,

$$\sqrt{4x^2} = 2x \quad \text{because} \quad (2x)^2 = (2x)(2x) = 4x^2$$

and

$$\sqrt{16x^6} = 4x^3 \quad \text{because} \quad (4x^3)^2 = (4x^3)(4x^3) = 16x^6.$$

EXERCISES 11.1

■ *Find the two square roots of the given number.*

Sample Problems

a. 25

b. $\dfrac{4}{9}$

Ans. 5, −5

Ans. $\dfrac{2}{3}, \dfrac{-2}{3}$

1. 81   **2.** 49   **3.** 121

4. 100   **5.** $\dfrac{16}{25}$ **6.** $\dfrac{4}{25}$

■ *Find each square root.*

Sample Problems

a. $\sqrt{49}$ b. $-\sqrt{\dfrac{4}{81}}$ c. $\pm\sqrt{\dfrac{4}{25}}$ d. $\sqrt{0}$

Ans. 7   *Ans.* $-\dfrac{2}{9}$ *Ans.* $\pm\dfrac{2}{5}$ *Ans.* 0

7. $\sqrt{16}$ **8.** $\sqrt{36}$ **9.** $-\sqrt{81}$ **10.** $-\sqrt{121}$

11. $\pm\sqrt{144}$ **12.** $\pm\sqrt{225}$ **13.** $\sqrt{289}$ **14.** $\sqrt{256}$

15. $\sqrt{\dfrac{1}{36}}$ **16.** $-\sqrt{\dfrac{1}{4}}$ **17.** $\pm\sqrt{\dfrac{4}{9}}$ **18.** $\pm\sqrt{\dfrac{9}{25}}$

■ *Write each of the following as an equivalent radical expression.*

Sample Problems

a. 5   b. −10   c. $\dfrac{2}{3}$

Ans. $\sqrt{25}$ *Ans.* $-\sqrt{100}$ *Ans.* $\sqrt{\dfrac{4}{9}}$

19. 3   **20.** 8   **21.** −7   **22.** −6

23. 13   **24.** 12   **25.** −5   **26.** −4

27. $\dfrac{1}{2}$ **28.** $\dfrac{8}{9}$ **29.** $-\dfrac{7}{8}$ **30.** $-\dfrac{2}{7}$

■ *Find each square root.*

Sample Problems

a. $\sqrt{y^6}$ b. $-\sqrt{49x^2y^6}$ c. $\pm\sqrt{(c+d)^2}$

$\sqrt{(y^3)^2}$ $-\sqrt{(7xy^3)^2}$

Ans. y^3 Ans. $-7xy^3$ Ans. $\pm(c+d)$

31. $\sqrt{x^2}$ 32. $\sqrt{y^4}$ 33. $\sqrt{4x^4}$
34. $\sqrt{9a^6}$ 35. $-\sqrt{a^2c^4}$ 36. $-\sqrt{x^6y^6}$
37. $\pm\sqrt{36a^6}$ 38. $\pm\sqrt{100x^{10}}$ 39. $\sqrt{121a^2b^8}$
40. $\sqrt{81x^4y^2}$ 41. $-\sqrt{169x^6y^2z^{10}}$ 42. $-\sqrt{225a^2b^8c^4}$
43. $\sqrt{64a^{64}b^8}$ 44. $\sqrt{100x^{100}y^{50}}$ 45. $\sqrt{(x+y)^2}$
46. $\sqrt{(a+3)^2}$ 47. $-\sqrt{4(x+y)^2}$ 48. $-\sqrt{25(b+c)^2}$
49. $\sqrt{49(x+2)^4}$ 50. $\sqrt{64(a-1)^6}$ 51. $\sqrt{\dfrac{a^2}{b^2}}$
52. $-\sqrt{\dfrac{b^4}{100}}$ 53. $\pm\sqrt{\dfrac{9}{x^2y^4}}$ 54. $\pm\sqrt{\dfrac{4x^2}{y^2}}$

■ Write each of the following as an equivalent radical expression.

Sample Problems

a. $3x^3y$ b. $-(x+y)$ c. $\pm\dfrac{3x}{4}$

$\sqrt{(3x^3y)^2}$ $\pm\sqrt{\left(\dfrac{3x}{4}\right)^2}$

Ans. $\sqrt{9x^6y^2}$ Ans. $-\sqrt{(x+y)^2}$ Ans. $\pm\sqrt{\dfrac{9x^2}{16}}$

55. x 56. y^2 57. xy^2 58. x^3y
59. $4y^3$ 60. $7x$ 61. $-8x^4$ 62. $-7x^7$
63. $\pm 9x^2y^3$ 64. $\pm 6xy^4$ 65. $-12x^{12}y^{12}$ 66. $-10x^{10}y^{10}$
67. $(x+y)$ 68. $-(x+y)$ 69. $\pm(2x+y)$ 70. $\pm(3x+2y)$
71. $\dfrac{1}{2}$ 72. $\dfrac{1}{3}$ 73. $\dfrac{3}{4}x$ 74. $\dfrac{2}{3}a$
75. $\pm\dfrac{4}{5}b^2$ 76. $\pm\dfrac{3}{2}ab$ 77. $\dfrac{1}{a}$ 78. $-\dfrac{1}{ab}$
79. $-\dfrac{3x}{y}$ 80. $-\dfrac{y}{4x}$ 81. $\dfrac{a+b}{a}$ 82. $\dfrac{b+1}{b}$

■ Simplify.

Sample Problems

a. $(\sqrt{10})^2$ b. $(\sqrt{5x})(\sqrt{5x})$

Ans. 10   Ans. $5x$

83. $(\sqrt{13})^2$ **84.** $(\sqrt{7})^2$

85. $(\sqrt{22})(\sqrt{22})$ **86.** $(\sqrt{31})(\sqrt{31})$

87. $(\sqrt{3ab})^2$ **88.** $(\sqrt{11xy})^2$

89. $(\sqrt{17x})(\sqrt{17x})$ **90.** $(\sqrt{2b})(\sqrt{2b})$

■ *State the restriction on the variable in each expression.*

Sample Problem

$\sqrt{x + 1}$

$x + 1$ must be nonnegative; thus,

$x + 1 \geq 0$ or $x \geq -1$.

Ans. We assume $x \geq -1$.

91. $\sqrt{x - 3}$ **92.** $\sqrt{x - 5}$ **93.** $\sqrt{x + 7}$

94. $\sqrt{x + 4}$ **95.** $\dfrac{1}{\sqrt{x - 6}}$ **96.** $\dfrac{1}{\sqrt{x - 4}}$

11.2 RATIONAL AND IRRATIONAL NUMBERS

RATIONAL NUMBERS

Any number that can be represented as the quotient of two integers $\dfrac{a}{b}$, where $b \neq 0$, is called a **rational number.** For example,

$$\frac{2}{3}, \qquad \frac{-4}{7}, \qquad \text{and} \qquad \frac{15}{8}$$

are rational numbers. Note that all integers are rational numbers, since any integer can be expressed as the quotient of itself and 1. That is,

$$-2 = \frac{-2}{1}, \qquad 6 = \frac{6}{1}, \qquad \text{and} \qquad 0 = \frac{0}{1}.$$

IRRATIONAL NUMBERS

As we saw in Section 6.1, we can associate rational numbers with points on the number line. But there are other numbers that can be associated with points on a number line that cannot be expressed as the quotient of two integers. Such numbers are called **irrational numbers.** A detailed study of irrational numbers is beyond the scope of this book. We will simply observe that any radical whose radicand is not the square of a rational number represents an irrational number.

Thus,

$$\sqrt{2}, \quad \sqrt{3}, \quad \sqrt{5}, \quad \text{and} \quad \sqrt{\frac{5}{7}}$$

are irrational numbers; but

$$\sqrt{4}, \quad \sqrt{\frac{9}{25}}, \quad \text{and} \quad \sqrt{16}$$

are rational numbers, since they can be written as 2, 3/5, and 4, respectively—that is, as the quotient of two integers.

As noted in Section 11.1, every positive number has two square roots. These square roots may be irrational. For example:

the two square roots of 7 are $\sqrt{7}$ and $-\sqrt{7}$;

the two square roots of 11 are $\sqrt{11}$ and $-\sqrt{11}$;

and

the two square roots of 15 are $\sqrt{15}$ and $-\sqrt{15}$.

SQUARE ROOT TABLE

We cannot represent irrational numbers exactly by common fractions or decimal fractions. However, we can approximate irrational numbers to any desired degree of accuracy. To do this, we can use either a table of square roots (see page 535) or a hand-held calculator that has square root capability.

By consulting the table of square roots or by using a calculator, we see that the square roots of 1, 4, 9, 16, 25, 36, 49, 64, 81, and 100 are the rational numbers 1, 2, 3, 4, 5, 6, 7, 8, 9, and 10, respectively. The square roots of all other integers between 1 and 100 are irrational numbers, and the entries shown for these numbers in the table are *only approximations* to their true value. For example, $\sqrt{2}$ is approximately equal to 1.414, and $\sqrt{3}$ is approximately equal to 1.732. The symbol $\approx$ is often used for the phrase "is approximately equal to." Thus,

$$\sqrt{2} \approx 1.414 \quad \text{and} \quad \sqrt{3} \approx 1.732.$$

In studying operations with radicals in this and the following sections, we assume that all laws valid for operations with rational numbers also hold for irrational numbers, and that the symbols for the fundamental operations are unchanged. For example, the expressions $2\sqrt{3}$ and $4 + \sqrt{7}$ both represent irrational numbers. Using the table of square roots on page 535 or a calculator, we find that $\sqrt{3} \approx 1.732$ and $\sqrt{7} \approx 2.646$. Thus

$$2\sqrt{3} \approx 2(1.732) \quad \text{and} \quad 4 + \sqrt{7} \approx 4 + 2.646$$

$$\approx 3.464 \qquad\qquad\qquad \approx 6.646.$$

REAL NUMBERS

Both rational and irrational numbers are called **real numbers.** The real numbers fill the number line completely. In Section 3.6 we used a number line to graph inequalities where the variable represented an integer. Now we can graph inequalities where the variable represents a real number. For example, if x represents a real number, the graph of $x > 3$ is shown in Figure. 11.1, and the graph of $x \geq 3$ is shown in Figure 11.2. The colored line on the number line represents an infinite set of points.

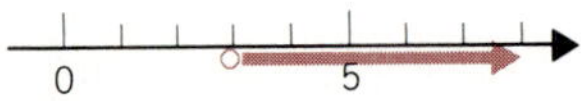

Figure 11.1 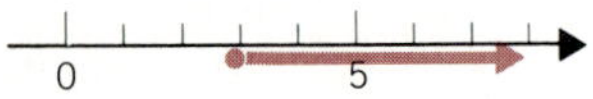   **Figure 11.2**

Notice that we use an open dot when we do not want to include the number in the graph (Figure 11.1), and a closed dot when we do (Figure 11.2).

EXERCISES 11.2

■ *Find the two square roots of the given number. Use radical notation for exact values.*

Sample Problems

a. 14   b. 18

Ans. $\sqrt{14}, -\sqrt{14}$ *Ans.* $\sqrt{18}, -\sqrt{18}$

1. 10	**2.** 8	**3.** 17
4. 19	**5.** 22	**6.** 27

■ *Which of the following numbers are rational and which are irrational? If the number is rational, express it as an integer or as a quotient of two integers.*

Sample Problems

a. $\sqrt{\dfrac{4}{9}}$ b. $\sqrt{3}$

Ans. Rational, since $\sqrt{\dfrac{4}{9}} = \dfrac{2}{3}$ *Ans.* Irrational

c. $5 + \sqrt{3}$ d. $3\sqrt{7}$

Ans. Irrational   *Ans.* Irrational

7. 6	**8.** 8	**9.** $\sqrt{2}$	**10.** $\sqrt{4}$
11. $\sqrt{6}$	**12.** $\sqrt{9}$	**13.** $\sqrt{25}$	**14.** $\sqrt{100}$

15. $3\sqrt{16}$ **16.** $\sqrt{7}$ **17.** $-\sqrt{13}$ **18.** $-2\sqrt{100}$

19. $\sqrt{\dfrac{4}{9}}$ **20.** $-\sqrt{\dfrac{16}{25}}$ **21.** $-\sqrt{\dfrac{2}{3}}$ **22.** $\sqrt{\dfrac{4}{5}}$

23. $1 + \sqrt{4}$ **24.** $3 + \sqrt{4}$ **25.** $2 + \sqrt{5}$ **26.** $1 + \sqrt{3}$

■ *Using the table of square roots on page 535 or a calculator, find a decimal approximation for each of the following. Round off answers to two decimal places. If the digit in the third decimal place is 5, round to the next higher digit in the second decimal place.*

Sample Problems

a. $\sqrt{23}$

4.796

Ans. 4.80

b. $-2\sqrt{46}$

$-2(6.782)$

-13.564

Ans. -13.56

c. $\dfrac{1}{2}\sqrt{62}$

$\dfrac{7.874}{2}$

3.937

Ans. 3.94

27. $\sqrt{57}$ **28.** $\sqrt{83}$ **29.** $\sqrt{3}$ **30.** $\sqrt{17}$

31. $\sqrt{5}$ **32.** $\sqrt{92}$ **33.** $-\sqrt{26}$ **34.** $-\sqrt{54}$

35. $2\sqrt{3}$ **36.** $3\sqrt{2}$ **37.** $-5\sqrt{3}$ **38.** $-6\sqrt{5}$

39. $\dfrac{1}{3}\sqrt{18}$ **40.** $\dfrac{1}{4}\sqrt{48}$ **41.** $-\dfrac{1}{5}\sqrt{75}$ **42.** $-\dfrac{2}{3}\sqrt{21}$

Sample Problems

a. $3 + 2\sqrt{2}$

$3 + 2(1.414)$

$3 + 2.828$

5.828

Ans. 5.83

b. $\dfrac{3 + \sqrt{3}}{2}$

$\dfrac{3 + 1.732}{2}$

$\dfrac{4.732}{2}$

2.366

Ans. 2.37

c. $1 - 2\sqrt{3}$

$1 - 2(1.732)$

$1 - 3.464$

-2.464

Ans. -2.46

43. $1 + \sqrt{3}$ **44.** $2 - \sqrt{5}$ **45.** $3 - \sqrt{2}$

46. $5 + \sqrt{5}$ **47.** $5 + 3\sqrt{7}$ **48.** $-3 + 2\sqrt{6}$

49. $-7 - 3\sqrt{28}$ **50.** $-6 + 2\sqrt{35}$ **51.** $\dfrac{3 + 2\sqrt{2}}{2}$

52. $\dfrac{7 - 5\sqrt{5}}{3}$ **53.** $\dfrac{6 - 2\sqrt{3}}{5}$ **54.** $\dfrac{7 + 3\sqrt{3}}{2}$

55. $\sqrt{3} - \sqrt{2}$ **56.** $\sqrt{5} - \sqrt{7}$ **57.** $3\sqrt{3} - 2\sqrt{5}$

58. $2\sqrt{2} - 5\sqrt{5}$

■ *Graph each set of numbers on a separate number line. Estimate the locations of graphs of numbers between integers.*

$\sqrt{13}$, 4, $\sqrt{7}$

From the table of square roots on page 535, or by using a calculator, $\sqrt{13} \approx 3.606$ and $\sqrt{7} \approx 2.646$.

Ans.
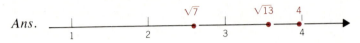

59. 7, 8, $\sqrt{55}$

61. $\sqrt{3}$, $\sqrt{5}$, $-\sqrt{7}$

63. $\sqrt{1}$, $-\sqrt{2}$, $\sqrt{3}$

65. $-\sqrt{4}$, $\sqrt{3}$, $-\sqrt{2}$

67. $-\sqrt{40}$, $\sqrt{30}$, $-\sqrt{20}$

69. $\sqrt{35}$, $-\sqrt{16}$, $-\sqrt{37}$

60. $\sqrt{21}$, $-\sqrt{25}$, 6

62. $\sqrt{9}$, $-\sqrt{16}$, $\sqrt{25}$

64. $-\sqrt{6}$, $\sqrt{8}$, $\sqrt{10}$

66. $\sqrt{2}$, 0, $-\sqrt{2}$

68. $-\sqrt{1}$, 0, $\sqrt{1}$

70. $\sqrt{21}$, $\sqrt{27}$, $\sqrt{30}$

■ *Graph the inequalities. All variables represent real numbers.*

a. $x > -3$

Ans.

b. $x \le 2$

Ans.

71. $x > 5$ **72.** $x > -2$ **73.** $x \le 4$ **74.** $x \le -2$

75. $x > -2$ **76.** $x > -4$ **77.** $x \le 1$ **78.** $x \le 3$

11.3 SIMPLIFYING RADICAL EXPRESSIONS I

We consider a square root to be in its *simplest form* if no prime factor of the radicand occurs more than once. We can determine whether a radical is in simplest form by examining the prime factors of the radicand. For example, $\sqrt{78}$ is in simplest form because none of the factors is repeated when we completely factor the radicand, $\sqrt{13 \cdot 3 \cdot 2}$. On the other hand, $\sqrt{20}$ is not in simplest form because the factor 2 occurs more than once when we factor the radicand completely, $\sqrt{2 \cdot 2 \cdot 5}$. We shall now consider a way to simplify a radical that has a repeated factor in the radicand.

First note that

$$\sqrt{4 \cdot 9} = \sqrt{36} = 6$$

and

$$\sqrt{4}\sqrt{9} = 2 \cdot 3 = 6.$$

Thus,

$$\sqrt{4 \cdot 9} = \sqrt{4} \cdot \sqrt{9}.$$

In general:

> **The square root of a product is equal to the product of the square roots of its factors.**

In symbols:

$$\sqrt{ab} = \sqrt{a}\,\sqrt{b} \qquad (a, b \geq 0).$$

To simplify radicals, we first express the radicand in completely factored form. For example,

$$\sqrt{216} = \sqrt{2 \cdot 2 \cdot 2 \cdot 3 \cdot 3 \cdot 3}.$$

Now, using the fact that $\sqrt{ab} = \sqrt{a}\,\sqrt{b}$ and $\sqrt{a^2} = a$, we can write $\sqrt{216}$ as

$$\sqrt{216} = \sqrt{2 \cdot 2 \cdot 2 \cdot 3 \cdot 3 \cdot 3}$$
$$= \sqrt{2^2}\,\sqrt{3^2}\,\sqrt{2 \cdot 3}$$
$$= 2 \cdot 3\sqrt{6}$$
$$= 6\sqrt{6}.$$

In practice, it is convenient to group the repeated factors by two's and then simplify the square roots directly. For example,

Group by two's.

$$\sqrt{216} = \sqrt{(2 \cdot 2) \cdot (3 \cdot 3) \cdot 2 \cdot 3}$$
$$= \sqrt{2 \cdot 2}\,\sqrt{3 \cdot 3}\,\sqrt{2 \cdot 3}$$
$$= 2 \cdot 3 \, \sqrt{2 \cdot 3}$$
$$= 6\sqrt{6}.$$

We can apply the same procedure when the radicand contains variables. For example,

Group by two's.

$$\sqrt{x^6} = \sqrt{(x \cdot x) \cdot (x \cdot x) \cdot (x \cdot x)}$$
$$= \sqrt{x \cdot x}\,\sqrt{x \cdot x}\,\sqrt{x \cdot x}$$
$$= x \cdot x \cdot x$$
$$= x^3,$$

and

Group by two's.

$$\sqrt{x^5} = \sqrt{(x \cdot x) \cdot (x \cdot x) \cdot x}$$
$$= \sqrt{x \cdot x} \ \sqrt{x \cdot x} \ \sqrt{x}$$
$$= x \cdot x \ \sqrt{x}$$
$$= x^2\sqrt{x}.$$

You may notice that when the radicand contains a variable with an even exponent, we obtain the square root by dividing the exponent by 2. Thus,

$$\overset{6 \div 2}{\sqrt{x^6} = x^3} \quad \text{and} \quad \overset{10 \div 2}{\sqrt{x^{10}} = x^5}.$$

Similarly, when the radicand has a variable with an odd exponent, we can simplify by factoring the variable factor into two factors, one having an even exponent and the other having an exponent of 1. For example,

$$\sqrt{x^7} = \sqrt{x^6}\sqrt{x} = x^3\sqrt{x}$$

and

$$\sqrt{x^{11}} = \sqrt{x^{10}}\sqrt{x} = x^5\sqrt{x}.$$

By simplifying radical expressions, we can extend the scope of the table of square roots. For example, $\sqrt{216}$, is not available in the table. But we can approximate $\sqrt{216}$ by observing that

$$\sqrt{216} = \sqrt{2 \cdot 2 \cdot 2 \cdot 3 \cdot 3 \cdot 3},$$
$$= 2 \cdot 3\sqrt{6}$$
$$\approx 6(2.449) = 14.694.$$

Common Error

Note that

$$\sqrt{16 + 9} \neq \sqrt{16} + \sqrt{9}$$

since

$$\sqrt{16 + 9} = \sqrt{25} = 5,$$

while

$$\sqrt{16} + \sqrt{9} = 4 + 3 = 7.$$

In general,

$$\sqrt{a + b} \neq \sqrt{a} + \sqrt{b}.$$

EXERCISES 11.3

■ *Simplify.*

Sample Problems

a. $\sqrt{24}$

b. $\sqrt{1575}$

Factor radicand completely and group factors by two's.

$\sqrt{(2 \cdot 2) \cdot 2 \cdot 3}$

$2\sqrt{2 \cdot 3}$

Ans. $2\sqrt{6}$

$\sqrt{(5 \cdot 5) \cdot (3 \cdot 3) \cdot 7}$

$5 \quad \cdot \quad 3\sqrt{7}$

Ans. $15\sqrt{7}$

1. $\sqrt{8}$ 2. $\sqrt{12}$ 3. $\sqrt{18}$ 4. $\sqrt{49}$

5. $-\sqrt{20}$ 6. $-\sqrt{27}$ 7. $-\sqrt{72}$ 8. $-\sqrt{24}$

9. $\sqrt{64}$ 10. $\sqrt{162}$ 11. $\sqrt{288}$ 12. $\sqrt{84}$

13. $\sqrt{125}$ 14. $\sqrt{450}$ 15. $-\sqrt{1080}$ 16. $-\sqrt{882}$

17. $\pm\sqrt{720}$ 18. $\pm\sqrt{588}$ 19. $\pm\sqrt{1944}$ 20. $\pm\sqrt{1125}$

Sample Problems

a. $\sqrt{x^4}$

Method I

$\sqrt{(x \cdot x) \cdot (x \cdot x)}$

$x \quad \cdot \quad x$

Ans. x^2

Method II

$\sqrt{x^4}$

$x^{4 \div 2}$

Ans. x^2

b. $-\sqrt{y^9}$

Method I

$-\sqrt{(y \cdot y) \cdot (y \cdot y) \cdot (y \cdot y) \cdot (y \cdot y) \cdot y}$

$- y \quad \cdot \quad y \quad \cdot \quad y \quad \cdot \quad y\sqrt{y}$

Ans. $-y^4\sqrt{y}$

Method II

$-\sqrt{y^8}\sqrt{y}$

$-y^{8 \div 2}\sqrt{y}$

Ans. $-y^4\sqrt{y}$

21. $\sqrt{x^3}$ 22. $\sqrt{y^5}$ 23. $\sqrt{y^7}$ 24. $\sqrt{x^{10}}$

25. $-\sqrt{x^{11}}$ 26. $-\sqrt{x^{13}}$ 27. $\sqrt{x^6}$ 28. $-\sqrt{x^4}$

29. $\pm\sqrt{x^8}$ 30. $\pm\sqrt{x^{15}}$ 31. $\sqrt{x^{12}}$ 32. $\sqrt{x^{14}}$

Sample Problems

a. $\sqrt{12y^3}$

b. $-\sqrt{20x^2y^3}$

Factor radicand completely and group factors by two's.

$\sqrt{(2 \cdot 2) \cdot 3 \cdot (y \cdot y) \cdot y}$

$2 \quad \cdot \quad y\sqrt{3 \cdot y}$

Ans. $2y\sqrt{3y}$

$-\sqrt{(2 \cdot 2) \cdot 5 \cdot (x \cdot x) \cdot (y \cdot y) \cdot y}$

$-2 \quad \cdot \quad x \quad \cdot \quad y\sqrt{5y}$

Ans. $-2xy\sqrt{5y}$

33. $\sqrt{4x^2}$ **34.** $\sqrt{8x^2}$ **35.** $\sqrt{9x^3}$ **36.** $\sqrt{12x^3}$

37. $-\sqrt{24y^5}$ **38.** $-\sqrt{121x^5}$ **39.** $\sqrt{64y^4}$ **40.** $\sqrt{36x^5}$

41. $\sqrt{49x^7}$ **42.** $\sqrt{16x^2}$ **43.** $\pm\sqrt{32x^3}$ **44.** $\pm\sqrt{72y^3}$

45. $\sqrt{80x^{12}}$ **46.** $\sqrt{98y^{13}}$ **47.** $-\sqrt{64x^9}$ **48.** $-\sqrt{3x^8}$

49. $\pm\sqrt{5y^3}$ **50.** $\pm\sqrt{7x^2}$ **51.** $\sqrt{48x^2y}$ **52.** $\sqrt{20x^2y^2}$

53. $\sqrt{25x^3y^2}$ **54.** $\sqrt{50xy^2}$ **55.** $-\sqrt{45a^4b^7}$ **56.** $-\sqrt{40x^5y^6}$

57. $\sqrt{\dfrac{9}{16}x^2y^2}$ **58.** $\sqrt{\dfrac{4}{9}x^3y}$ **59.** $\pm\sqrt{\dfrac{25}{36}y^2}$ **60.** $\pm\sqrt{\dfrac{1}{4}ab^2c^3}$

Sample Problems

a. $3\sqrt{4x^3}$

b. $\pm\dfrac{3x}{y}\sqrt{18x^3y^3}$

$3\sqrt{(2 \cdot 2) \cdot (x \cdot x) \cdot x}$

$3 \cdot 2 \cdot x\sqrt{x}$

Ans. $6x\sqrt{x}$

$\pm\dfrac{3x}{y}\sqrt{(3 \cdot 3) \cdot 2 \cdot (x \cdot x) \cdot x \cdot (y \cdot y) \cdot y}$

$\pm\dfrac{3x}{y} \cdot 3 \cdot x \cdot y\sqrt{2xy}$

Ans. $\pm 9x^2\sqrt{2xy}$

61. $2\sqrt{x^2}$ **62.** $3\sqrt{y^2}$ **63.** $3\sqrt{4x}$

64. $4\sqrt{5x}$ **65.** $-7\sqrt{49y^3}$ **66.** $-4\sqrt{16x^3}$

67. $2x\sqrt{x^2y}$ **68.** $3x\sqrt{9xy^2}$ **69.** $-\dfrac{1}{3}\sqrt{9a^3b^5}$

70. $\dfrac{1}{5}\sqrt{25y^3z^5}$ **71.** $\pm\dfrac{1}{2}x\sqrt{16x^4y^3}$ **72.** $\pm\dfrac{2}{3}a\sqrt{36a^3b^3}$

73. $-\dfrac{3a^2}{4b}\sqrt{144a^6b^9}$ **74.** $-\dfrac{4x}{3y^2}\sqrt{81x^8y^5}$

■ *Use the table of square roots (page 535) to approximate each expression. Round off answers to two decimal places.*

Sample Problems

a. $\sqrt{243}$

b. $3 + 2\sqrt{200}$

$$\sqrt{\boxed{3 \cdot 3} \cdot \boxed{3 \cdot 3} \cdot 3}$$

$$3 \quad \cdot \quad 3\sqrt{3}$$

$$9(1.732)$$

$$15.588$$

Ans. 15.59

$$3 + 2\sqrt{\boxed{2 \cdot 2} \cdot \boxed{5 \cdot 5} \cdot 2}$$

$$3 + 2 \cdot 2 \cdot 5\sqrt{2}$$

$$3 + 20(1.414)$$

$$3 + 28.28$$

Ans. 31.28

75. $\sqrt{108}$ **76.** $\sqrt{162}$ **77.** $\sqrt{275}$

78. $\sqrt{207}$ **79.** $3 + \sqrt{125}$ **80.** $24 - \sqrt{176}$

81. $5 - \sqrt{300}$ **82.** $11 - \sqrt{242}$ **83.** $-2\sqrt{243}$

84. $-5\sqrt{120}$ **85.** $6 + 3\sqrt{104}$ **86.** $1 + 2\sqrt{128}$

87. Use a numerical example to show that $\sqrt{x} + \sqrt{9}$ is not equivalent to $\sqrt{x + 9}$.

88. Use a numerical example to show that $\sqrt{y} + \sqrt{16}$ is not equivalent to $\sqrt{y + 16}$.

11.4 SIMPLIFYING RADICAL EXPRESSIONS II

SUMS AND DIFFERENCES

Recall from Section 3.1 that we add like terms by adding their numerical coefficients; that is,

$$2r + 3r = 5r,$$

where r represents any number. In particular, if r represents an irrational number, say $\sqrt{2}$, we have

$$2\sqrt{2} + 3\sqrt{2} = 5\sqrt{2}.$$

Thus, we may add radical expressions by adding their numerical coefficients, provided the radicands involved are identical. If the radicands differ, we can only indicate addition; for example, $3\sqrt{2} + 4\sqrt{3}$ cannot be written as a single term. As before, if there is no numerical coefficient before a radical, it is understood that the coefficient is 1. Thus,

$$4\sqrt{3} + \sqrt{3} = 4\sqrt{3} + 1\sqrt{3}$$

$$= 5\sqrt{3}.$$

These same ideas apply to differences. For example,

$$5\sqrt{2} - 2\sqrt{2} = 3\sqrt{2}$$

and

$$\sqrt{7} - 4\sqrt{7} = 1\sqrt{7} - 4\sqrt{7}$$

$$= -3\sqrt{7}$$

It is a good idea to write radicals in simplest form before attempting to combine like terms. For example,

$$\sqrt{20} + 2\sqrt{45} = \sqrt{2 \cdot 2 \cdot 5} + 2\sqrt{3 \cdot 3 \cdot 5}$$
$$= 2\sqrt{5} + 2 \cdot 3\sqrt{5}$$
$$= 2\sqrt{5} + 6\sqrt{5}$$
$$= 8\sqrt{5},$$

and

$$3\sqrt{4x} - 2\sqrt{x} = 3\sqrt{2 \cdot 2x} - 2\sqrt{x}$$
$$= 3 \cdot 2\sqrt{x} - 2\sqrt{x}$$
$$= 6\sqrt{x} - 2\sqrt{x}$$
$$= 4\sqrt{x}.$$

Common Error

Note that

$$5 - 3\sqrt{2} \neq 2\sqrt{2}$$

since 5 and $3\sqrt{2}$ are not like terms. Similarly

$$2 + 7\sqrt{5} \neq 9\sqrt{5}.$$

PRODUCTS AND FACTORS

Products involving radicals can also be written in equivalent forms. For example, using the distributive property,

$$6(\sqrt{3} + \sqrt{2}) = 6\sqrt{3} + 6\sqrt{2}.$$

We can also express binomials that contain radicals in factored form. For example,

$$6\sqrt{3} + 18 = 6\sqrt{3} + 6 \cdot 3$$
$$= 6(\sqrt{3} + 3).$$

FRACTIONS

The properties of fractions we considered in Chapters 6 and 7 also apply to fractions that contain radical expressions. For example, to reduce

$$\frac{8 - \sqrt{80}}{4},$$

we first simplify the radical expression to get

$$\frac{8 - 4\sqrt{5}}{4}.$$

We then factor the numerator to obtain

$$\frac{4(2 - \sqrt{5})}{4} = 2 - \sqrt{5}.$$

To write the sum

$$\frac{\sqrt{2}}{3} + \frac{x}{2}$$

as a single fraction, we find the L.C.D. 6 and build each fraction to obtain

$$\frac{(2)\sqrt{2}}{(2)3} + \frac{(3)x}{(3)2} = \frac{2\sqrt{2} + 3x}{6}.$$

Common Error

Note that

$$\frac{4 + \sqrt{20}}{4} \neq \sqrt{20}.$$

Since 4 is a term of the numerator, not a factor, we cannot reduce the fraction as shown above. However if we first simplify the radical as

$$\frac{4 + \sqrt{4 \cdot 5}}{4} = \frac{4 + 2\sqrt{5}}{4},$$

we can then factor 2 from the numerator and reduce the fraction as follows:

$$\frac{4 + 2\sqrt{5}}{4} = \frac{2(2 + \sqrt{5})}{\overset{4}{2}}$$

$$= \frac{2 + \sqrt{5}}{2}.$$

The fraction cannot be reduced further.

EXERCISES 11.4

■ *Simplify.*

Sample Problems

a. $\sqrt{5} + 3\sqrt{5}$

Ans. $4\sqrt{5}$

b. $\sqrt{5} - 3\sqrt{5}$

Ans. $-2\sqrt{5}$

1. $\sqrt{3} + 2\sqrt{3}$

2. $\sqrt{7} - 3\sqrt{7}$

3. $3\sqrt{5} - 2\sqrt{5}$

4. $8\sqrt{5} - 2\sqrt{5} + 3\sqrt{5}$

5. $2\sqrt{3} - 4\sqrt{3} + 2\sqrt{3}$

6. $\sqrt{5} - 3\sqrt{5} + 7\sqrt{5}$

Sample Problems

a. $5\sqrt{2} - \sqrt{8} + \sqrt{12}$ b. $2\sqrt{3a} + \sqrt{27a} - 2\sqrt{12a}$

Simplify radicals.

$5\sqrt{2} - 2\sqrt{2} + 2\sqrt{3}$ $2\sqrt{3a} + 3\sqrt{3a} - 4\sqrt{3a}$

Combine like terms.

Ans. $3\sqrt{2} + 2\sqrt{3}$ *Ans.* $\sqrt{3a}$

7. $2\sqrt{3} + \sqrt{27}$ **8.** $\sqrt{8} + \sqrt{18}$
9. $\sqrt{12} + 2\sqrt{27} - 3\sqrt{48}$ **10.** $\sqrt{20} + \sqrt{45} - 2\sqrt{80}$
11. $3\sqrt{2} - 4\sqrt{3} + \sqrt{2}$ **12.** $2\sqrt{3} - \sqrt{4} + 3\sqrt{3}$
13. $\sqrt{3} + 2\sqrt{12} + \sqrt{18}$ **14.** $\sqrt{36} - 2\sqrt{32} + \sqrt{49}$
15. $3\sqrt{144} - 4\sqrt{49} + 3\sqrt{24}$ **16.** $\sqrt{12} - \sqrt{27} + 2\sqrt{8}$
17. $\sqrt{4a} + \sqrt{9a} - 7\sqrt{a}$ **18.** $\sqrt{12a} - \sqrt{3a} + 3\sqrt{3a}$
19. $2\sqrt{x} + 2\sqrt{25x} - 6\sqrt{5x}$ **20.** $3\sqrt{2x} - \sqrt{8x} + \sqrt{4x}$
21. $\sqrt{16b^3} - b\sqrt{25b} + 3b\sqrt{b}$ **22.** $\sqrt{xy^2} + 2\sqrt{xy^2} - \sqrt{4xy^2}$

■ *Express without parentheses.*

Sample Problems

a. $\overset{\frown}{2(3 + 4\sqrt{2})}$ b. $\overset{\frown}{a(4\sqrt{3} - 6\sqrt{a})}$

Apply distributive property.

Ans. $6 + 8\sqrt{2}$ *Ans.* $4a\sqrt{3} - 6a\sqrt{a}$

23. $4(\sqrt{3} + 1)$ **24.** $2(3 - \sqrt{2})$
25. $-5(6 + \sqrt{7})$ **26.** $-2(\sqrt{6} - 3)$
27. $4(\sqrt{2} - 2\sqrt{3})$ **28.** $3(\sqrt{3} + 3\sqrt{7})$
29. $2x(\sqrt{5} + 3\sqrt{x})$ **30.** $5y(2\sqrt{y} - \sqrt{2})$
31. $6xy(x^2\sqrt{2} - 3y\sqrt{3x})$ **32.** $4ab(2ab\sqrt{3} + b^2\sqrt{a})$

■ *Simplify radical expressions where possible and factor.*

Sample Problems

a. $8\sqrt{3} - 10$ b. $\sqrt{12} + 4$ c. $y\sqrt{x} - y^2\sqrt{y}$

$2\sqrt{3} + 4$

Ans. $2(4\sqrt{3} - 5)$ *Ans.* $2(\sqrt{3} + 2)$ *Ans.* $y(\sqrt{x} - y\sqrt{y})$

33. $2 + 2\sqrt{3}$ **34.** $5 + 5\sqrt{7}$ **35.** $6 - 3\sqrt{2}$
36. $4\sqrt{2} - 12$ **37.** $8 + 32\sqrt{5}$ **38.** $6 + 24\sqrt{2}$
39. $3 + \sqrt{18}$ **40.** $4 + \sqrt{32}$ **41.** $4 - 2\sqrt{8}$

42. $6 - 2\sqrt{27}$ **43.** $3\sqrt{50} - 2\sqrt{75}$ **44.** $4\sqrt{45} - 3\sqrt{72}$

45. $3x\sqrt{x} - 6x\sqrt{y}$ **46.** $2y^2\sqrt{x} - 8y\sqrt{y}$ **47.** $3\sqrt{x^2y} - 9x\sqrt{x}$

48. $4\sqrt{x^3y} + 6x\sqrt{x}$ **49.** $2\sqrt{y^3} - 6\sqrt{y^4}$ **50.** $3\sqrt{x^4} + 6\sqrt{x^5}$

■ *Simplify fractions.*

Sample Problems

a. $\dfrac{6 - 3\sqrt{7}}{3}$ b. $\dfrac{-2 - \sqrt{72}}{4}$

Simplify radicals.

$$\dfrac{-2 - 6\sqrt{2}}{4}$$

Factor numerator and simplify.

$$\dfrac{3(2 - \sqrt{7})}{3}$$ $$\dfrac{\overset{1}{2}(-1 - 3\sqrt{2})}{\underset{2}{4}}$$

Ans. $2 - \sqrt{7}$ *Ans.* $\dfrac{-1 - 3\sqrt{2}}{2}$

51. $\dfrac{4 + 6\sqrt{3}}{2}$ **52.** $\dfrac{3 - 3\sqrt{2}}{3}$ **53.** $\dfrac{6 - 2\sqrt{5}}{2}$

54. $\dfrac{9 - 3\sqrt{5}}{3}$ **55.** $\dfrac{-2 + \sqrt{8}}{2}$ **56.** $\dfrac{-6 + \sqrt{54}}{3}$

57. $\dfrac{2 + 3\sqrt{12}}{4}$ **58.** $\dfrac{5 - \sqrt{75}}{10}$ **59.** $\dfrac{2x - \sqrt{8}}{2x}$

60. $\dfrac{6x - \sqrt{18}}{6x}$ **61.** $\dfrac{4y - 2\sqrt{y^3}}{2y}$ **62.** $\dfrac{6y + 3\sqrt{2y^3}}{6y}$

■ *Write each sum or difference as a single fraction.*

Sample Problems

a. $\dfrac{2}{3} - \dfrac{\sqrt{7}}{3}$ b. $\dfrac{2}{3} + \dfrac{5\sqrt{7}}{6}$

In problem b, build $\frac{2}{3}$ to a fraction with denominator 6.

$$\dfrac{(2)2}{(2)3} + \dfrac{5\sqrt{7}}{6}$$

Add or subtract numerators.

Ans. $\dfrac{2 - \sqrt{7}}{3}$ *Ans.* $\dfrac{4 + 5\sqrt{7}}{6}$

63. $\dfrac{2}{3} + \dfrac{\sqrt{2}}{3}$ **64.** $\dfrac{5}{2} - \dfrac{\sqrt{3}}{2}$ **65.** $\dfrac{\sqrt{3}}{5} - \dfrac{1}{5}$

66. $\dfrac{\sqrt{2}}{7} + \dfrac{1}{7}$ **67.** $\dfrac{2\sqrt{10}}{3} - \dfrac{\sqrt{3}}{3}$ **68.** $\dfrac{\sqrt{17}}{5} - \dfrac{3\sqrt{7}}{5}$

69. $\dfrac{\sqrt{11}}{a} + \dfrac{1}{a}$ **70.** $\dfrac{\sqrt{5}}{b} - \dfrac{3}{b}$ **71.** $\dfrac{5}{4} + \dfrac{3\sqrt{2}}{2}$

72. $\dfrac{1}{10} - \dfrac{2\sqrt{3}}{5}$ **73.** $\dfrac{1}{2a} + \dfrac{\sqrt{3}}{6a}$ **74.** $\dfrac{\sqrt{5}}{3b} - \dfrac{5}{6b}$

Sample Problems

a. $\dfrac{1}{2} - \dfrac{\sqrt{3}}{3}$ b. $4 - \dfrac{2\sqrt{3}}{5}$

Find L.C.D. and build fractions.

$\dfrac{(3)1}{(3)2} - \dfrac{\sqrt{3}(2)}{3(2)}$ $\dfrac{(5)4}{(5)1} - \dfrac{2\sqrt{3}}{5}$

Ans. $\dfrac{3 - 2\sqrt{3}}{6}$ *Ans.* $\dfrac{20 - 2\sqrt{3}}{5}$

75. $\dfrac{2}{5} + \dfrac{\sqrt{3}}{3}$ **76.** $\dfrac{3}{7} - \dfrac{\sqrt{2}}{2}$ **77.** $\dfrac{2\sqrt{3}}{3} - \dfrac{\sqrt{2}}{2}$

78. $\dfrac{3\sqrt{5}}{4} + \dfrac{\sqrt{3}}{5}$ **79.** $4 + \dfrac{3\sqrt{2}}{2}$ **80.** $\dfrac{3\sqrt{3}}{2} + 3$

81. $\dfrac{\sqrt{2}}{x} + \dfrac{1}{2x}$ **82.** $\dfrac{\sqrt{3}}{x} - \dfrac{1}{4x}$ **83.** $\dfrac{3}{4} - 2\sqrt{y}$

84. $\dfrac{4}{5} + 3\sqrt{y}$ **85.** $\dfrac{\sqrt{x}}{2} + \dfrac{\sqrt{y}}{3}$ **86.** $\dfrac{2\sqrt{y}}{3} - \dfrac{\sqrt{x}}{4}$

11.5   PRODUCTS OF RADICAL EXPRESSIONS

In Section 11.3, we saw that

$$\sqrt{ab} = \sqrt{a}\sqrt{b}.$$

By the symmetric property of equality,

$$\sqrt{a}\sqrt{b} = \sqrt{ab} \qquad (a, b \geq 0)$$

Stated in words:

The product of two square roots is equal to the square root of the product of the radicands.

We can use this property to simplify products involving radicals. For example, to simplify

$$\sqrt{6x}\,\sqrt{2xy},$$

we write the product as

$$\sqrt{6x \cdot 2xy}$$

and then we simplify as in Section 11.3 to obtain

$$\sqrt{6x \cdot 2xy} = \sqrt{(2 \cdot 2) \cdot 3 \cdot (x \cdot x) \cdot y}$$
$$= 2 \quad \cdot \quad x\sqrt{3y}$$
$$= 2x\sqrt{3y}.$$

   We can use the property above together with the distributive property to simplify radical expressions containing parentheses. For example,

$$\sqrt{3}(\sqrt{3} - 2) = \sqrt{3} \cdot \sqrt{3} - \sqrt{3} \cdot 2$$
$$= 3 - 2\sqrt{3},$$

and

$$(\sqrt{3} - 1)(\sqrt{3} + 2) = \sqrt{3} \cdot \sqrt{3} + \sqrt{3} \cdot 2 - 1 \cdot \sqrt{3} - 1 \cdot 2$$
$$= 3 + 2\sqrt{3} - \sqrt{3} - 2$$
$$= 1 + \sqrt{3}.$$

EXERCISES 11.5

■ *Simplify.*

Sample Problems

a. $\sqrt{2}\sqrt{3}$

   $\sqrt{2 \cdot 3}$

Ans. $\sqrt{6}$

b. $\sqrt{2x}\,\sqrt{10xy}$

   $\sqrt{2x \cdot 10xy}$

   $\sqrt{(2 \cdot 2) \cdot 5 \cdot (x \cdot x) \cdot y}$

   $2 \quad \cdot \quad x\sqrt{5y}$

Ans. $2x\sqrt{5y}$

1. $\sqrt{3}\sqrt{5}$ 2. $\sqrt{2}\sqrt{7}$ 3. $\sqrt{3}\sqrt{10}$

4. $\sqrt{5}\sqrt{13}$ 5. $\sqrt{3}\sqrt{6}$ 6. $\sqrt{2}\sqrt{10}$

7. $\sqrt{8}\sqrt{2}$ 8. $\sqrt{27}\sqrt{3}$ 9. $\sqrt{2x}\sqrt{3x}$

10. $\sqrt{5a}\sqrt{3a}$ 11. $\sqrt{2xy}\sqrt{6xy^2}$ 12. $\sqrt{6x^2}\sqrt{3x^2y}$

13. $\sqrt{8a}\sqrt{18a}$ 14. $\sqrt{12b}\sqrt{32b}$ 15. $\sqrt{10x^2}\sqrt{15y}$

16. $\sqrt{18a^2}\sqrt{6b}$

Sample Problems

a. $2\sqrt{6}\sqrt{8}$

$2\sqrt{48}$

$2\sqrt{(2\cdot2)\cdot(2\cdot2)\cdot3}$

$2\cdot2\cdot2\sqrt{3}$

Ans. $8\sqrt{3}$

b. $(\sqrt{x})(2\sqrt{xy})$

$2\sqrt{(x\cdot x)\cdot y}$

$2\cdot x\sqrt{y}$

Ans. $2x\sqrt{y}$

17. $\sqrt{2}\sqrt{5}\sqrt{3}$ 18. $\sqrt{5}\sqrt{3}\sqrt{7}$

19. $\sqrt{5}\sqrt{10}\sqrt{2}$ 20. $\sqrt{6}\sqrt{3}\sqrt{2}$

21. $(2\sqrt{3})(\sqrt{2})(\sqrt{9})$ 22. $(5\sqrt{5})(3\sqrt{10})(\sqrt{4})$

23. $(2\sqrt{x})(3\sqrt{x})(\sqrt{x})$ 24. $(x\sqrt{2})(x\sqrt{3})(\sqrt{6})$

25. $(a\sqrt{b})(b\sqrt{c})(c\sqrt{a})$ 26. $(b\sqrt{a})(a\sqrt{b})(a\sqrt{ab})$

27. $(x\sqrt{x})(\sqrt{x^2})(\sqrt{x^3})$ 28. $(a^2\sqrt{a})(2a\sqrt{a})(a\sqrt{a^2})$

Sample Problems

a. $\sqrt{3}(2 + \sqrt{2})$

$(\sqrt{3})(2) + (\sqrt{3})(\sqrt{2})$

Ans. $2\sqrt{3} + \sqrt{6}$

b. $\sqrt{3}(\sqrt{6} - \sqrt{15})$

Apply distributive property.

$(\sqrt{3})(\sqrt{6}) - (\sqrt{3})(\sqrt{15})$

$\sqrt{18} - \sqrt{45}$

Simplify radicals.

Ans. $3\sqrt{2} - 3\sqrt{5}$

29. $\sqrt{2}(3 + \sqrt{3})$ 30. $\sqrt{3}(5 + \sqrt{5})$ 31. $\sqrt{3}(\sqrt{6} + 2)$

32. $\sqrt{2}(\sqrt{6} + 3)$ 33. $\sqrt{5}(4 + \sqrt{10})$ 34. $\sqrt{3}(2 - \sqrt{15})$

35. $\sqrt{3}(\sqrt{2} + \sqrt{6})$ 36. $\sqrt{5}(\sqrt{3} - \sqrt{10})$ 37. $\sqrt{3}(\sqrt{3} + \sqrt{2})$

38. $\sqrt{5}(\sqrt{5} + \sqrt{3})$ 39. $\sqrt{2}(\sqrt{10} - \sqrt{2})$ 40. $\sqrt{3}(\sqrt{3} + \sqrt{15})$

Sample Problem

$(2 + \sqrt{3})(1 - 2\sqrt{3})$

Apply distributive property.

$2 - 4\sqrt{3} + \sqrt{3} - 2\sqrt{3}\sqrt{3}$

Simplify.

$2 - 3\sqrt{3} - 6$

Ans. $-4 - 3\sqrt{3}$

41. $(3 + \sqrt{2})(1 - \sqrt{2})$ **42.** $(2 - \sqrt{2})(3 + \sqrt{2})$

43. $(\sqrt{5} - 1)(\sqrt{5} + 3)$ **44.** $(\sqrt{7} + 3)(\sqrt{7} - 5)$

45. $(2 + \sqrt{3})(2 - \sqrt{3})$ **46.** $(3 + \sqrt{2})(3 - \sqrt{2})$

47. $(2 + \sqrt{3})^2$ **48.** $(3 - \sqrt{2})^2$

49. $(3 - 2\sqrt{5})(3 + 2\sqrt{5})$ **50.** $(4 - 3\sqrt{6})(4 + 3\sqrt{6})$

51. $(2\sqrt{3} + \sqrt{5})(\sqrt{3} - 2\sqrt{5})$ **52.** $(2\sqrt{5} - 3\sqrt{2})(\sqrt{5} + \sqrt{2})$

53. $(3\sqrt{7} - 2\sqrt{5})(2\sqrt{7} + 3\sqrt{5})$ **54.** $(5\sqrt{6} - 2\sqrt{3})(\sqrt{6} - \sqrt{3})$

55. $(2\sqrt{7} - 3)^2$ **56.** $(3\sqrt{5} - 2)^2$

57. $(4\sqrt{3} + 3\sqrt{5})^2$ **58.** $(5\sqrt{2} + 2\sqrt{7})^2$

11.6 QUOTIENTS OF RADICAL EXPRESSIONS

Observe that

$$\sqrt{\frac{36}{9}} = \sqrt{4} = 2$$

and

$$\frac{\sqrt{36}}{\sqrt{9}} = \frac{6}{3} = 2.$$

Thus,

$$\frac{\sqrt{36}}{\sqrt{9}} = \sqrt{\frac{36}{9}}.$$

Stated in words:

The quotient of two square roots is equal to the square root of the quotient of the radicands.

In general,

$$\frac{\sqrt{a}}{\sqrt{b}} = \sqrt{\frac{a}{b}}, \qquad (a \geq 0, \, b > 0)$$

For example,

$$\frac{\sqrt{6}}{\sqrt{3}} = \sqrt{\frac{6}{3}} = \sqrt{2}$$

and

$$\frac{\sqrt{4x^2}}{\sqrt{2x}} = \sqrt{\frac{4x^2}{2x}} = \sqrt{2x}.$$

RATIONALIZING THE DENOMINATOR

It is often convenient to express a fraction whose denominator contains a radical as an equivalent fraction whose denominator is free of radicals. This process is called **rationalizing the denominator.** We rationalize the denominator of a fraction by building to a fraction with a perfect square in the denominator and then simplifying. For example, we can rationalize the denominator of the fraction $\frac{\sqrt{2}}{\sqrt{3}}$ by multiplying the numerator and the denominator by $\sqrt{3}$ to obtain

$$\frac{\sqrt{2}\sqrt{3}}{\sqrt{3}\sqrt{3}} = \frac{\sqrt{6}}{\sqrt{9}} = \frac{\sqrt{6}}{3}.$$

Alternatively, by applying the property above, we can represent the fraction $\frac{\sqrt{2}}{\sqrt{3}}$ in the form $\sqrt{\frac{2}{3}}$ and then multiply the numerator and the denominator of the radicand by 3 to get

$$\sqrt{\frac{2 \cdot 3}{3 \cdot 3}} = \sqrt{\frac{6}{9}}$$
$$= \frac{\sqrt{6}}{\sqrt{9}} = \frac{\sqrt{6}}{3}.$$

The result is the same, so use whichever approach you find simpler.

To see one advantage of the rationalized form, we will compute a decimal approximation for $\frac{\sqrt{2}}{\sqrt{3}}$. If we approach this problem directly, we obtain

$$\frac{\sqrt{2}}{\sqrt{3}} \approx \frac{1.414}{1.732}$$

and arrive at a problem in long division. But if we first rationalize the denominator, we obtain

$$\frac{\sqrt{2}}{\sqrt{3}} = \frac{\sqrt{6}}{3} \approx \frac{2.449}{3}$$

and arrive at a simple division process, one that can even be done mentally.

We can also rationalize the denominator of a fraction when the radical expression involves variables. For example,

$$\frac{3}{\sqrt{x}} = \frac{3\sqrt{x}}{\sqrt{x}\sqrt{x}} = \frac{3\sqrt{x}}{x},$$

and

$$\frac{\sqrt{2}}{\sqrt{3a}} = \frac{\sqrt{2}\sqrt{3a}}{\sqrt{3a}\sqrt{3a}} = \frac{\sqrt{6a}}{3a}.$$

BINOMIAL DENOMINATORS

To rationalize a binomial denominator that contains radicals, we use the fact that

$$(a + b)(a - b) = a^2 - b^2.$$

For example,

$$(3 + \sqrt{5})(3 - \sqrt{5}) = (3)^2 - (\sqrt{5})^2 = 9 - 5 = 4.$$

Notice that the product, 4, contains no radicals. The expressions $3 + \sqrt{5}$ and $3 - \sqrt{5}$ are called *conjugates*. In general the conjugate of $a + b$ is $a - b$. Thus to rationalize the denominator of

$$\frac{7}{3 - \sqrt{5}},$$

we multiply numerator and denominator by the conjugate of $3 - \sqrt{5}$ to obtain

$$\frac{7(3 + \sqrt{5})}{(3 - \sqrt{5})(3 + \sqrt{5})} = \frac{7(3 + \sqrt{5})}{(3)^2 - (\sqrt{5})^2}$$

$$= \frac{21 + 7\sqrt{5}}{9 - 5}$$

$$= \frac{21 + 7\sqrt{5}}{4}.$$

EXERCISES 11.6

■ *Rewrite each fraction so that no radical appears in the denominator and no radicand contains a fraction.*

Sample Problems

a. $\dfrac{\sqrt{12}}{\sqrt{3}}$ b. $\dfrac{2\sqrt{30}}{\sqrt{6}}$ c. $\dfrac{3\sqrt{50a}}{\sqrt{8a}}$

$$\sqrt{\frac{12}{3}} \qquad\qquad 2\sqrt{\frac{30}{6}} \qquad\qquad 3\sqrt{\frac{50a}{8a}}$$

$$Ans.\ 2 \qquad\qquad Ans.\ 2\sqrt{5} \qquad\qquad 3\sqrt{\frac{25}{4}}$$

$$Ans.\ \frac{15}{2}$$

1. $\dfrac{\sqrt{18}}{\sqrt{2}}$ 2. $\dfrac{\sqrt{8}}{\sqrt{2}}$ 3. $\dfrac{\sqrt{75x^3}}{\sqrt{3x}}$ 4. $\dfrac{\sqrt{80a^3}}{\sqrt{5a}}$

5. $\dfrac{\sqrt{32b}}{\sqrt{18b}}$ 6. $\dfrac{\sqrt{48a}}{\sqrt{27a}}$ 7. $\dfrac{\sqrt{5a^3}}{\sqrt{125a^5}}$ 8. $\dfrac{\sqrt{2x}}{\sqrt{32x^5}}$

9. $\dfrac{2\sqrt{14bc}}{\sqrt{2c}}$ 10. $\dfrac{\sqrt{15abc}}{5\sqrt{3b}}$ 11. $\dfrac{2\sqrt{50a^6b^5}}{b\sqrt{98a^3b}}$ 12. $\dfrac{x\sqrt{20x^3y}}{3\sqrt{45x^2y^5}}$

13. $\dfrac{\sqrt{6a}\sqrt{8b}}{\sqrt{12ab}}$ 14. $\dfrac{\sqrt{3a}\sqrt{10a}}{\sqrt{6a^2}}$ 15. $\dfrac{2x\sqrt{81xy}\sqrt{30y}}{\sqrt{10x^3y^4}}$ 16. $\dfrac{y\sqrt{21x^3}\sqrt{35y}}{2\sqrt{15xy^3}}$

■ *Rationalize each denominator.*

Sample Problems

a. $\dfrac{\sqrt{3}}{\sqrt{a}}$ or $\dfrac{\sqrt{3}}{\sqrt{a}}$ b. $\sqrt{\dfrac{1}{2}}$ or $\sqrt{\dfrac{1}{2}}$

$\dfrac{\sqrt{3}\,\sqrt{a}}{\sqrt{a}\,\sqrt{a}}$ $\sqrt{\dfrac{3\cdot a}{a\cdot a}}$ $\sqrt{\dfrac{1\cdot 2}{2\cdot 2}}$ $\dfrac{\sqrt{1}\,\sqrt{2}}{\sqrt{2}\,\sqrt{2}}$

$\dfrac{\sqrt{3a}}{\sqrt{a^2}}$ $\dfrac{\sqrt{3a}}{\sqrt{a^2}}$ $\dfrac{\sqrt{2}}{\sqrt{4}}$ $\dfrac{\sqrt{2}}{\sqrt{4}}$

$Ans.\ \dfrac{\sqrt{3a}}{a}$ $Ans.\ \dfrac{\sqrt{2}}{2}$

17. $\dfrac{5}{\sqrt{2}}$ 18. $\dfrac{5}{\sqrt{3}}$ 19. $\dfrac{2}{\sqrt{x}}$ 20. $\dfrac{5}{\sqrt{x}}$

21. $\sqrt{\dfrac{1}{3}}$ 22. $\sqrt{\dfrac{1}{5}}$ 23. $\sqrt{\dfrac{3a}{b}}$ 24. $\sqrt{\dfrac{5b}{a}}$

Sample Problems

a. $\dfrac{\sqrt{12}}{\sqrt{3x}}$ or $\dfrac{\sqrt{12}}{\sqrt{3x}}$ b. $\dfrac{2\sqrt{5}}{\sqrt{8}}$ or $\dfrac{2\sqrt{5}}{\sqrt{8}}$

$\sqrt{\dfrac{12}{3x}}$ $\dfrac{\sqrt{12}\,\sqrt{3x}}{\sqrt{3x}\,\sqrt{3x}}$ $2\sqrt{\dfrac{5\cdot 2}{8\cdot 2}}$ $\dfrac{2\sqrt{5}\,\sqrt{2}}{\sqrt{8}\,\sqrt{2}}$

$$\sqrt{\frac{4 \cdot x}{x \cdot x}} \qquad \frac{\sqrt{36x}}{\sqrt{9x^2}} \qquad\qquad 2\sqrt{\frac{10}{16}} \qquad \frac{2\sqrt{10}}{\sqrt{16}}$$

$$\frac{\sqrt{4}\sqrt{x}}{\sqrt{x^2}} \quad \text{or} \quad \frac{6\sqrt{x}}{3x} \qquad\qquad \frac{2\sqrt{10}}{4} \quad \text{or} \quad \frac{2\sqrt{10}}{4}$$

$$Ans. \ \frac{2\sqrt{x}}{x} \qquad\qquad\qquad Ans. \ \frac{\sqrt{10}}{2}$$

25. $\dfrac{\sqrt{18}}{\sqrt{2x}}$ **26.** $\dfrac{\sqrt{8}}{\sqrt{2y}}$ **27.** $\dfrac{\sqrt{75}}{\sqrt{3y}}$ **28.** $\dfrac{\sqrt{80}}{\sqrt{5x}}$

29. $\dfrac{a\sqrt{2}}{\sqrt{a}}$ **30.** $\dfrac{b\sqrt{3}}{\sqrt{b}}$ **31.** $\dfrac{4\sqrt{3x}}{\sqrt{8}}$ **32.** $\dfrac{9\sqrt{5x}}{\sqrt{27}}$

33. $\dfrac{a\sqrt{32}}{\sqrt{2a}}$ **34.** $\dfrac{b\sqrt{21}}{\sqrt{3b}}$ **35.** $\dfrac{4y\sqrt{3x}}{\sqrt{4y}}$ **36.** $\dfrac{9x\sqrt{2y}}{\sqrt{27x}}$

Sample Problems   **a.** $\sqrt{\dfrac{20}{3}}$ or $\sqrt{\dfrac{20}{3}}$ **b.** $\sqrt{\dfrac{4}{3x}}$ or $\sqrt{\dfrac{4}{3x}}$

$$\frac{\sqrt{2 \cdot 2 \cdot 5}}{\sqrt{3}} \qquad \sqrt{\frac{2 \cdot 2 \cdot 5 \cdot 3}{3 \cdot 3}} \qquad \frac{2}{\sqrt{3x}} \qquad \sqrt{\frac{2 \cdot 2 \cdot 3x}{3x \cdot 3x}}$$

$$\frac{2\sqrt{5}\sqrt{3}}{\sqrt{3}\sqrt{3}} \qquad \frac{2\sqrt{15}}{\sqrt{9}} \qquad \frac{2\sqrt{3x}}{\sqrt{3x}\sqrt{3x}} \qquad \frac{2\sqrt{3x}}{\sqrt{9x^2}}$$

$$Ans. \ \frac{2\sqrt{15}}{3} \qquad\qquad\qquad Ans. \ \frac{2\sqrt{3x}}{3x}$$

37. $\sqrt{\dfrac{8}{3}}$ **38.** $\sqrt{\dfrac{18}{5}}$ **39.** $\sqrt{\dfrac{9}{8}}$ **40.** $\sqrt{\dfrac{7}{12}}$

41. $\sqrt{\dfrac{72}{5}}$ **42.** $\sqrt{\dfrac{98}{3}}$ **43.** $\sqrt{\dfrac{50}{2x}}$ **44.** $\sqrt{\dfrac{75}{3y}}$

45. $\sqrt{\dfrac{24}{3x}}$ **46.** $\sqrt{\dfrac{32}{4y}}$ **47.** $\sqrt{\dfrac{x^3}{xy}}$ **48.** $\sqrt{\dfrac{y^5}{xy^2}}$

Sample Problems   **a.** $\dfrac{3}{\sqrt{3}+2}$ **b.** $\dfrac{4+\sqrt{2}}{3-\sqrt{5}}$

$$\frac{3(\sqrt{3}-2)}{(\sqrt{3}+2)(\sqrt{3}-2)} \qquad\qquad \frac{(4+\sqrt{2})(3+\sqrt{2})}{(3-\sqrt{5})(3+\sqrt{5})}$$

$$\frac{3\sqrt{3} - 6}{(\sqrt{3})^2 - (2)^2}$$

$$\frac{12 + 4\sqrt{2} + 3\sqrt{2} + 2}{(3)^2 - (\sqrt{5})^2}$$

$$\frac{3\sqrt{3} - 6}{3 - 4}$$

$$\frac{14 + 7\sqrt{2}}{9 - 5}$$

Ans. $6 - 3\sqrt{3}$

Ans. $\dfrac{14 + 7\sqrt{2}}{4}$

49. $\dfrac{1}{\sqrt{2} - 1}$ **50.** $\dfrac{1}{3 - \sqrt{3}}$ **51.** $\dfrac{4 - \sqrt{3}}{\sqrt{3} + 1}$

52. $\dfrac{\sqrt{5} - 3}{2 - \sqrt{5}}$ **53.** $\dfrac{\sqrt{3}}{\sqrt{7} - \sqrt{3}}$ **54.** $\dfrac{2\sqrt{2}}{\sqrt{3} + \sqrt{5}}$

■ *Rationalize each denominator and find a decimal approximation (round off answers to two decimal places).*

Sample Problems

a. $\sqrt{\dfrac{5}{3}}$ b. $4\sqrt{\dfrac{1}{3}}$

$\dfrac{\sqrt{5}\sqrt{3}}{\sqrt{3}\sqrt{3}}$ $4\sqrt{\dfrac{1 \cdot 3}{3 \cdot 3}}$

$\dfrac{\sqrt{15}}{3}$ ($\sqrt{15} \approx 3.873$) $\dfrac{4}{3}\sqrt{3}$ ($\sqrt{3} \approx 1.732$)

$\dfrac{\sqrt{15}}{3} \approx \dfrac{3.873}{3} = 1.291$ $\dfrac{4}{3}\sqrt{3} \approx \dfrac{4}{3}(1.732) = 2.309$

Ans. 1.29   *Ans.* 2.31

55. $\sqrt{\dfrac{1}{7}}$ **56.** $\sqrt{\dfrac{3}{5}}$ **57.** $\dfrac{3}{\sqrt{2}}$ **58.** $\dfrac{2}{\sqrt{3}}$

59. $\dfrac{3}{\sqrt{5}}$ **60.** $\dfrac{2}{\sqrt{6}}$ **61.** $3\sqrt{\dfrac{1}{3}}$ **62.** $5\sqrt{\dfrac{1}{5}}$

11.7 Nth ROOTS: RADICAL NOTATION (OPTIONAL SECTION)*

The work in the preceding sections of this chapter involved only square roots of positive numbers. In this section, and the next, we briefly consider other roots of numbers and symbols that are used to represent them.

* The material in this section may be omitted without loss of continuity. The Cumulative Reviews do not include exercises on this material.

CUBE ROOTS

The **cube root** of a number a is a number that when raised to the third power equals a. For example,

the cube root of 8 is 2 because $2^3 = 8$

and

the cube root of 27 is 3 because $3^3 = 27$

Recall that a negative number such as -4 does not have a square root because there is no real number whose square is 4. However, negative numbers do have cube roots. For example,

the cube root of -8 is -2 because $(-2)^3 = -8$

and

the cube root of -27 is -3 because $(-3)^3 = -27$.

Each real number a has *one and only one* cube root, which we denote $\sqrt[3]{a}$. Thus $\sqrt[3]{8} = 2$ and $\sqrt[3]{-27} = -3$.

Nth ROOTS

Higher roots of numbers are defined in a way similar to that in which square and cube roots are defined.

> The nth root of a number a is a number that, when raised to the nth power, equals a.

We consider two distinct cases.

Case 1   If n is even (as in the square root of a or the fourth root of a), we make the restriction that $a > 0$. Then there are two nth roots of a, one positive and one negative. For example, as we have noted,

the square roots of 4 are 2 and -2 because

$$2^2 = 4 \quad \text{and} \quad (-2)^2 = 4.$$

Also,

the fourth roots of 81 are 3 and -3 because

$$3^4 = 81 \quad \text{and} \quad (-3)^4 = 81.$$

Case 2   If n is odd (as in the cube root of a or the fifth root of a), and a is any real number, then there is *exactly one* nth root. For example,

the cube root of -64 is -4 because $(-4)^3 = -64$.

Also,

the fifth root of 32 is 2 because $2^5 = 32$.

The symbol $\sqrt[n]{a}$ is used to represent the positive nth root of a if n is even, and the one and only nth root of a if n is odd. The number n is called the **index** of the radical. Where $n = 2$, we continue to use $\sqrt{a}$; the index 2 is omitted. Thus we write

$$\sqrt{4} = 2 \quad \text{and} \quad \sqrt[4]{81} = 3,$$
$$\sqrt[5]{32} = 2 \quad \text{and} \quad \sqrt[3]{-64} = -4.$$

EXPRESSIONS INVOLVING VARIABLES

Algebraic expressions that involve radicals with indexes greater than 2 can sometimes be written in simpler form. First consider a number of examples.

$$\sqrt[3]{x^6} = x^2 \quad \text{because} \quad (x^2)^3 = x^6,$$
$$\sqrt[4]{x^{20}} = x^5 \quad \text{because} \quad (x^5)^4 = x^{20},$$
$$\sqrt[5]{x^{10}} = x^2 \quad \text{because} \quad (x^2)^5 = x^{10}.$$

In each example the exponent of the radicand was a multiple of the index, and we obtained the root by dividing the index into the exponent. Thus

$$\sqrt[3]{x^6} = x^{6 \div 3} = x^2.$$

If the exponent of the radicand is not divisible by the index we may still be able to simplify the radical by the following techniques.

SIMPLIFYING Nth ROOTS

In Section 11.3 we saw that $\sqrt{ab} = \sqrt{a}\sqrt{b}$. A similar relationship holds for nth roots; that is

$$\sqrt[n]{ab} = \sqrt[n]{a}\sqrt[n]{b}$$

if all the roots are defined.

We can use this principle to simplify radicals of index greater than 2. For example, to simplify $\sqrt[3]{54}$ we factor 54 into the product of a perfect cube and another factor to get

$$\sqrt[3]{54} = \sqrt[3]{27 \cdot 2}$$
$$= \sqrt[3]{27}\sqrt[3]{2} = 3\sqrt[3]{2}.$$

Similarly, to simplify $\sqrt[5]{96}$ we factor 96 into the product of a fifth power and another factor to obtain

$$\sqrt[5]{96} = \sqrt[5]{32 \cdot 3}$$
$$= \sqrt[5]{32}\sqrt[5]{3} = 2\sqrt[5]{3}.$$

We use the same method when the radicand contains variables. For example,

$$\sqrt[4]{x^{15}} = \sqrt[4]{x^{12} \cdot x^3}$$
$$= \sqrt[4]{x^{12}}\sqrt[4]{x^3} = x^3\sqrt[4]{x^3},$$

where we factored the radicand into two powers, one of which is the largest possible multiple of the index. As another example,

$$\sqrt[3]{24x^6y^4} = \sqrt[3]{8x^6y^3 \cdot 3y}$$
$$= \sqrt[3]{8x^6y^3}\sqrt[3]{3y} = 2x^2y\sqrt[3]{3y}.$$

EXERCISES 11.7

■ *Simplify.*

Sample Problems

a. $\sqrt[3]{64}$ b. $\sqrt[4]{16}$ c. $\sqrt[3]{-8}$ d. $-\sqrt[4]{16}$

Ans. 4   *Ans.* 2   *Ans.* -2 *Ans.* -2

1. $\sqrt[3]{1}$ 2. $\sqrt[4]{1}$ 3. $\sqrt[3]{-1}$ 4. $\sqrt[5]{-1}$

5. $\sqrt[4]{81}$ 6. $\sqrt[3]{125}$ 7. $\sqrt[3]{-125}$ 8. $\sqrt[3]{-64}$

9. $\sqrt[5]{-32}$ 10. $\sqrt[3]{-1000}$ 11. $-\sqrt[4]{256}$ 12. $-\sqrt[5]{243}$

13. $\sqrt[4]{625}$ 14. $\sqrt[3]{343}$ 15. $-\sqrt[3]{-729}$ 16. $-\sqrt[3]{-216}$

Sample Problems

a. $\sqrt[4]{x^{12}}$

$x^{12 \div 4}$

Ans. x^3

b. $\sqrt[5]{a^{23}}$

$\sqrt[5]{a^{20}}\sqrt[5]{a^3}$

Ans. $a^4\sqrt[5]{a^3}$

c. $\sqrt[4]{16x^7y^9}$

$\sqrt[4]{16x^4y^8}\sqrt[4]{x^3y}$

Ans. $2xy^2\sqrt[4]{x^3y}$

17. $\sqrt[6]{x^{12}}$ 18. $\sqrt[4]{x^8}$ 19. $\sqrt[3]{a^{27}}$

20. $\sqrt[4]{a^{16}}$ 21. $\sqrt[5]{-32x^{10}z^{25}}$ 22. $\sqrt[3]{-64a^6b^{15}}$

23. $\sqrt[4]{y^{11}}$ 24. $\sqrt[5]{y^{29}}$ 25. $\sqrt[6]{a^{21}b^7}$

26. $\sqrt[6]{a^{13}b^{11}}$ 27. $\sqrt[3]{250x^{20}y^{19}}$ 28. $\sqrt[3]{81x^{17}y^{22}}$

29. $\sqrt[4]{8x^9y^3}$ 30. $\sqrt[3]{25x^{13}y^5}$ 31. $\sqrt[5]{-3000x^2y^{18}z^{20}}$

32. $\sqrt[5]{-64x^{32}y^3z^9}$ 33. $3b\sqrt[3]{40a^{24}b^{20}}$ 34. $4a^2\sqrt[3]{54a^{18}b^{16}}$

Sample Problems

a. $(\sqrt[3]{7})^3$

b. $(\sqrt[3]{-2})(\sqrt[3]{-2})(\sqrt[3]{-2})$

c. $2(\sqrt[5]{5})^3(\sqrt[5]{5})(3\sqrt[5]{5})$

$2 \cdot 3 \cdot 5$

Ans. 7

Ans. -2

Ans. 30

35. $(\sqrt[3]{-8})^3$

36. $(\sqrt[4]{81})^4$

37. $(\sqrt[4]{9})(\sqrt[4]{9})(\sqrt[4]{9})(\sqrt[4]{9})$

38. $(\sqrt[3]{-4})(\sqrt[3]{-4})(\sqrt[3]{-4})$

39. $(\sqrt[6]{10})^3(\sqrt[6]{10})^3$

40. $(\sqrt[6]{4})^2(\sqrt[6]{4})^2(\sqrt[6]{4})^2$

41. $3(\sqrt[3]{12})^2(5\sqrt[3]{12})$

42. $4(2\sqrt[3]{6})(\sqrt[3]{6})^2$

43. $(2\sqrt[4]{3})^4$

44. $(3\sqrt[4]{2})^4$

45. $(2\sqrt[3]{5})^2(2\sqrt[3]{5})$

46. $(2\sqrt[4]{2})^3(2\sqrt[4]{2})$

11.8   *N*th ROOTS: FRACTIONAL EXPONENTS (OPTIONAL SECTION)*

In some work in mathematics and science, powers with fractional exponents are used instead of radical notation to represent certain numbers. Thus, we make the following definition:

$$a^{1/n} = \sqrt[n]{a}$$

where it is understood that the restrictions on a in $a^{1/n}$ are the same restrictions made for $\sqrt[n]{a}$. Thus

$$9^{1/2} = \sqrt{9} \qquad \text{and} \qquad (-8)^{1/3} = \sqrt[3]{-8},$$

$$\left(\frac{4}{25}\right)^{1/2} = \sqrt{\frac{4}{25}} \qquad \text{and} \qquad (16)^{1/4} = \sqrt[3]{16}.$$

We evaluate fractional exponents just as we evaluate radicals. For example,

$$9^{1/2} = 3 \qquad \text{because} \qquad 3^2 = 9,$$

$$\left(\frac{4}{25}\right)^{1/2} = \frac{2}{5} \qquad \text{because} \qquad \left(\frac{2}{5}\right)^2 = \frac{4}{25},$$

$$(-8)^{1/3} = -2 \qquad \text{because} \qquad (-2)^3 = -8,$$

$$(16)^{1/4} = 2 \qquad \text{because} \qquad 2^4 = 16.$$

Numbers of the form $\sqrt[n]{a}$ or $a^{1/n}$ may be rational or irrational. In the exam-

---

* The material in this section may be omitted without loss of continuity. The Cumulative Reviews do not include exercises on this material.

ples above, the powers $9^{1/2}$, $\left(\dfrac{4}{25}\right)^{1/2}$, $(-8)^{1/3}$ and $16^{1/4}$ are *rational numbers* that can be written as 3, $\dfrac{2}{5}$, -2, and 2, respectively; but the powers $5^{1/3}$, $(-4)^{1/3}$, and $7^{1/4}$ are *irrational numbers*. Decimal approximations for such irrational numbers can be obtained; however, the mathematical procedures are beyond the scope of this text.

The laws of exponents presented in Chapters 4 and 6 also apply to fractional exponents. For example, by applying law II on page 194, we have

$$(8^{1/3})^2 = 8^{2(1/3)} = 8^{2/3}.$$

But since $8^{1/3} = \sqrt[3]{8}$, we also have

$$8^{2/3} = (8^{1/3})^2 = (\sqrt[3]{8})^2,$$

or

$$8^{2/3} = (8^2)^{1/3} = \sqrt[3]{8^2}.$$

These examples suggest the following rule.

$$a^{m/n} = \sqrt[n]{a^m} = (\sqrt[n]{a})^m, \qquad a \geq 0, n \neq 0$$

Thus

$$8^{2/3} = (\sqrt[3]{8})^2 = 2^2 = 4,$$

$$81^{3/4} = (\sqrt[4]{81})^3 = 3^3 = 27,$$

$$16^{5/4} = (\sqrt[4]{16})^5 = 2^5 = 32.$$

Many calculations are simplified by using the laws of exponents. For example,

$$\frac{(5^{1/3})(5^{4/3})}{5^{-2/3}} = 5^{1/3+4/3-(-2/3)} = 5^{7/3},$$

and

$$x^{3/4}(x^{5/4})^2 = x^{3/4}(x^{10/4}) = x^{13/4}.$$

Also, some radicals can be simplified by first writing them in exponential form. For example,

$$\sqrt[4]{6^2} = 6^{2/4} = 6^{1/2} = \sqrt{6},$$

and

$$\sqrt[6]{x^4} = x^{4/6} = x^{2/3} = \sqrt[3]{x^2}.$$

EXERCISES 11.8

■ *Write in radical notation.*

Sample Problems

a. $(8)^{1/4}$

Ans. $\sqrt[4]{8}$

b. $3x^{2/3}$

Ans. $3\sqrt[3]{x^2}$

1. $(7)^{1/3}$ **2.** $(6)^{1/5}$ **3.** $(13)^{4/3}$ **4.** $(20)^{5/2}$

5. $(2a)^{3/4}$ **6.** $(5b)^{5/3}$ **7.** $\left(\dfrac{2x}{3}\right)^{2/5}$ **8.** $\left(\dfrac{3x}{4}\right)^{3/4}$

9. $4x^{5/6}$ **10.** $8y^{4/3}$ **11.** $xy^{3/2}$ **12.** $xy^{2/7}$

■ *Write with fractional exponents.*

Sample Problems

a. $\sqrt[3]{x^7}$

Ans. $x^{7/3}$

b. $\sqrt[5]{3x^4}$

Ans. $3^{1/5}x^{4/5}$

13. $\sqrt[5]{x^3}$ **14.** $\sqrt[4]{x^3}$ **15.** $\sqrt[4]{8^3}$ **16.** $\sqrt[3]{5^2}$

17. $\sqrt[3]{(3x)^4}$ **18.** $\sqrt[5]{(7x)^6}$ **19.** $\sqrt[5]{2x^3}$ **20.** $\sqrt[6]{4x^5}$

21. $\sqrt[3]{x^2y^4}$ **22.** $\sqrt[3]{a^5b^2}$ **23.** $\sqrt[4]{xy^9}$ **24.** $\sqrt[4]{ab^{11}}$

■ *Simplify.*

Sample Problems

a. $(-125)^{1/3}$

b. $\left(\dfrac{8}{27}\right)^{1/3}$

Ans. -5

Ans. $\dfrac{2}{3}$

25. $(64)^{1/3}$ **26.** $(16)^{1/4}$ **27.** $(625)^{1/4}$ **28.** $(125)^{1/3}$

29. $(-243)^{1/5}$ **30.** $(-64)^{1/3}$ **31.** $(256)^{1/4}$ **32.** $(81)^{1/4}$

33. $\left(\dfrac{-27}{125}\right)^{1/3}$ **34.** $\left(\dfrac{-64}{125}\right)^{1/3}$ **35.** $\left(\dfrac{16}{81}\right)^{1/4}$ **36.** $\left(\dfrac{16}{625}\right)^{1/4}$

Sample Problems

a. $(-125x^9)^{1/3}$

b. $(16x^8)^{1/4}$

Ans. $-5x^3$

Ans. $2x^2$

37. $(8x^6)^{1/3}$ **38.** $(27x^9)^{1/3}$ **39.** $(16x^{12})^{1/4}$ **40.** $(81x^{16})^{1/4}$

41. $(64x^6y^{12})^{1/3}$ **42.** $(125x^{15}y^9)^{1/3}$ **43.** $\left(\dfrac{y^6}{-27}\right)^{1/3}$ **44.** $\left(\dfrac{x^3y^6}{125}\right)^{1/3}$

45. $\left(\dfrac{x^4y^{16}}{z^8}\right)^{1/4}$ **46.** $\left(\dfrac{x^4y^8}{z^{12}}\right)^{1/4}$ **47.** $\left(\dfrac{x^9y^{12}}{z^6}\right)^{1/3}$ **48.** $\left(\dfrac{x^{15}y^9}{z^{12}}\right)^{1/3}$

■ *For the given n express the nth root of each number using fractional exponents and state whether the root is rational or irrational.*

Sample Problems   a. 256; $n = 4$

b. 17; $n = 3$

Ans. $256^{1/4}$; rational because $256^{1/4} = 4$

Ans. $17^{1/3}$; irrational

49. 30; $n = 3$ **50.** 33; $n = 3$ **51.** 64; $n = 3$ **52.** 125; $n = 3$
53. 18; $n = 4$ **54.** 22; $n = 4$ **55.** 81; $n = 4$ **56.** 625; $n = 4$
57. -27; $n = 3$ **58.** -125; $n = 3$ **59.** 81; $n = 3$ **60.** 49; $n = 3$

■ *Simplify.*

Sample Problems   a. $(2^{2/3})^{3/4}$

$2^{(2/3)(3/4)} = 2^{2/4}$

b. $\dfrac{a^{5/4} \cdot a^{-1/4}}{a^{3/4}}$

$a^{5/4-1/4-3/4}$

Ans. $2^{1/2}$

Ans. $a^{1/4}$

61. $3^{5/4} \cdot 3^{3/4}$ **62.** $8^{1/5} \cdot 8^{-3/5}$ **63.** $(6^{5/2})^{2/3}$ **64.** $(7^{3/5})^{5/2}$
65. $\dfrac{x^{-2/5}}{x^{-3/5}}$ **66.** $\dfrac{x^{3/4}}{x^{-5/4}}$ **67.** $a^{1/3}(a^{2/3})^3$ **68.** $a^{3/2}(a^{1/2})^4$
69. $(x^{2/3}y^{4/5})^{3/2}$ **70.** $(a^{3/4}b^{6/5})^{2/3}$ **71.** $\dfrac{4^{-3/2} \cdot 4^{5/2}}{4^{-9/2}}$ **72.** $\dfrac{8^{11/3} \cdot 8^{-2/3}}{8^{5/3}}$

Sample Problems   a. $\sqrt[10]{3^5}$

$3^{5/10} = 3^{1/2}$

b. $\sqrt[12]{9^6}$

$9^{6/12} = 9^{1/2}$

Ans. $\sqrt{3}$

Ans. 3

73. $\sqrt[12]{5^4}$ **74.** $\sqrt[12]{6^3}$ **75.** $\sqrt[6]{8^2}$ **76.** $\sqrt[18]{27^6}$
77. $\sqrt[8]{x^4}$ **78.** $\sqrt[16]{x^4}$ **79.** $\sqrt[4]{25}$ **80.** $\sqrt[4]{49}$

CHAPTER SUMMARY

[11.1]   A **square root** of a non-negative number a is a number whose square is a. A positive number has two square roots, the positive or **principal square root** and the negative square root.

For positive numbers a, $\sqrt{a}$ denotes the positive square root of the positive number a, $-\sqrt{a}$ denotes the negative square root, and $\pm\sqrt{a}$ denotes both. Also, $\sqrt{0} = 0$. For all nonnegative numbers a,

$$\sqrt{a}\,\sqrt{a} = (\sqrt{a})^2 = a.$$

In the symbol $\sqrt{a}$, $\sqrt{}$ is called the **radical sign** and a is called the **radicand.**

[11.2]   Numbers that can be expressed as the quotient of two integers are called **rational numbers.** Numbers that can be associated with points on the number line but cannot be expressed as the quotient of two integers are called **irrational numbers. Real numbers** consist of the rational and irrational numbers. Real numbers fill the number line completely.

[11.3-11.5] We can simplify some radical expressions using the following property:

The square root of a product is equal to the product of the square roots of its factors.

This property is useful in either of these two forms:

$$\sqrt{ab} = \sqrt{a}\,\sqrt{b} \quad \text{or} \quad \sqrt{a}\,\sqrt{b} = \sqrt{ab}, \quad (a,b \geq 0).$$

[11.6]   Some radical expressions can be simplified using the following property:

The quotient of two square roots is equal to the square root of the quotient of the radicands.

This property is useful in either of these two forms:

$$\sqrt{\frac{a}{b}} = \frac{\sqrt{a}}{\sqrt{b}} \quad \text{or} \quad \frac{\sqrt{a}}{\sqrt{b}} = \sqrt{\frac{a}{b}}, \quad (a \geq , b > 0).$$

The process of expressing a fraction whose denominator contains a radical as an equivalent fraction whose denominator is free of radicals is called **rationalizing the denominator.**

[11.7]   The **nth root** of a number a is a number whose nth power is a. If n is even and a is positive, then a has two nth roots. The symbol $\sqrt[n]{a}$ is used to represent the positive nth root of a if n is even, and the one and only one nth root of a if n is odd. Some nth roots can be simplified using the following property:

$$\sqrt[n]{ab} = \sqrt[n]{a}\,\sqrt[n]{b}.$$

[11.8]   Fractional exponents can be used to represent nth roots where

$$a^{1/n} = \sqrt[n]{a},$$

and

$$a^{m/n} = \sqrt[n]{a^m} = (\sqrt[n]{a})^m.$$

The laws of exponents apply to powers with fractional exponents.

■ *The symbols introduced in this chapter appear on the inside of the front cover.*

CHAPTER REVIEW

1. Which of the following are irrational numbers?

$$\sqrt{6}, -\sqrt{9}, \sqrt{4}, \sqrt{\frac{9}{25}}, -\sqrt{\frac{2}{3}}$$

2. Using the table of square roots, find a decimal approximation for:

a. $\sqrt{93}$ b. $2\sqrt{47}$ c. $\frac{1}{4}\sqrt{32}$

3. a. Locate the numbers $8, \sqrt{80}, \frac{19}{2}$ on a number line.

b. Graph $x \leq 3$ where x represents a real number.

■ *Simplify each expression in exercises 4–8.*

4. a. $\sqrt{72}$ b. $-\sqrt{90}$ c. $\sqrt{175}$

5. a. $\sqrt{y^4}$ b. $\sqrt{x^{15}}$ c. $\sqrt{x^3y^7}$

6. a. $x\sqrt{27}$ b. $3\sqrt{3x^2}$ c. $\frac{1}{3}\sqrt{27x^2}$

7. a. $\sqrt{7} + 3\sqrt{7}$ b. $4\sqrt{6} - 3\sqrt{6} + \sqrt{6}$ c. $7\sqrt{2} - 3\sqrt{2} + \sqrt{2}$

8. a. $2\sqrt{12} - 4\sqrt{27}$

b. $\sqrt{9x} - \sqrt{4x}$

c. $\sqrt{x^2y} - 3\sqrt{4x^2y} + 7x\sqrt{y}$

9. Express without parentheses.

a. $3(\sqrt{y} - 6)$ b. $y(\sqrt{xy} - 2y)$ c. $\sqrt{3}(\sqrt{2} - \sqrt{6})$

10. Simplify radicals where possible and factor.

a. $2 - \sqrt{8}$ b. $\sqrt{27} - 3\sqrt{5}$ c. $3x^2 - \sqrt{5x^4}$

■ *Simplify each expression.*

11. a. $\dfrac{3 - 2\sqrt{27}}{3}$ b. $\dfrac{6 - 3\sqrt{12}}{3}$ c. $\dfrac{4 - \sqrt{32}}{4}$

12. a. $\dfrac{3}{2} - \dfrac{\sqrt{3}}{2}$ b. $\dfrac{\sqrt{15}}{5} + \dfrac{2\sqrt{15}}{5}$ c. $\dfrac{2}{3} - \dfrac{\sqrt{3}}{6}$

13. a. $\dfrac{\sqrt{5}}{3} - \dfrac{3}{2}$ b. $\dfrac{2\sqrt{3}}{5} - 1$ c. $\dfrac{3\sqrt{6}}{4} + 2$

14. a. $\sqrt{5}\sqrt{3}$ b. $\sqrt{16}\sqrt{32}$ c. $\sqrt{3x^2}\sqrt{6x}$

15. a. $\sqrt{5}\sqrt{2}\sqrt{15}$ b. $(2\sqrt{6})(\sqrt{3})(\sqrt{2})$ c. $(x\sqrt{3})(\sqrt{4x})$

16. a. $\sqrt{3}(2 - \sqrt{2})$ b. $\sqrt{7}(\sqrt{2} - \sqrt{14})$ c. $\sqrt{8}(\sqrt{2} - \sqrt{3})$

17. a. $(2 - \sqrt{3})(2 + \sqrt{3})$

     b. $(\sqrt{5} - \sqrt{7})(\sqrt{5} + \sqrt{7})$

     c. $(2 - \sqrt{3})(3 - 2\sqrt{3})$

■ *Rationalize the denominator and simplify.*

18. a. $\dfrac{\sqrt{27}}{\sqrt{3}}$ b. $\dfrac{\sqrt{12a^2}}{\sqrt{2b}}$ c. $\dfrac{\sqrt{3a}\sqrt{ab}}{\sqrt{b}}$

19. a. $\sqrt{\dfrac{3}{5}}$ b. $\sqrt{\dfrac{12}{5}}$ c. $\sqrt{\dfrac{3}{2x}}$

20. a. $\dfrac{3}{1 + \sqrt{2}}$ b. $\dfrac{\sqrt{2}}{\sqrt{3} - \sqrt{2}}$ c. $\dfrac{\sqrt{5} + \sqrt{3}}{2 - \sqrt{3}}$

EXERCISES FROM OPTIONAL SECTIONS

■ *Simplify.*

21. a. $\sqrt[4]{625}$ b. $\sqrt[3]{-729}$ c. $\sqrt[5]{243}$

22. a. $\sqrt[3]{375x^5y}$ b. $\sqrt[6]{x^{28}y^{36}z^{13}}$ c. $\sqrt[4]{80a^{16}b^3}$

23. a. $\dfrac{x^3}{2}\sqrt[5]{64x^{10}}$ b. $3b\sqrt[3]{81a^9b^{27}}$ c. $\dfrac{2}{xy}\sqrt[4]{8x^8y^{18}}$

24. a. $(\sqrt[3]{-2})^3$ b. $(\sqrt[5]{7})^3(\sqrt[5]{7})^2$ c. $(\sqrt[3]{11})(\sqrt[3]{11})(\sqrt[3]{11})$

25. Write with fractional exponents.

     a. $\sqrt[4]{x^9}$ b. $\sqrt[3]{x^2y^7}$ c. $\sqrt[6]{xy^5}$

26. Write with radical notation.

    a. $3x^{5/3}$ b. $x^{2/5}y^{3/5}$ c. $(2xy^2)^{4/3}$

27. State whether each number is rational or irrational.

    a. $\sqrt[5]{625}$ b. $8^{3/4}$ c. $343^{1/3}$

■ *Simplify.*

28. a. $(125x^{15})^{1/3}$ b. $(-64x^3y^{81})^{1/3}$ c. $(81x^{16}y^8)^{1/4}$

29. a. $8^{1/4}(8^{3/4})^2$ b. $\dfrac{x^{2/5} \cdot x^{-3/5}}{x^{-6/5} \cdot x^{3/5}}$ c. $(8x^{3/2}y^3)^{2/3}$

30. a. $\sqrt[12]{5^8}$ b. $\sqrt[9]{x^3}$ c. $\sqrt[6]{36}$

CUMULATIVE REVIEW

1. A positive number has _?_ square roots.

2. If $a = -4$, $b = 2$, and $c = 0$, find the value of $\dfrac{2a - 4}{3 - bc}$.

3. Multiply: $(-3ab)(2c)(b^2)$.

4. Represent $\dfrac{a - b}{2} + \dfrac{a - 3b}{3}$ as a single fraction.

5. Reduce to lowest terms: $\dfrac{3a^2x - 3ax}{24a^2x^2 + 60ax^2}$.

6. Divide: $(x^4 + 3x^3 - x^2)$ by (x^2).

7. Factor: $4x^2 - 12x + 9$.

8. Solve: $3(x - 2) - 7 = 5(2x + 5) - 3$.

9. Solve: $\dfrac{y - 6}{5} = \dfrac{12y + 3}{5} - 3y + 3$.

10. The sum of two numbers is 146. If n represents the smaller of the two numbers, represent the larger number in terms of n.

11. Write in terms of x, the number of cents in x quarters.

12. Solve the system: $\dfrac{x}{2} - \dfrac{y}{3} = 1$

$$x - \dfrac{2y}{3} = 2.$$

13. Write in scientific notation.

    a. 3,270,000   b. 0.0000649

14. Solve: $2x^2 = 5x + 3$.

15. The sum of two numbers is 154. If the larger number is eight less than twice the smaller, what are the numbers?

16. What is the slope of the line with the equation $3x + 2y = 5$?

17. Simplify: $\sqrt{18} - 2\sqrt{5} - \sqrt{20} + \sqrt{50}$.

18. Simplify: $(2 - \sqrt{3})(2 + \sqrt{3})$.

19. Rationalize the denominator and simplify: $\dfrac{\sqrt{4x}\,\sqrt{9x}}{\sqrt{6x}}$.

20. Rationalize the denominator and simplify: $\dfrac{3\sqrt{2y}}{\sqrt{3y}}$.

■ *True or false.*

21. $\dfrac{x^2 + x^4}{x^2}$ is equivalent to $1 + x^2$.

22. $\sqrt{25 + 81} = \sqrt{25} + \sqrt{81}$

23. $(4xy^{-2})^0 = 0$

24. $4^{-2} = \dfrac{1}{8}$

25. x^{-2} is equivalent to $-2x$.

12 SOLVING QUADRATIC EQUATIONS BY OTHER METHODS

In Chapter 10, we solved quadratic equations whose solutions are rational numbers by using factoring methods. Now that we have studied irrational numbers, we are ready to examine additional methods of solving quadratic equations.

12.1 EXTRACTION OF ROOTS

We can solve an equation of the form $x^2 - a = 0$ by first writing it in the form

$$x^2 = a. \tag{1}$$

We see that for a greater than or equal to zero, x is a number which, when multiplied by itself, yields a. Therefore, by the definition of a square root, x must be a square root of a. Since a has two square roots, we have $\sqrt{a}$ and $-\sqrt{a}$ as solutions for Equation (1). That is,

$$(\sqrt{a})^2 = \sqrt{a}\,\sqrt{a} = a,$$

and

$$(-\sqrt{a})^2 = (-\sqrt{a})(-\sqrt{a}) = a.$$

For example, if

$$x^2 = 5, \quad \text{then} \quad x = \pm\sqrt{5}.$$

As another example, consider the equation

$$\frac{2}{3}x^2 - 4 = 0,$$

which is equivalent to

$$\frac{2}{3} x^2 = 4,$$

$$x^2 = 6.$$

Therefore, $x = \pm \sqrt{6}$.

This method of solving equations is sometimes referred to as **extraction of roots.** Note that to apply this method, we must get the squared term on one side of the equation and the constant term on the other.

We can also use extraction of roots to solve quadratic equations of the form

$$(x + k)^2 = d.$$

For example, from

$$(x - 2)^2 = 9,$$

we obtain

$$x - 2 = \pm \sqrt{9} = \pm 3.$$

Solving for x yields the equations:

$$x - 2 = 3, \qquad \text{or} \qquad x - 2 = -3,$$

$$x = 3 + 2, \qquad\qquad x = -3 + 2,$$

$$x = 5, \qquad\qquad x = -1.$$

The solutions are $x = 5$, $x = -1$.

EXERCISES 12.1

■ *Solve.*

Sample Problem

$3y^2 = 48$

      Write with y^2 as left-hand member.

$y^2 = 16$

      Extract square root of each member.

$y = \pm \sqrt{16}$

      Simplify.

$y = \pm 4$

Ans. $y = 4$; $y = -4$

1. $x^2 = 4$	**2.** $y^2 = 9$	**3.** $x^2 - 16 = 0$
4. $x^2 - 25 = 0$	**5.** $98 = 2x^2$	**6.** $12 = 3x^2$
7. $x^2 - 3 = 0$	**8.** $7 - z^2 = 0$	**9.** $y^2 - 10 = 0$
10. $3x^2 - 15 = 0$	**11.** $4x^2 - 24 = 0$	**12.** $7x^2 = 42$

13. $12 = z^2$

14. $24 = b^2$

15. $x^2 - 18 = 0$

16. $3x^2 - 24 = 0$

17. $3x^2 - 54 = 0$

18. $5x^2 - 100 = 0$

Sample Problem

$5x^2 - 3 = 2x^2 + 33$

$\qquad$ Write with x^2 as left-hand member.

$3x^2 = 36$

$x^2 = 12$

$\qquad$ Extract square root of each member.

$x = \pm\sqrt{12}$

$\qquad$ Simplify radical expression.

Ans. $x = 2\sqrt{3}; \quad x = -2\sqrt{3}$

19. $2x^2 - 4 = x^2$

20. $6x^2 + 3 = 4x^2 + 11$

21. $3 = x^2 - 2$

22. $4t^2 - 16 = 16$

23. $4y^2 - 10 = y^2 - 10$

24. $s^2 - 2 = 4 - s^2$

Sample Problem

$$\frac{1}{3}x^2 - \frac{3}{4} = 0$$

$\qquad$ Multiply each member by L.C.D. 12.

$$(\cancel{12})\overset{4}{\frac{1}{\cancel{3}}}x^2 - (\cancel{12})\overset{3}{\frac{3}{\cancel{4}}} = (12)0$$

$\qquad$ Write with x^2 as left-hand member.

$$4x^2 - 9 = 0$$

$$4x^2 = 9$$

$$x^2 = \frac{9}{4}$$

$\qquad$ Extract square root of each member.

$$x = \pm\frac{3}{2}$$

Ans. $x = \frac{3}{2}; \quad x = \frac{-3}{2}$

25. $\dfrac{1}{4}x^2 = 5$

26. $\dfrac{2}{3}y^2 - 4 = 0$

27. $\dfrac{2}{3}x^2 = 6$

28. $\dfrac{2x^2}{3} - 4 = \dfrac{x^2}{3}$

29. $\dfrac{5x^2}{2} - 4 = 2x^2$

30. $\dfrac{1}{2}x^2 - 4 = \dfrac{3}{2}$

■ *Given* $x^2 + y^2 = z^2$ *and values for two of the variables, find values for the third variable.*

Sample Problem

$y = 4, z = 5$

Substitute 4 for y and 5 for z in $x^2 + y^2 = z^2$.

$x^2 + (4)^2 = (5)^2$

Simplify.

$x^2 + 16 = 25$

Write with x^2 as left-hand member.

$x^2 = 9$

Extract square root of each member.

$x = \pm 3$

Ans. $x = 3;\quad x = -3$

31. $x = 15, y = 20$ **32.** $x = 10, y = 24$ **33.** $x = 6, z = 10$
34. $y = 9, z = 41$ **35.** $x = 5, y = 5$ **36.** $x = 11, z = 61$

■ *Solve for x.*

Sample Problem

$(x + 3)^2 = 25$

Extract square root of each member.

$x + 3 = \pm 5$

Solve resulting first-degree equations.

$x = \pm 5 - 3$

$x = +5 - 3;\quad x = -5 - 3$

Ans. $x = 2,\quad x = -8$

37. $(x - 1)^2 = 4$ **38.** $(x + 3)^2 = 9$ **39.** $(x - 2)^2 = 25$
40. $(x + 1)^2 = 36$ **41.** $(x - 5)^2 = 1$ **42.** $(x + 7)^2 = 1$
43. $(x - a)^2 = 25$ **44.** $(x + b)^2 = 4$ **45.** $(x - 3)^2 = a^2$
46. $(x + 5)^2 = b^2$ **47.** $(x - a)^2 = b^2$ **48.** $(x + a)^2 = b^2$

Sample Problem

$(x - 2)^2 = 20$

Extract square root of each member.

$x - 2 = \pm\sqrt{20}$

Solve resulting first-degree equations.

$x - 2 = \pm 2\sqrt{5}$

$x = 2 \pm 2\sqrt{5}$

Ans. $x = 2 + 2\sqrt{5};\quad x = 2 - 2\sqrt{5}$

49. $(x + 3)^2 = 2$ **50.** $(x - 2)^2 = 3$ **51.** $(x + 5)^2 = 5$

52. $(x - 6)^2 = 7$ **53.** $(x + 10)^2 = 8$ **54.** $(x - 1)^2 = 12$

55. $(x - 5)^2 = a$ **56.** $(x + 2)^2 = a$ **57.** $(x + 1)^2 = b$

58. $(x - 7)^2 = b$ **59.** $(x - b)^2 = a$ **60.** $(x + b)^2 = a$

Sample Problem

$(2x + 3)^2 = 49$

Extract square root of each member.

$2x + 3 = \pm 7$

Solve resulting first-degree equations.

$2x = -3 \pm 7$

$2x = -3 + 7; \quad 2x = -3 - 7$

$2x = 4; \quad 2x = -10$

Ans. $x = 2; \quad x = -5$

61. $(3x + 4)^2 = 25$ **62.** $(2x - 7)^2 = 9$ **63.** $(5x - 2)^2 = 121$

64. $(4x + 3)^2 = 144$ **65.** $(2x + 1)^2 = 8$ **66.** $(3x - 2)^2 = 12$

67. $(6x - 2)^2 = 13$ **68.** $(5x + 4)^2 = 17$ **69.** $3(4x - 3)^2 = 75$

70. $2(2x - 3)^2 = 128$ **71.** $3(2x - 8)^2 = 60$ **72.** $2(3x - 6)^2 = 96$

■ *Solve each of the following equations for the indicated variable.*

Sample Problem

$K = \dfrac{v^2}{64},$ for v

Multiply each member by 64.

$(64)K = (64)\dfrac{v^2}{64}$

$64K = v^2$

$v^2 = 64K$

Extract the square root of each member and simplify.

$v = \pm\sqrt{64K}$

$v = \pm\sqrt{64}\,\sqrt{K}$

$v = \pm 8\sqrt{K}$

Ans. $v = 8\sqrt{K}; \quad v = -8\sqrt{K}$

73. $x^2 - a = 0,$ for x **74.** $b = x^2,$ for x

75. $\dfrac{x^2}{3} - b^2a^3 = 0,$ for x **76.** $\dfrac{ax^2}{2} - b = 0,$ for x

77. $\dfrac{2y^2}{5} = \dfrac{b}{3}$, for y

78. $\dfrac{y^2}{2} + a = \dfrac{2a}{3} + 2y^2$, for y

79. $s = \dfrac{1}{2}gt^2$, for t

80. $V = \dfrac{1}{3}\pi r^2 h$, for r

81. $A = 4\pi r^2$, for r

82. $C = bh^2 r$, for h

83. $I = \dfrac{3k}{d^2}$, for d

84. $F = \dfrac{k}{d^2}$, for d

12.2 COMPLETING THE SQUARE

So far, the methods we have used to solve quadratic equations apply to special cases only. Let us now develop a method that is applicable to any quadratic equation.

First, we examine the squares of several binomials expressed as trinomials.

$$(x + 1)^2 = x^2 + x + x + 1 = x^2 + 2x + 1;$$

$$(x - 2)^2 = x^2 - 2x - 2x + 4 = x^2 - 4x + 4;$$

$$\left(x + \frac{1}{3}\right)^2 = x^2 + \frac{1}{3}x + \frac{1}{3}x + \frac{1}{9} = x^2 + \frac{2}{3}x + \frac{1}{9};$$

$$\left(x - \frac{2}{5}\right)^2 = x^2 - \frac{2}{5}x - \frac{2}{5}x + \frac{4}{25} = x^2 - \frac{4}{5}x + \frac{4}{25}.$$

In general, we have

$$(x + k)^2 = x^2 + kx + kx + k^2 = x^2 + 2kx + k^2.$$

The coefficient of x is twice the second term of the binomial.

The constant term is the square of the second term of the binomial.

We use these observations to perform a process called **completing the square.** For example, if we have the expression $x^2 + 6x$, we want to find a value k^2, such that when we add k^2 to $x^2 + 6x$, we get the square of a binomial. That is, we want to find a value of k^2 such that

$$x^2 + 6x + k^2 = (x + k)^2.$$

From the examples above, we know that 6 (the coefficient of x) must be twice k. Therefore, k must be $\dfrac{1}{2} \cdot 6$ or 3. Now, since k is 3, k^2 must be 3^2 or 9, and we can write

$$x^2 + 6x + 9 = (x + 3)^2.$$

In general, if we are given an expression of the form $x^2 + bx$, we can *complete the square* by adding $\left[\dfrac{1}{2} b\right]^2$.

$$x^2 + bx + \left[\dfrac{1}{2} b\right]^2$$

Add to complete
the square.

We can now write this trinomial as the square of a binomial:

$$\left(x + \dfrac{1}{2} b\right)^2.$$

We can now write any quadratic equation in the form $(x + k)^2 = d$, which we learned to solve in the preceding section. For example, consider the equation

$$x^2 - 6x - 7 = 0.$$

We first write the equation in the form

$$x^2 - 6x \quad = 7.$$

Now, by adding 9 [the *square* of $\dfrac{1}{2}(-6)$] to each member, we have

$$x^2 - 6x + 9 = 7 + 9,$$

or

$$(x - 3)^2 = 16.$$

We can now solve by extraction of roots to obtain

$$x - 3 = \pm\sqrt{16} = \pm 4$$

$$x - 3 = 4 \quad \text{or} \quad x - 3 = -4$$

$$x = 7 \quad \text{or} \quad x = -1.$$

Thus the solutions are 7 or -1.

In the event the coefficient on the second-degree term is not 1, we must divide each term in the equation by that coefficient before proceeding. Thus, to solve

$$2x^2 - 3x - 9 = 0,$$

we begin by dividing each term by 2 to obtain

$$\dfrac{2x^2}{2} - \dfrac{3x}{2} - \dfrac{9}{2} = \dfrac{0}{2},$$

or

$$x^2 - \dfrac{3}{2}x - \dfrac{9}{2} = 0.$$

Now, we will add $\dfrac{9}{2}$ to each member to obtain

$$x^2 - \frac{3}{2}x = \frac{9}{2}.$$

We can complete the square by adding $\dfrac{9}{16}$ [the *square* of $\dfrac{1}{2}\left(\dfrac{-3}{2}\right)$] to each member to get

$$x^2 - \frac{3}{2}x + \frac{9}{16} = \frac{9}{2} + \frac{9}{16}.$$

We can now express the left-hand member as the square of a binomial and simplify the right-hand member. We obtain

$$\left(x - \frac{3}{4}\right)^2 = \frac{(8)9}{(8)2} + \frac{9}{16}$$

$$\left(x - \frac{3}{4}\right)^2 = \frac{81}{16}.$$

Extracting the roots, we obtain

$$x - \frac{3}{4} = \pm\sqrt{\frac{81}{16}} = \pm\frac{9}{4}$$

$$x - \frac{3}{4} = \frac{9}{4} \quad \text{or} \quad x - \frac{3}{4} = \frac{-9}{4}$$

$$x = \frac{9}{4} + \frac{3}{4} \qquad\qquad x = \frac{-9}{4} + \frac{3}{4}$$

$$= \frac{12}{4} = 3 \qquad\qquad = \frac{-6}{4} = \frac{-3}{2}.$$

Thus, the solutions are $x = 3$ or $x = \dfrac{-3}{2}$.

The method of completing the square is summarized as follows.

To Solve an Equation by Completing the Square:

1. Write the equation in standard form.
2. If the coefficient of the second-degree term is not 1, divide each term in the equation by this coefficient.
3. Write the equation with the constant term in the right-hand member.
4. Add to each member the square of one half the coefficient of the first-degree term.

5. Rewrite the equation with the left-hand member expressed as a square of a binomial; simplify the right-hand member.
6. Solve by extraction of roots.

To become familiar with the method of completing the square, solve all equations in the following exercises by this method even though many can be solved by factoring, which is usually easier.

EXERCISES 12.2

■ *Find the number that must be added to each binomial to form an expression that is a perfect square trinomial. Write the expression in the form $(x + b)^2$.*

Sample Problems

a. $x^2 + 8x$

b. $x^2 - 9x$

Square one half the coefficient of x.

$\left(\dfrac{8}{2}\right)^2 = 16;$

$\left(\dfrac{-9}{2}\right)^2 = \dfrac{81}{4};$

Ans. $x^2 + 8x + 16 = (x + 4)^2$

Ans. $x^2 - 9x + \dfrac{81}{4} = \left(x - \dfrac{9}{2}\right)^2$

1. $x^2 + 2x$ **2.** $x^2 + 4x$ **3.** $x^2 - 6x$

4. $x^2 - 8x$ **5.** $x^2 - 10x$ **6.** $x^2 - 14x$

7. $x^2 - 12x$ **8.** $x^2 + 20x$ **9.** $x^2 + 3x$

10. $x^2 - 7x$ **11.** $x^2 + 11x$ **12.** $x^2 + \dfrac{2}{3}x$

13. $x^2 - \dfrac{4}{3}x$ **14.** $x^2 + \dfrac{3}{2}x$

■ *Solve by completing the square.*

Sample Problem

$x^2 - x - 2 = 0$

Rewrite equation with constant term in the right-hand member.

$x^2 - x \qquad = 2$

Square one half the coefficient of x and add to each member; $\left[\dfrac{1}{2}(-1)\right]^2 = \dfrac{1}{4}.$

$x^2 - x + \dfrac{1}{4} = 2 + \dfrac{1}{4}$

Rewrite left-hand member as a perfect square; simplify right-hand member.

$$\left(x - \frac{1}{2}\right)^2 = \frac{9}{4}$$

Extract square root of each member.

$$x - \frac{1}{2} = \pm\frac{3}{2}$$

Solve resulting first-degree equations.

$$x - \frac{1}{2} = \frac{3}{2} \qquad \text{or} \qquad x - \frac{1}{2} = \frac{-3}{2}$$

$$x = \frac{3}{2} + \frac{1}{2} \qquad\qquad x = \frac{-3}{2} + \frac{1}{2}$$

Ans. $x = 2; \quad x = -1$

15. $x^2 + 4x - 12 = 0$ 16. $x^2 - 2x - 15 = 0$ 17. $z^2 - 2z + 1 = 0$
18. $y^2 - y - 6 = 0$ 19. $y^2 + y - 20 = 0$ 20. $x^2 - x - 20 = 0$
21. $x^2 + 3x + 2 = 0$ 22. $u^2 + 5u + 6 = 0$ 23. $z^2 - 3z - 4 = 0$
24. $y^2 + 9y + 20 = 0$ 25. $r^2 - 3r - 10 = 0$ 26. $p^2 - 5p + 4 = 0$
27. $x^2 - 2x - 1 = 0$ 28. $y^2 + 4y = 4$ 29. $z^2 = 3z + 3$
30. $s^2 + 1 = -3s$ 31. $t^2 = 3 - t$ 32. $-x^2 - 6x = 1$

Sample Problem

$3x^2 - 2x - 5 = 0$

Divide each term by 3, the coefficient of x^2, and rewrite with constant in right-hand member.

$$x^2 - \frac{2}{3}x \quad = \frac{5}{3}$$

Square one half the coefficient of x and add to each member; $\left[\frac{1}{2}\left(\frac{-2}{3}\right)\right]^2 = \frac{1}{9}$.

$$x^2 - \frac{2}{3}x + \frac{1}{9} = \frac{5}{3} + \frac{1}{9}$$

Rewrite left-hand member as a perfect square; simplify right-hand member.

$$\left(x - \frac{1}{3}\right)^2 = \frac{(3)5}{(3)3} + \frac{1}{9} = \frac{16}{9}$$

Extract square root of each member.

$$x - \frac{1}{3} = \pm\sqrt{\frac{16}{9}} = \pm\frac{4}{3}$$

Solve resulting first-degree equations.

$$x - \frac{1}{3} = \frac{4}{3} \qquad \text{or} \qquad x - \frac{1}{3} = \frac{-4}{3}$$

$$x = \frac{4}{3} + \frac{1}{3} \qquad\qquad x = \frac{-4}{3} + \frac{1}{3}$$

Ans. $x = \frac{5}{3}; \quad x = -1$

33. $4x^2 + 4x - 3 = 0$ **34.** $4y^2 - 4y = 3$ **35.** $2x^2 = 2 - 3x$

36. $6z^2 + 6 = 13z$ **37.** $2t^2 - t - 15 = 0$ **38.** $1 - r = 6r^2$

12.3   QUADRATIC FORMULA

If we solve the general quadratic equation

$$ax^2 + bx + c = 0, \quad a \neq 0$$

by completing the square, we can obtain a formula expressing the solutions of the equation in terms of a, b, and c, where a is the coefficient of the squared term, b is the coefficient of the linear term, and c is the constant term. We can then solve any quadratic equation by simply substituting the numerical coefficients of the terms in the formula and evaluating the result.

We complete the square in the general quadratic equation as follows:

$$ax^2 + bx + c = 0$$

$$ax^2 + bx = -c,$$

$$x^2 + \frac{b}{a}x = -\frac{c}{a},$$

$$x^2 + \frac{b}{a}x + \frac{b^2}{4a^2} = -\frac{c}{a} + \frac{b^2}{4a^2},$$

$$\left(x + \frac{b}{2a}\right)^2 = \frac{-c(4a)}{a(4a)} + \frac{b^2}{4a^2},$$

$$\left(x + \frac{b}{2a}\right)^2 = \frac{b^2 - 4ac}{4a^2},$$

$$x + \frac{b}{2a} = \pm \sqrt{\frac{b^2 - 4ac}{4a^2}},$$

$$x = \frac{-b}{2a} \pm \frac{\sqrt{b^2 - 4ac}}{2a},$$

$$x = \frac{-b \pm \sqrt{b^2 - 4ac}}{2a}.$$

The last equation is called the **quadratic formula.** Since this formula was developed from the quadratic equation in standard form, we should write any quadratic equation in standard form before attempting to determine values for a, b, and c to substitute in the formula. Furthermore, the sign on the coefficient must be substituted with the coefficient. For example, we first write the equation

$$-x + 3x^2 = 4$$

in the form

$$3x^2 - x - 4 = 0,$$

from which

$$a = 3, \qquad b = -1, \qquad \text{and} \qquad c = -4.$$

We can now solve for x by substituting these values into the quadratic formula.

$$x = \frac{-b \pm \sqrt{b^2 - 4ac}}{2a},$$

$$x = \frac{-(-1) \pm \sqrt{(-1)^2 - 4(3)(-4)}}{2(3)}$$

$$= \frac{1 \pm \sqrt{1 - (-48)}}{6} = \frac{1 \pm \sqrt{49}}{6} = \frac{1 \pm 7}{6}.$$

Simplifying, we obtain

$$x = \frac{1 + 7}{6} \qquad \text{or} \qquad x = \frac{1 - 7}{6},$$

$$x = \frac{8}{6} = \frac{4}{3} \qquad\qquad x = \frac{-6}{6} = -1.$$

Common Error

Note that the denominator $2a$ in the quadratic formula must be divided into *both* terms of the numerator $-b \pm \sqrt{b^2 - 4ac}$. Thus in the example above it is *incorrect* to write

$$x = 1 \pm \frac{\sqrt{49}}{6}.$$

If a quadratic equation has fractional coefficients, it is generally helpful to clear the equation of fractions before proceeding. For example, if we have

$$\frac{2}{3} - \frac{1}{2}x = -2x^2,$$

we multiply each term by the L.C.D. 6 to get

$$(6)\frac{2}{3} - (6)\frac{1}{2}x = (6)(-2x^2),$$

$$4 - 3x = -12x^2,$$

from which

$$12x^2 - 3x + 4 = 0$$

and

$$a = 12, \qquad b = -3, \qquad \text{and} \qquad c = 4.$$

If we apply the quadratic formula to the equation

$$x^2 + 2x + 2 = 0, \tag{1}$$

we find that

$$x = \frac{-b \pm \sqrt{b^2 - 4ac}}{2a}$$

$$= \frac{-2 \pm \sqrt{(2)^2 - 4(1)(2)}}{2(1)}$$

$$= \frac{-2 \pm \sqrt{4 - 8}}{2} = \frac{-2 \pm \sqrt{-4}}{2}.$$

Recall that the square root of a negative number, such as $\sqrt{-4}$, is not a meaningful expression for us, since there is no real number whose square is -4. Since $\sqrt{-4}$ is not a real number, we say that Equation (1) has no real number solutions.

In actual practice, we should use the quadratic formula only when easier methods (factoring or extraction of roots) fail. Many of the following exercises are easier to solve by other methods, but they are included to indicate the complete generality of the quadratic formula. These exercises should be solved by use of the formula.

EXERCISES 12.3

■ *Indicate the values for a, b, and c to be substituted in the quadratic formula.*

Sample Problems

a. $x^2 = x + 2$

$$x^2 - x - 2 = 0$$

Ans. $a = 1, b = -1, c = -2$

b. $2x^2 = x$

Write in standard form.

$$2x^2 - x = 0$$

Ans. $a = 2, b = -1, c = 0$

1. $x^2 - 3x + 2 = 0$
2. $y^2 + 5y + 4 = 0$
3. $x^2 - x - 30 = 0$
4. $y^2 + 3y - 4 = 0$
5. $x^2 - 2x = 0$
6. $y^2 = 5y$
7. $4y^2 - 3 = 0$
8. $2y^2 - 1 = 0$
9. $2x^2 = 7x - 6$
10. $6x^2 + x = 1$
11. $6x^2 = 5x - 1$
12. $3x^2 - 5 = 0$
13. $y^2 + 4 = 8y$
14. $x^2 = 7x$

Sample Problem

$$\frac{x}{3} = 4 - \frac{x^2}{2}$$

Multiply each term by L.C.D. 6.

$$(\overset{2}{\cancel{6}})\frac{x}{\cancel{3}} = (6)4 - (\overset{3}{\cancel{6}})\frac{x^2}{\cancel{2}}$$

$$2x = 24 - 3x^2$$

Write in standard form.

$$3x^2 + 2x - 24 = 0$$

Ans. $a = 3, b = 2, c = -24$

15. $x^2 = x + \dfrac{1}{2}$ **16.** $x^2 = \dfrac{15}{4} - x$

17. $2x^2 - 1 + \dfrac{7}{3}x = 0$ **18.** $y^2 + 1 = \dfrac{13}{6}y$

19. $\dfrac{9}{4}y^2 + \dfrac{3}{2}y - 2 = 0$ **20.** $\dfrac{x^2}{3} = \dfrac{x}{2} + \dfrac{3}{2}$

■ *Solve by use of the quadratic formula.*

Sample Problem

$$x^2 - x - 3 = 0$$

$$x = \frac{-b \pm \sqrt{b^2 - 4ac}}{2a}$$

Substitute 1 for a, -1 for b, and -3 for c.

$$x = \frac{-(-1) \pm \sqrt{(-1)^2 - 4(1)(-3)}}{2(1)}$$

Perform indicated operations.

$$x = \frac{1 \pm \sqrt{1 + 12}}{2}$$

$$x = \frac{1 \pm \sqrt{13}}{2}$$

Ans. $x = \dfrac{1 + \sqrt{13}}{2}$; $x = \dfrac{1 - \sqrt{13}}{2}$

21. $x^2 - 3x + 2 = 0$ **22.** $y^2 + 5y + 4 = 0$ **23.** $z^2 - 4z - 12 = 0$

24. $x^2 - x - 30 = 0$ **25.** $x^2 + 2x - 15 = 0$ **26.** $y^2 + 3y - 4 = 0$

27. $x^2 + 3x - 1 = 0$ **28.** $y^2 + 5y + 5 = 0$ **29.** $y^2 - 3y - 2 = 0$

30. $x^2 + x - 1 = 0$

Sample Problem

$$x^2 - 9 = 0$$

$$x = \frac{-b \pm \sqrt{b^2 - 4ac}}{2a}$$

Substitute 1 for a, 0 for b, and -9 for c.

$$x = \frac{-(0) \pm \sqrt{(0)^2 - 4(1)(-9)}}{2(1)}$$

Perform indicated operations.

$$x = \frac{\pm \sqrt{36}}{2}$$

Simplify.

$$x = \frac{+6}{2}; \quad x = \frac{-6}{2}$$

Ans. $x = 3; \quad x = -3$

31. $x^2 - 2x = 0$ (*Hint.* $c = 0$)   **32.** $x^2 - 4 = 0$ (*Hint.* $b = 0$)

33. $y^2 = 5y$ **34.** $z^2 = 9$

35. $7x = x^2$ **36.** $16 = y^2$

37. $z^2 - 3z = 0$ **38.** $x^2 = 1$

39. $4y^2 - 3 = 0$ **40.** $2y^2 - 1 = 0$

Sample Problem $2x^2 = 2 - 3x$

Write in standard form.

$$2x^2 + 3x - 2 = 0$$

$$x = \frac{-b \pm \sqrt{b^2 - 4ac}}{2a}$$

Substitute 2 for a, 3 for b, and -2 for c.

$$x = \frac{-(3) \pm \sqrt{(3)^2 - 4(2)(-2)}}{2(2)}$$

Perform indicated operations.

$$x = \frac{-3 \pm \sqrt{9 + 16}}{4}$$

$$x = \frac{-3 \pm \sqrt{25}}{4}$$

Simplify.

$$x = \frac{-3 + 5}{4}; \quad x = \frac{-3 - 5}{4}$$

Ans. $x = \frac{1}{2}; \quad x = -2$

41. $2x^2 = 7x - 6$ **42.** $5 = 6y - y^2$ **43.** $6x^2 + x = 1$

44. $-z = 3 - 2z^2$ **45.** $6x^2 - 13x - 5 = 0$ **46.** $6x^2 = 5x - 1$

47. $x^2 = 2x + 1$ **48.** $x^2 = 2x + 4$ **49.** $y^2 - 4y - 2 = 0$

50. $z^2 + 4 = 8z$ **51.** $2x^2 - 3x - 1 = 0$ **52.** $3x^2 - x - 1 = 0$

Sample Problem
$$\frac{x^2}{3} = \frac{1}{3} - \frac{x}{2}$$

Multiply by L.C.D. 6.

$$(6)\overset{2}{\frac{x^2}{\cancel{3}}} = (6)\overset{2}{\frac{1}{\cancel{3}}} - (6)\overset{3}{\frac{x}{\cancel{3}}}$$

$$2x^2 = 2 - 3x$$

Write in standard form. Solve as shown in preceding sample problem.

53. $x^2 = \dfrac{15}{4} - x$ **54.** $2x^2 - 1 + \dfrac{7}{3}x = 0$

55. $y^2 + 1 = \dfrac{13}{6}y$ **56.** $\dfrac{9}{4}y^2 + \dfrac{3}{2}y - 2 = 0$

57. $\dfrac{1}{3}x^2 = \dfrac{1}{2}x + \dfrac{3}{2}$ **58.** $\dfrac{3}{5}x^2 - x - \dfrac{2}{5} = 0$

12.4   GRAPHING QUADRATIC EQUATIONS IN TWO VARIABLES

In Chapter 8, we learned how to graph first-degree equations in two variables. We can now use similar procedures to graph second-degree equations of the form

$$y = ax^2 + bx + c.$$

The graph of such an equation is called a **parabola** (see Figure 12.1).

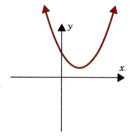 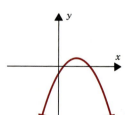

Figure 12.1

GRAPHING WITH ORDERED PAIRS

Since the graph of a second-degree equation of this form is not a straight line, we must plot more than two points to determine the graph. For example, consider the equation

$$y = x^2 - 2x - 3. \tag{1}$$

For x-components, we use all integers between -2 and 4. To find the second component for each ordered pair, we substitute the given value of x into the expression $x^2 - 2x - 3$.

For $x = -2$, $y = (-2)^2 - 2(-2) - 3 = 5$; $(-2, 5)$

For $x = -1$, $y = (-1)^2 - 2(-1) - 3 = 0$; $(-1, 0)$

For $x = 0$, $y = (0)^2 - 2(0) - 3 = -3$; $(0, -3)$

For $x = 1$, $y = (1)^2 - 2(1) - 3 = -4$; $(1, -4)$

For $x = 2$, $y = (2)^2 - 2(2) - 3 = -3$; $(2, -3)$

For $x = 3$, $y = (3)^2 - 2(3) - 3 = 0$; $(3, 0)$

For $x = 4$, $y = (4)^2 - 2(4) - 3 = 5$; $(4, 5)$

Each of these ordered pairs is a solution to the equation. For convenience, we can list the solutions in a table. We then graph the ordered pairs and connect them with a smooth curve, as shown in Figure 12.2.

x	y
-2	5
-1	0
0	-3
1	-4
2	-3
3	0
4	5

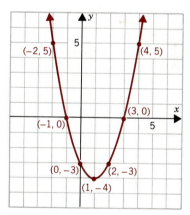

Figure 12.2

As another example, consider the equation

$$y = -x^2 - 4x - 5. \tag{2}$$

For x-components, we use all integers between -5 and 1. To find the y-components, we substitute each x-value into the expression $-x^2 - 4x - 5$. You should verify the results of these calculations, which are listed in the table on page 400. The graph of the parabola is illustrated in Figure 12.3.

x	y
-5	-10
-4	-5
-3	-2
-2	-1
-1	-2
0	-5
1	-10

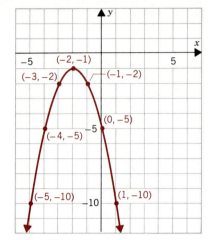

Figure 12.3

SKETCHING PARABOLAS

The graphs of all parabolas of the form $y = ax^2 + bx + c$ have a number of features in common. If we can identify these features from the equation we can use them to sketch the graph quickly and easily. First notice that each of the graphs in the foregoing examples has a maximum (high) point or a minimum (low) point called the **vertex**. If the coefficient of x^2 is *positive*, as in Equation (1) on page 399, the graph opens *upwards* from a minimum point. If the coefficient of x^2 is *negative*, as in Equation (2) on page 399, the graph opens *downwards* from a maximum point.

INTERCEPTS

The point where a parabola crosses the y-axis is called its **y-intercept**. Since the x-coordinate of the y-intercept is zero, we can find the y-intercept by setting x equal to zero in the equation of the parabola. For our first example

$$y = x^2 - 2x - 3, \tag{1}$$

we find that when $x = 0$

$$y = (0)^2 - 2(0) - 3 = -3,$$

so the y-intercept is -3. (Notice that in general for $y = ax^2 + bx + c$, the y-intercept of the parabola will be c.)

A parabola may cross the x-axis in zero, one, or two points called **x-intercepts**. Since the y-coordinate of any point on the x-axis is zero, we can find the x-intercepts by solving the equation $0 = ax^2 + bx + c$ obtained by setting y equal to zero. For the previous example, the x-intercepts are the solutions to the equation

$$0 = x^2 - 2x - 3,$$

or

$$0 = (x + 1)(x - 3).$$

Thus the parabola crosses the x-axis at $(-1, 0)$ and $(3, 0)$.

VERTEX

To find the coordinates of the vertex, we use the technique of completing the square to write the equation in a more useful form. First consider our example

$$y = x^2 - 2x - 3. \tag{1}$$

Since the graph opens upwards, the vertex is the low point; that is, its y-coordinate will be the minimum for all the points on the graph. We rewrite the equation as follows.

$$y = (x^2 - 2x \quad) - 3,$$

$$y = (x^2 - 2x + 1) - 3 - 1,$$

$$y = (x - 1)^2 - 4. \tag{1'}$$

We obtained the $+1$ in the parentheses by completing the square on the binomial $x^2 - 2x$. By adding $+1$ and -1 we do not change the original equation. However, the minimum value for y can now be determined easily from Equation (1'). Since the squared term $(x - 1)^2$ is never negative, the smallest value it can have is zero, which occurs when $x = 1$. Therefore $y = (1 - 1)^2 - 4 = -4$ is the smallest y-value that can occur. Thus the minimum point or vertex of the parabola is the point $(1, -4)$.

   Just as we obtained the quadratic formula by completing the square on the general quadratic, $y = ax^2 + bx + c$, we can obtain a formula for the coordinates of the vertex. Thus,

$$y = a\left(x^2 + \frac{b}{a}x \quad\right) + c$$

$$= a\left(x^2 + \frac{b}{a}x + \frac{b^2}{4a^2}\right) + c - \frac{b^2}{4a^2}$$

$$= a\left(x + \frac{b}{2a}\right)^2 + c - \frac{b^2}{4a^2}.$$

The minimum value for y occurs when $(x + b/2a)^2 = 0$, or when $x = -b/2a$. If a is negative, we find the maximum value for y when $x = -b/2a$.

> **The vertex of the graph of $y = ax^2 + bx + c$ has x-coordinate given by $x = -b/2a$.**

We find the y-coordinate of the vertex by substituting this value for x into the equation. For our example

$$y = x^2 - 2x - 3, \tag{1}$$

the x-coordinate of the vertex is

$$x = \frac{-b}{2a} = \frac{-(-2)}{2(1)} = 1,$$

and the y-coordinate is

$$y = (1)^2 - 2(1) - 3 = -4,$$

so the vertex is the point $(1, -4)$.

AXIS OF SYMMETRY

Once we have found the intercepts of the parabola and its vertex, we can use them as a guide to sketch the graph. Every parabola of the form

$$y = ax^2 + bx + c$$

is symmetric about a vertical line called the **axis of symmetry,** which passes through the vertex. In particular, the x-intercepts, if they exist, lie on either side of the axis, at equal distances from it.

The significant features of the graph of our example $y = x^2 - 2x - 3$ are shown in Figure 12.4.

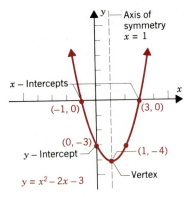

Figure 12.4

We summarize our procedure for sketching parabolas as follows.

To graph $y = ax^2 + bx + c$:
1. The parabola opens upwards if $a > 0$ and downwards if $a < 0$.
2. a. The x-coordinate of the vertex is given by $x = -b/2a$.

4. $(2, ?)$ **5.** $(-2, ?)$ **6.** $(3, ?)$

■ *Find a second component such that each of the ordered pairs satisfies the equation* $y = x^2 + x - 2$.

7. $(0, ?)$ **8.** $(-1, ?)$ **9.** $(1, ?)$

10. $(2, ?)$ **11.** $(-2, ?)$ **12.** $(-3, ?)$

■ *Find a second component such that each of the ordered pairs satisfies the equation* $y = x^2 - 7x + 12$.

13. $(0, ?)$ **14.** $(1, ?)$ **15.** $(2, ?)$

16. $(3, ?)$ **17.** $(4, ?)$ **18.** $(5, ?)$

19. Graph the ordered pairs obtained in exercises 1–6 and connect the points with a smooth curve.

20. Graph the ordered pairs obtained in exercises 7–12 and connect the points with a smooth curve.

■ *Graph each quadratic equation by following the five steps on page 402. Label the significant features of the graph.*

Sample Problem

$y = x^2 - 2x + 1$

Step 1   Since the coefficient of x^2 is 1, the parabola opens upwards.

Step 2   The x-coordinate of the vertex is

$$x = \frac{-b}{2a} = \frac{-(-2)}{2(1)} = 1,$$

and the y-coordinate is

$$y = (1)^2 - 2(1) + 1 = 0,$$

so the vertex is the point $(1, 0)$.

Step 3   When $x = 0$,

$$y = (0)^2 - 2(0) + 1 = 1,$$

so the y-intercept is 1.

Step 4   To find the x-intercepts, set y equal to zero.

$$0 = x^2 - 2x + 1.$$

Solve for x.

$$0 = (x - 1)(x - 1).$$

There is only one x-intercept, 1.

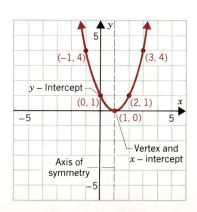

Step 5   Since the *x*-intercept is the same point as the vertex, we will find a third point to aid us in graphing. The point $(0, 1)$ lies one unit to the left of the axis of symmetry, the line $x = 1$. So there must be another point on the parabola with *y*-coordinate 1 and one unit to the right of the axis. This is the point $(2, 1)$. If we wish, we can find additional ordered pairs from the equation:

$$\text{when } x = -1, \quad y = (-1)^2 - 2(-1) + 1 = 4; \quad (-1, 4).$$

$$\text{when } x = 3, \quad y = (3)^2 - 2(3) + 1 = 4; \quad (3, 4).$$

Graph these ordered pairs and connect them with a smooth curve.

21. $y = x^2 - 2x$ **22.** $y = x^2 + 2x$ **23.** $y = x^2 - 4$

24. $y = x^2 - 1$ **25.** $y = 9 - x^2$ **26.** $y = 4 - x^2$

27. $y = x^2 + 1$ **28.** $y = x^2 + 3$ **29.** $y = x^2$

30. $y = -x^2$ **31.** $y = -2x^2$ **32.** $y = 2x^2$

33. $y = x^2 + 4x + 4$ **34.** $y = x^2 - 6x + 9$ **35.** $y = -x^2 + 4x - 3$

36. $y = 8 + 2x - x^2$ **37.** $y = x^2 + 6x + 11$ **38.** $y = x^2 - 2x + 4$

39. $y = 12 - x - x^2$ **40.** $y = 10 + 3x - x^2$

12.5   THE PYTHAGOREAN THEOREM

A particularly useful application of quadratic equations is illustrated in problems involving the sides of a right triangle. The early Greeks proved that in any right triangle, the sum of the squares of the lengths of the shorter sides (called **legs**) of a right triangle is equal to the square of the length of the longest side (called the **hypotenuse**). Thus, in Figure 12.6,

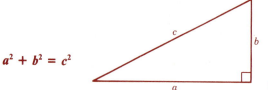

$$a^2 + b^2 = c^2$$

Figure 12.6

This relationship is known as the **Pythagorean theorem,** in honor of the Greek mathematician Pythagoras.

We can use the theorem to help us solve word problems that in some way concern right triangles. Again, we will follow the six steps outlined on page 84.

> Find the length of the diagonal of a rectangle whose length is 6 meters and whose width is 4 meters.

Steps 1–2   We represent what we want to find as a word phrase and in terms of a variable.

Length of diagonal of rectangle: c

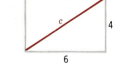

Step 3   A sketch is useful.

Step 4   Using the Pythagorean theorem and substituting 4 for a and 6 for b in $c^2 = a^2 + b^2$, we obtain

$$c^2 = (4)^2 + (6)^2$$

Step 5   Solving yields

$$c^2 = 16 + 36$$
$$c = \pm \sqrt{52}$$
$$c = \pm 2\sqrt{13}$$

Step 6 $2\sqrt{13}$ is the only meaningful answer. The length of the diagonal is $2\sqrt{13}$ meters.

EXERCISES 12.5

■ *Solve.*

Sample Problem

Find the width of a rectangle whose length is 8 centimeters and whose diagonal is 10 centimeters long.

Steps 1–2   Width of rectangle: b

Step 3   A sketch is helpful.

Step 4   Substitute 8 for a and 10 for c in $a^2 + b^2 = c^2$.

$$(8)^2 + b^2 = (10)^2$$

Step 5 $64 + b^2 = 100$

$$b^2 = 100 - 64$$
$$b^2 = 36$$
$$b = \pm 6$$

Step 6   **Ans.** 6 is the only meaningful solution. The width is 6 centimeters.

1. Find the length of the diagonal of a rectangle whose length is 4 centimeters and whose width is 3 centimeters.

2. Find the length of the diagonal of a rectangle whose length is 12 meters and whose width is 5 meters.

3. Find the length of the diagonal of a square whose side is 3 kilometers in length.

4. Find the length of the diagonal of a square whose side is 2 meters in length.

5. A baseball diamond is a square whose sides are 90 feet in length. Find the straight-line distance from home plate to second base. (Use the table of square roots and find the length to the nearest foot.)

6. Find the length of the diagonal of a square whose side is a millimeters in length.

7. Find the length of a rectangle whose width is 5 inches and whose diagonal is 13 inches long.

8. Find the length of a rectangle whose diagonal is 20 meters long and whose width is 12 meters.

9. Find the width of a rectangle whose diagonal is 11 yards long and whose length is 9 yards.

10. Find the width of a rectangle whose length is 5 feet and whose diagonal is 6 feet long.

Sample Problem

The length of a rectangle is 2 meters greater than the width. The diagonal is 10 meters long. Find the dimensions of the rectangle.

Steps 1–2   Width of rectangle: x

Length of rectangle: $x + 2$

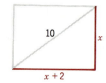

Step 3   A sketch is helpful.

Step 4   Substitute x for a, $x + 2$ for b, and 10 for c in $a^2 + b^2 = c^2$.

$$(x)^2 + (x + 2)^2 = (10)^2$$

Step 5 $x^2 + x^2 + 4x + 4 = 100$

$2x^2 + 4x - 96 = 0$

$2(x - 6)(x + 8) = 0$

$x - 6 = 0 \qquad x + 8 = 0$

$x = 6 \qquad\quad x = -8$

Step 6   **Ans.** 6 is the only meaningful solution. The width is 6 meters, and the length is $6 + 2 = 8$ meters.

11. The length of a rectangle is 3 meters greater than the width, and the diagonal is 15 meters in length. Find the dimensions of the rectangle.

12. The width of a rectangle is 7 centimeters less than the length, and the diagonal is 13 centimeters in length. Find the dimensions of the rectangle.

13. The width of a rectangle is 3 millimeters less than the length, and the square of the length of the diagonal is 29 millimeters. Find the dimensions of the rectangle.

14. The length of a rectangle is 5 inches greater than the width, and the square of the length of the diagonal is 73 inches. Find the dimensions of the rectangle.

15. The length of a rectangle is twice the width, and the diagonal is $3\sqrt{5}$ yards in length. Find the dimensions of the rectangle.

16. The length of a rectangle is three times the width, and the diagonal is $2\sqrt{10}$ feet in length. Find the dimensions of the rectangle.

Sample Problem   A 25-foot ladder is placed against a wall so that its foot is 7 feet from the foot of the wall. How far up the wall does the ladder extend?

Steps 1–2   Height up the wall: a

Step 3   A sketch is helpful.

Step 4   Substitute 7 for b and 25 for c in $a^2 + b^2 = c^2$.

$$a^2 + (7)^2 = (25)^2$$

Step 5 $a^2 + 49 = 625$

$a^2 = 576$

$a = \pm\sqrt{576}$

$a = \pm 24$

Step 6   **Ans.** 24 is the only meaningful solution. The ladder extends 24 feet up the wall.

17. How long must a wire be to stretch from the top of a 40-meter telephone pole to a point on the ground 30 meters from the foot of the pole?

18. How high on a building will a 25-foot ladder reach if its foot is 15 feet from the wall against which the ladder is to be placed?

19. If a 30-meter pine tree casts a shadow of 30 meters, how far is it from the tip of the shadow to the top of the tree?

20. One leg of a right triangle is 6 centimeters longer than the other, and the hypotenuse has a length of 6 centimeters less than twice that of the shorter leg. Find the lengths of the sides of the right triangle.

12.6   RADICAL EQUATIONS (OPTIONAL SECTION)*

A **radical equation** is one in which the variable appears in the radicand of some radical expression. For example,

$$\sqrt{x - 3} = 4. \tag{1}$$

If the radical involved is a square root, we can solve the equation by squaring both sides to produce an equation without radicals. Thus, for Equation (1),

$$(\sqrt{x - 3})^2 = (4)^2,$$

$$x - 3 = 16,$$

$$x = 19.$$

The technique of squaring both sides of an equation does not always produce an equivalent equation. For example, consider the equation

$$\sqrt{x + 2} = -3. \tag{2}$$

Squaring both sides gives us

$$(\sqrt{x + 2})^2 = (-3)^2,$$

$$x + 2 = 9,$$

$$x = 7.$$

However, substituting 7 for x into Equation (2) shows that it is not a solution:

$$\sqrt{7 + 2} \stackrel{?}{=} -3,$$

$$\sqrt{9} \stackrel{?}{=} -3,$$

$$3 \neq -3.$$

The value $x = 7$ is a solution to the squared equation but *not* to the original equation. In this case, $x = 7$ is called an **extraneous solution;** Equation (2) does not have a solution. Thus whenever we square both sides of an equation, we must check the solutions in the original equation.

---

* The material in this section may be omitted without loss of continuity. The Cumulative Reviews do not include exercises on this material.

If a radical equation involves several terms, it is easiest to isolate the radical term in one member of the equation before squaring both sides. To solve the equation

$$4 + \sqrt{8 - 2x} = x, \tag{3}$$

we first isolate the radical by subtracting 4 from each member to get

$$4 + \sqrt{8 - 2x} - 4 = x - 4,$$
$$\sqrt{8 - 2x} = x - 4,$$

and then square both sides to obtain

$$(\sqrt{8 - 2x})^2 = (x - 4)^2,$$
$$8 - 2x = x^2 - 8x + 16,$$
$$0 = x^2 - 6x + 8,$$
$$0 = (x - 4)(x - 2).$$

Thus $x = 4$ or $x = 2$. Checking these solutions in Equation (3), we find that for $x = 4$,

$$4 + \sqrt{8 - 2(4)} \overset{?}{=} 4$$
$$4 + \sqrt{0} \overset{?}{=} 4$$
$$4 = 4,$$

and for $x = 2$,

$$4 + \sqrt{8 - 2(2)} \overset{?}{=} 2$$
$$4 + \sqrt{4} \overset{?}{=} 2$$
$$6 \neq 2.$$

Thus $x = 2$ is an extraneous solution, so the solution to Equation (3) is $x = 4$.

Some radical equations must be squared twice in order to eliminate the square roots. For example,

$$\sqrt{x} - \sqrt{2x - 2} = 1. \tag{4}$$

It is usually best to begin by isolating the more complicated radical; in this case we write

$$\sqrt{x} - 1 = \sqrt{2x - 2}.$$

When squaring both sides, we must be careful to square the *entire* left-hand member, using binomial multiplication. It is *not* correct to square each term separately. Thus

$$(\sqrt{x} - 1)^2 = (\sqrt{2x - 2})^2,$$
$$x - 2\sqrt{x} + 1 = 2x - 2.$$

The equation still contains a radical term, $-2\sqrt{x}$, so we isolate this radical and square again.

$$x - 2\sqrt{x} + 1 - x - 1 = 2x - 2 - x - 1$$

$$-2\sqrt{x} = x - 3$$

$$(-2\sqrt{x})^2 = (x - 3)^2$$

$$4x = x^2 - 6x + 9$$

$$0 = x^2 - 10x + 9$$

$$0 = (x - 9)(x - 1).$$

Thus $x = 9$ or $x = 1$. Checking these solutions in Equation (4), we find that for $x = 9$,

$$\sqrt{9} - \sqrt{2(9) - 2} \stackrel{?}{=} 1$$

$$\sqrt{9} - \sqrt{16} \stackrel{?}{=} 1$$

$$3 - 4 \neq 1,$$

and for $x = 1$,

$$\sqrt{1} - \sqrt{2(1) - 2} \stackrel{?}{=} 1$$

$$\sqrt{1} - \sqrt{0} \stackrel{?}{=} 1$$

$$1 - 0 = 1.$$

Thus $x = 9$ is an extraneous solution, and the solution to Equation (4) is $x = 1$.

EXERCISES 12.6

■ *Solve.*

Sample Problem

$2\sqrt{x + 1} = 6$

Isolate the radical in one member.

$\sqrt{x + 1} = 3$

Square both sides.

$(\sqrt{x + 1})^2 = 3^2$

$x + 1 = 9$

Ans. $x = 8$

Check. $2\sqrt{8 + 1} \stackrel{?}{=} 6$

$2(3) = 6$

1. $\sqrt{x + 4} = 5$ **2.** $\sqrt{x - 6} = 2$

3. $\sqrt{x} - 4 = 5$ **4.** $\sqrt{x} + 3 = 9$

5. $6 - \sqrt{x} = 8$ **6.** $5 - \sqrt{x} = 6$

7. $2 + 3\sqrt{x - 1} = 8$ **8.** $1 + 2\sqrt{x + 3} = 7$

9. $2\sqrt{3x + 1} - 3 = 5$ **10.** $3\sqrt{5x - 1} - 5 = 16$

Sample Problem

$\sqrt{x - 3} = x - 5$

Square both sides.

$(\sqrt{x - 3})^2 = (x - 5)^2$

$x - 3 = x^2 - 10x + 25$

Write in standard form and solve.

$0 = x^2 - 11x + 28$

$0 = (x - 7)(x - 4)$

$x = 7, x = 4$

Check. $\sqrt{7 - 3} \overset{?}{=} 7 - 5$ $\sqrt{4 - 3} \overset{?}{=} 4 - 5$

$\sqrt{4} \overset{?}{=} 2$ $\sqrt{1} \overset{?}{=} -1$

$2 = 2$ $1 \neq -1$

Ans. $x = 7$

11. $\sqrt{x} = 3 - 2x$ **12.** $\sqrt{x} = 2x - 6$

13. $\sqrt{x + 4} + 2 = x$ **14.** $\sqrt{x + 8} - 2 = x$

15. $x + \sqrt{2x + 7} = -2$ **16.** $\sqrt{3x + 10} - 2x = 6$

17. $\sqrt{x + 7} = 2x + 4$ **18.** $\sqrt{x + 10} = 2x - 1$

19. $6 + \sqrt{5x - 4} - x = 4$ **20.** $x + \sqrt{2x + 3} - 7 = 9$

Sample Problem

$\sqrt{2x + 4} = \sqrt{x + 3} + 1$

Square both sides.

$(\sqrt{2x + 4})^2 = (\sqrt{x + 3} + 1)^2$

$2x + 4 = x + 3 + 2\sqrt{x + 3} + 1$

Isolate the radical in the right-hand member.

$2x + 4 - x - 4 = x + 3 + 2\sqrt{x + 3} + 1 - x - 4$

$x = 2\sqrt{x + 3}$

Square both sides.

$x^2 = (2\sqrt{x + 3})^2$

$x^2 = 4(x + 3)$

Write in standard form and solve.

$$x^2 - 4x - 12 = 0$$

$$(x - 6)(x + 2) = 0$$

$$x = 6, \quad x = -2$$

Check. $\sqrt{2(6) + 4} \overset{?}{=} \sqrt{6 + 3} + 1$ $\quad$ $\sqrt{2(-2) + 4} \overset{?}{=} \sqrt{-2 + 3} + 1$

$$\sqrt{16} \overset{?}{=} \sqrt{9} + 1 \qquad\qquad \sqrt{0} \overset{?}{=} \sqrt{1} + 1$$

$$4 = 3 + 1 \qquad\qquad\qquad 0 \neq 1 + 1$$

Ans. $x = 6$

21. $\sqrt{x - 7} = 7 - \sqrt{x}$ $\qquad\qquad$ **22.** $\sqrt{x - 3} = 3 - \sqrt{x}$

23. $\sqrt{4x + 5} - \sqrt{x} = 2$ $\qquad\quad$ **24.** $\sqrt{3x - 2} - \sqrt{x} = 2$

25. $\sqrt{2x - 5} = 1 + \sqrt{x - 3}$ $\qquad$ **26.** $\sqrt{3x - 5} = 2 + \sqrt{x - 3}$

27. $2 + \sqrt{x} + \sqrt{x + 3} = 0$ $\qquad$ **28.** $1 + \sqrt{x + 1} + \sqrt{x - 1} = 0$

29. $\sqrt{3x + 1} - \sqrt{2x - 6} = 2$ $\quad$ **30.** $\sqrt{3x + 1} - \sqrt{x - 4} = 3$

CHAPTER SUMMARY

[12.1] The equation $x^2 = a$ has two solutions: $\sqrt{a}$ and $-\sqrt{a}$. This direct method of solving equations of this form is called **extraction of roots.**

[12.2] We can write any quadratic equation in the form $(x + k)^2 = d$ by the method of **completing the square** (see page 390).

[12.3] The **quadratic formula**

$$x = \frac{-b \pm \sqrt{b^2 - 4ac}}{2a}$$

can be used to solve any quadratic equation, where a, b, and c are the coefficients in the quadratic equation $ax^2 + bx + c = 0$.

[12.4] The graph of an equation of the form

$$y = ax^2 + bx + c$$

is called a **parabola.** To graph the equation, we locate the **vertex** and the **x-** and **y-intercepts.** The parabola is symmetric about a vertical line called the **axis of symmetry.**

[12.5] In any right triangle, the sum of the squares of the lengths of the shorter sides, a and b, is equal to the square of the length of the longest side, c, called the

hypotenuse. This relationship, called the **Pythagorean theorem,** is given in symbols by the equation

$$a^2 + b^2 = c^2.$$

[12.6] To solve a **radical equation,** we square both sides of the equation to eliminate the radicals. We check for **extraneous solutions** which do not satisfy the original equation.

CHAPTER REVIEW

■ *Solve for x, y, or z by any method.*

1. a. $x^2 - 25 = 0$ b. $3x^2 - 27 = 0$

2. a. $6x^2 - 42 = 0$ b. $\dfrac{2y^2}{3} = 4$

3. a. $(z - 2)^2 = 9$ b. $(z + 3)^2 = 1$

4. a. $(x - 7)^2 = 16$ b. $(x - a)^2 = c^2$

5. a. $(x + 3)^2 = a$ b. $(y + a)^2 = 4$

6. a. $x^2 + 3x - 4 = 0$ b. $y^2 = 3y + 3$

7. a. $2x^2 + 4x + 2 = 0$ b. $y^2 - y = 2$

8. Solve by completing the square: $y^2 - 4y = 5$.

■ *In exercises 9–11, solve each equation using the quadratic formula.*

9. $\dfrac{x^2}{4} = \dfrac{15}{4} - \dfrac{x}{2}$ **10.** $x^2 + 3x + 1 = 0$ **11.** $\dfrac{x^2}{4} + 1 = \dfrac{13}{12}x$

12. Graph: $y = x^2 + 3x$.

13. Graph: $y = x^2 + 3x - 4$.

14. Graph: $y = -x^2 + x + 6$.

15. If $x^2 + y^2 = z^2$, and $x = 12$, $z = 20$ find y.

16. If $a^2 + b^2 = c^2$, and $a = 6$, $b = 6$, find c.

17. Find the length of the diagonal of a rectangle whose width is 7 meters and whose length is 9 meters.

18. Find the length of the diagonal of a square with sides 8 centimeters long.

19. A cable is to be stretched from a point 18 meters up a pole to a point on the ground 24 meters from the base of the pole. If the pole is mounted on level ground, what is the length of the stretched cable?

20. The length of a rectangle is 3 times the width and the diagonal is $5\sqrt{10}$ meters in length. Find the dimensions of the rectangle.

EXERCISES FROM OPTIONAL SECTION

■ *Solve.*

21. $\sqrt{x+2} - 1 = 2$ **22.** $\sqrt{x-3} + 4 = 2$ **23.** $\sqrt{2x+1} = x - 7$

24. $2 + \sqrt{4x+1} = \sqrt{9x+7}$ **25.** $\sqrt{x-1} + \sqrt{x+4} = 5$

CUMULATIVE REVIEW

1. Write $24x^3y^2$ in completely factored form.

2. If $x = -2$ and $y = -4$, find the value of $\dfrac{x^2 + 2x - 4}{y}$.

3. For what value of x is $\dfrac{3x-2}{x-1}$ meaningless?

4. Simplify: $\dfrac{6a^3b^2}{3ab^2}$.

5. Solve for x: $3x - a = \dfrac{2ax - a^2}{a}$.

6. Solve the system: $3x + 4y = 11$
$2x - y = 0$.

7. Represent $\dfrac{3}{x+y} + \dfrac{4}{x^2+xy}$ as a single fraction.

8. Find three ordered pairs that are solutions of $2x - y = 4$.

9. Simplify: $3x\sqrt{x^2y} + 2\sqrt{x^4y}$.

10. The sum of two numbers is a. If one number is x, the other is __?__.

11. How many solutions has the equation $2x^2 - 8 = 0$?

12. What conclusions can be drawn concerning a and b if $ab = 0$?

13. The sum of two numbers is 18. If one number is -9, the other is __?__.

14. One number is four times another. If their sum is 40, find the numbers.

15. The sum of two numbers is 16, and their product is 63. Find the numbers.

16. The first angle of a triangle is one half of the second, and the third is equal to the sum of the other two. Find the size of each angle.

17. The area of a triangle is 18 square centimeters. If the length of the base is four times the length of the altitude, find the base and altitude of the triangle.

18. The perimeter of a triangle is 30 inches. If the second side is 2 inches longer than the first, and the first is two thirds of the third, find the length of each side.

19. The length of a rectangle is twice the width, and its area is 242 square centimeters. Find the dimensions of the rectangle.

20. The base of a triangle is three times its altitude. If the triangle contains the same area as a rectangle whose length and width are 4 centimeters and 6 centimeters, respectively, find the base and altitude of the triangle.

■ *True or false.*

21. $\sqrt{36 + 25} = 6 + 5$ **22.** 8 divided by $\dfrac{1}{2}$ is 4.   **23.** $\dfrac{1}{\sqrt{2}} = \dfrac{\sqrt{2}}{2}$

24. $(\sqrt{x} + \sqrt{y})^2$ is equivalent to $x + y$.   **25.** $\dfrac{4 + x^2}{4x}$ is equivalent to x.

FINAL CUMULATIVE REVIEWS

Review I

1. List all prime numbers between 70 and 100.

2. If a and b are natural numbers, which of the following statements are true for all values of a and b?

$$a + b = b + a \qquad a - b = b - a \qquad ab = ba \qquad \frac{a}{b} = \frac{b}{a}$$

3. Arrange the numbers $6, -2, -4, 5, 3, -1$ in order, from smallest to largest.

4. Write an equation expressing the word sentence: "The volume (V) of a cylindrical silo is equal to π times the square of the radius (r) times the height (h)."

5. If a 30-meter rope is cut into two pieces and x represents one of the pieces, how could the second piece be represented in terms of x?

6. Simplify: $2a(a - 1) + a(a + 2) - a^2$.

7. Divide $(6x^2 + 3x - 1)$ by $3x$.

8. Find two consecutive positive odd integers whose product is 195.

9. Simplify: $\dfrac{2ab - 4b^2}{a^2 - 4b^2}$.

10. Represent $\dfrac{3}{4} + \dfrac{3}{4a}$ using a single fraction.

11. Simplify: $\dfrac{3xy^2}{4a} \cdot \dfrac{2xa^2}{5y}$.

12. Solve: $\dfrac{3}{x} = \dfrac{6}{12 - x}$.

13. Graph the following ordered pairs on a rectangular coordinate system.

$$(2, 6); (3, -4); (-5, -5); (-5, 0).$$

14. Solve the system: $y = x$

$$3x + 2y = 5.$$

15. Where does the graph of $3x + 4y = 12$ cross the x-axis?

16. Simplify: $\dfrac{\sqrt{4^4 y^3}}{8^2 y}$.

17. For what values of y will the fraction $\dfrac{5(y-2)}{y+3}$ equal 0?

18. An approximate value of $\sqrt{3}$ is 1.73. Find an approximation for $\sqrt{12}$ without using the table of square roots.

19. Simplify: $(\sqrt{2} - \sqrt{5})(\sqrt{2} + \sqrt{5})$.

20. Side b of a triangle is 2 centimeters longer than a second side a. Side c is twice as long as side a. How long is each side of the triangle if its perimeter is 34 centimeters?

Review II

1. Simplify: $\dfrac{5^2 + 5}{5} - \dfrac{3^2 + 3}{3}$.

2. Division by __?__ is meaningless.

3. State whether $(\frac{2}{3}, 1)$ is a solution of $3x - 4y = 6$.

4. The product of two numbers is 24. If one of the numbers is b, represent the other number in terms of b.

5. Simplify: $(a + b) - (2a - b) + (a + 3b)$.

6. Solve: $7(a + 3) < a - 9$.

7. Two trucks deliver 47 tons of gravel to a building site. The larger truck carries 5 tons more than the smaller, and makes only 1 trip while the smaller truck makes 5 trips. What is the capacity of each truck?

8. Represent $\dfrac{a}{b} + \dfrac{b}{a}$ using a single fraction.

9. Express $\dfrac{3}{x + 3}$ as an equivalent fraction with denominator $x^2 - 9$.

10. Simplify: $\dfrac{2x}{3y} \div \dfrac{x^2}{y}$.

11. A statement that two ratios are equal is called a __?__.

12. Graph $2x + 3y \geq 6$.

13. Which equation has a straight line for a graph?

$$y = x^2 - 1 \qquad x + y = 1 \qquad x = y^2 - 1$$

14. Solve the system: $2y - x = 7$
$$y + 2x = 1.$$

15. Solve: $(x - 3)^2 = 25$.

16. Which of the following numbers are irrational?

$$\sqrt{4} \quad \sqrt{5} \quad \sqrt{6} \quad \sqrt{7} \quad \sqrt{8} \quad \sqrt{9} \quad \sqrt{10}$$

17. Simplify: $\sqrt{160x^3y^3z}$.

18. Simplify: $\dfrac{8 - \sqrt{32}}{8}$.

19. Solve: $x^2 + 5x + 2 = 0$.

20. How far from the foot of a vertical pole will a 14-meter wire reach if the other end of the wire is tied to the pole at a height of 6 meters?

Review III

1. Express 360 in completely factored form.

2. If $a = -2$, $b = 1$, $c = 3$, find the value of $a^2bc - abc^2$.

3. Simplify: $a(a + 1) - 2(a^2 + a) - a$.

4. State whether -2 is a solution of $4x + 4 = 2x - 4$.

5. What is the area of a square whose perimeter is 100 millimeters?

6. Factor completely: $abc - ab$.

7. Factor completely: $6x^2 - 3x - 9$.

8. In a collection of coins, there are 6 more dimes than nickels. If n represents the number of nickels, represent the total value of the collection (in cents) in terms of n.

9. Represent $\dfrac{a - 3b}{5} - \dfrac{a + 3b}{5}$ using a single fraction.

10. Represent $(2x)^{-1} + (3x)^{-1}$ using a single fraction.

11. Divide $(x^2 - 6x - 16)$ by $(x + 2)$.

12. Simplify: $\dfrac{ab}{a^2 - b^2} \div \dfrac{ab}{2a - 2b}$.

13. The denominator of a certain fraction is four more than the numerator and the fraction is equivalent to $\frac{5}{6}$. Find the numerator of the fraction.

14. If 228 bricks are required for 4 meters of a wall, how many bricks will be required for 10 meters?

15. First-degree equations are called $\underline{\ ?\ }$ equations because their graphs are straight lines.

16. Solve for y explicitly in terms of x: $x = \dfrac{3y - 2}{4}$.

17. What value would be substituted for b in the quadratic formula when solving the equation $x^2 + x - 5 = 0$?

18. Simplify: $\sqrt{36a^6b^4c^5}$.

19. Simplify: $4\sqrt{a} + \sqrt{4a} - \sqrt{9a}$.

20. How far from the foot of a vertical pole will a 14-meter wire reach if the other end of the wire is tied to the pole at a height of 6 meters?

Review IV

1. Graph the first ten prime numbers on a line graph.

2. If r and s represent two numbers: (a) What is their sum? (b) What is their product?

3. If $4a$ and $-2b$ are two factors of $24ab$, what is the third factor?

4. If $a = 3$, $b = 2$, $c = -1$, find the value of $\dfrac{ab}{c^2} - \dfrac{a+c}{b}$.

5. Find three consecutive even integers whose sum is -78.

6. Simplify: $(x - b)(x - 2b) - (x^2 + 2b^2)$.

7. Factor completely: $2x^2 - 8$.

8. At a baseball game 260 tickets were sold. Adults paid $4 each for their tickets, and children paid $2 each. If the total receipts for the game were $820, how many tickets of each kind were sold?

9. Simplify: $\dfrac{2x - 2}{2}$.

10. Represent $\dfrac{3}{x + 2} - \dfrac{1}{(x + 2)^2}$ using a single fraction.

11. Divide $12x^2$ by $\dfrac{3x}{4}$.

12. Solve: $\dfrac{13}{x} = 1 + \dfrac{4}{x}$.

13. If $\frac{1}{4}$ inch on a map represents 8 miles, how many miles does 6 inches represent?

14. Find the slope and y-intercept of the line $4x - 3y = 5$.

15. What is the value of s when r equals 6 if $2r + 6s = 20$?

16. Supply the missing components so that the ordered pairs (0,), (2,), and $(-3,)$ satisfy the equation $3x - y = 6$.

17. Divide $\dfrac{a^2\sqrt{3}}{4}$ by $\dfrac{3a}{\sqrt{3}}$.

18. Solve: $8r(r - 1) = -6r$.

19. Where does the graph of $y = x^2 + x - 30$ intersect the x-axis?

20. If the perimeter of a rectangle is 56 centimeters and its area is 192 square centimeters, what are the dimensions of the rectangle?

Review V

1. Simplify: $\dfrac{6 + 2^2}{5} - \dfrac{3^3 - 5^2}{2}$.

2. The signed whole numbers together with zero are called _?_.

3. Solve: $\dfrac{3x}{2} - 9 = 6$.

4. Factor: $y^2 - 12y + 11$.

5. Simplify: $(x + 3b)(x - 3b) - (x^2 - b^2)$.

6. The length of a rectangle is 8 feet more than its width w. Represent two-thirds of the length in terms of w.

7. Write $-\dfrac{1 - a}{3}$ as a positive fraction with a positive denominator.

8. Divide $(x^3 - 2x^2 + x)$ by x.

9. Divide $\dfrac{2\pi rh}{6 + h}$ by $\dfrac{2\pi r}{h}$.

10. A plane travels 875 miles in t hours. Represent the distance the plane can travel in 1 hour in terms of t.

11. Solve: $\dfrac{x - 9}{2} = \dfrac{x + 12}{9}$.

12. In a proportion, the product of the _?_ equals the product of the _?_.

13. Graph the set of all points for which $x = 5$ in a rectangular coordinate system.

14. Solve the system: $y = 2x + 5$

$$x - 3y = -20.$$

15. For what value(s) of a will the product $(a + 3)(a - 7)$ equal zero?

16. Graph $-\sqrt{4}, \sqrt{7}, \sqrt{21}, \sqrt{29}$ on a line graph.

17. Simplify the product of $\sqrt{3}$ and $\sqrt{27}$.

18. Solve: $x^2 + 5x + 2 = 0$.

19. Solve: $x^2 - 2 + \dfrac{7}{3}x = 0$.

20. Write in scientific notation.

a. 48,700,000   b. 0.0000632

Review VI

1. If $x = -1$, $y = -2$, $z = -3$, find the value of $x^{-3} - y^3 + z^3$.

2. Simplify: $(3^3 - 2^2 - 1^2) \cdot 2^{-1}$.

3. Represent $\dfrac{3x^2 - x^2}{x} - \dfrac{x^3}{x^2}$ using a single fraction.

4. The temperature drops $23°$ from a reading of $6°$. What is the new temperature?

5. Simplify: $\dfrac{a^2 - 3a - 4}{a^2 - 1} \cdot \dfrac{a + 2}{a - 4}$.

6. At a recent election, the winning candidate received 62 votes more than his opponent. If there were 3626 votes cast in all, how many did each candidate receive?

7. Simplify: $3(y + 3)^2 - (18y + 27)$.

8. Factor completely: $a^2 - 25b^2$.

9. One typist averages 8 words-per-minute fewer than a second typist. If the rate of the second typist is r words per minute, represent the output of the first typist in 5 minutes in terms of r.

10. Find the lowest common denominator of $\frac{1}{2}, \frac{1}{3}, \frac{1}{4}, \frac{1}{5},$ and $\frac{1}{6}$.

11. Simplify: $\dfrac{3 + \dfrac{2}{3}}{1 - \dfrac{1}{3}}$.

12. Represent $\dfrac{x - 1}{6} - \dfrac{2x + 5}{3}$ using a single fraction.

13. Solve: $\dfrac{7}{8} = \dfrac{21}{y + 2}$.

14. Solve: $\dfrac{3}{5} x + \dfrac{3}{10} = x - \dfrac{1}{2}$.

15. Solve for m: $E = \frac{1}{2}mv^2$

16. An approximate value for $\sqrt{2}$ is 1.41. Find an approximation for $\sqrt{72}$.

17. Solve the system $5x - 3y = -1$ using graphical methods.
$$3x + 3y = 9$$

18. Solve: $3w(2w + 3) = 0$.

19. Solve: $(x + 1)(x + 3) = 1$.

20. Given $a^2 + b^2 = c^2$.   Find b, if $a = 5$ and $c = 13$.

Review VII

1. What is the numerical coefficient of $-x^3$?

2. Solve: $\dfrac{x^2}{2} - x = \dfrac{5}{2}$.

3. Simplify: $(a + b - 2c) - (3a - b + 2c) + (2a + 2b - c)$.

4. Which of the following equations are true for all values of x?
$$x^3 \cdot x^2 = x^5 \qquad x^3 \cdot x^2 = x^6 \qquad 3x^2 = 9x^2$$

5. Factor completely: $4y^2 + 16y + 15$.

6. A 22-meter cable is divided into 2 parts. If y represents the longer piece, represent six times the shorter piece in terms of y.

7. Represent $\dfrac{4}{3y} + \dfrac{7}{3y} - \dfrac{2}{3y}$ using a single fraction.

8. Solve for P: $A = P + Prt$.

9. Simplify: $\dfrac{1 - \dfrac{a}{b}}{1 + \dfrac{a}{b}}$.

10. A car travels 90 miles in the same time that a slower car travels 60 miles. If the first car goes 10 miles per hour faster than the second, find the rate of each.

11. Find second components for the ordered pairs $(3, \ \ )$, $(-2, \ \ )$, $(0, \ \ )$, and $(6, \ \ )$ so that each ordered pair satisfies $y = 3x + 4$.

12. Solve the system $\ 2x + 3y = -5 \ $ by graphical methods.
$$2y + 8 = x$$

13. Solve the system $\ x - 2y = 7 \ $ by algebraic methods.
$$2x + y = 4$$

14. Solve: $14x^2 = 28x$.

15. Solve: $\dfrac{4}{x - 2} - \dfrac{7}{x - 3} = \dfrac{2}{15}$.

16. What term must be added to the expression $x^2 - 8x$ in order to make the expression a perfect square?

17. Find the equation of a line with slope -3 and passing through the point $(2, -1)$.

18. Simplify: $\dfrac{\sqrt{12a^2b^3}}{ab}$.

19. Solve for x: $b^2x^2 - c = 0$.

20. A man rowed 9 miles downstream and back again in 6 hours. The rate of the current was 2 miles per hour. Find the rate of the boat in still water.

Review VIII

1. Simplify: $\dfrac{-3^2 + (-3)^2 - 2^3}{3}$.

2. For what value of x is the expression $\dfrac{x + 2}{x - 5}$ meaningless?

3. If n is an odd integer, represent the next three odd integers in terms of n.

4. Factor completely: $a^2 - 6ab + 8b^2$.

5. Simplify: $(x - 2y)^2 - (x^2 + 4y^2)$.

6. Where should a 64-meter cable be cut so that twice the length of the longer piece equals five times that of the shorter?

7. Simplify: $\dfrac{x^2 - x - 20}{x^2 - 7x + 10} \cdot \dfrac{x^2 + 9x + 18}{x^2 + 7x + 12}$.

8. Divide $\dfrac{ab}{a^2 - b^2}$ by $\dfrac{ab}{2a - 2b}$.

9. Simplify: $\dfrac{\dfrac{1}{a} - \dfrac{1}{b}}{\dfrac{1}{ab}}$.

10. Divide $(x^2 + 3x - 7)$ by $(x - 5)$.

11. Solve the system $\quad 3y - x - 1 = 0$ algebraically.
$$y + 6x + 6 = 0$$

12. Solve the system $\quad y - 2x - 5 = 0$ by graphical methods.
$$x - 3y + 20 = 0$$

13. Two packages weighed together total 146 kilograms. One of the packages weighs 12 kilograms more than the other. How much does each package weigh?

14. Solve: $x^2 - 5x - 14 = 0$.

15. Simplify: $\sqrt{200x^3y^2}$.

16. Simplify: $\sqrt{125} + 2\sqrt{5}$.

17. Graph: $y = x^2 + x - 6$.

18. Simplify: $5\sqrt{\dfrac{2}{5}}$

19. Solve: $2x^2 - 2x = 7$.

20. The length of a rectangle is 4 centimeters less than twice its width, and the area is 240 square centimeters. Find the dimensions of the rectangle.

Review IX

1. Simplify: $\dfrac{4^2 + 2}{2} - \dfrac{3^3 - 7}{4}$.

2. If $a = 1$, $b = -2$, $c = 2$, find the value of $abc - b^2 + c^2$.

3. Simplify: $2a(a - b) - b(a + b) - (a^2 - b^2)$.

4. Find three consecutive even integers whose sum is 78.

5. Factor completely: $2x^2 - 24x + 22$.

6. The difference of two numbers is 28. If n represents the smaller number, represent the larger in terms of n.

7. Solve: $2(a - 1) < a + 3$.

8. Represent $\dfrac{2}{x} - \dfrac{3}{y}$ using a single fraction.

9. Divide $\dfrac{3b}{4a}$ by $\dfrac{12b^2}{a}$.

10. Divide $(2x^2 + 3x - 1)$ by $(x + 2)$.

11. Solve for R: $\dfrac{1}{R} = \dfrac{1}{A} + \dfrac{1}{B}$.

12. Graph: $x - 3y \geq 6$.

13. Graph: $y = -2x^2 + 3$.

14. Solve the system $3y - 5x = 1$ by algebraic methods.
$$x + y = 3$$

15. The value of a collection of coins is $3.15. There are three more dimes than nickels and two more quarters than dimes. How many of each kind of coin is there in the collection?

16. Simplify the product of $\sqrt{6x}$ and $\sqrt{15x}$.

17. Simplify: $\dfrac{2 + 3\sqrt{12}}{2}$.

18. An approximate value for $\sqrt{2}$ is 1.41. Find an approximation for $\sqrt{18}$ without using the table of square roots.

19. A plane flies directly east for 12 kilometers, turns and flies south for 16 kilometers. How far is the plane from its starting point?

20. Where does the graph of $4x - 5y = 5$ intersect the y axis?

Review X

1. Graph the prime numbers between 20 and 40 on a line graph.

2. Express $240x^2y$ in completely factored form.

3. Arrange the numbers 5, -2, 3, -5, 1, 7, -1 in order, from smallest to largest.

4. Simplify: $(a - b) - 2(a + b) - (a + 2b)$.

5. Simplify: $(a - b)(a + b) - (a - b)^2$.

6. Factor completely: $3x^2 - 18xy + 24y^2$.

7. Simplify: $\dfrac{2a}{6a^2 + 8a}$.

8. Represent $\dfrac{a + 3b}{7} - \dfrac{a + 2b}{7}$ using a single fraction.

9. Represent $\dfrac{1}{a - 3} - \dfrac{2}{(a - 3)^2}$ using a single fraction.

10. Solve: $\dfrac{3}{5}x = x - \dfrac{4}{5}$.

11. Graph: $y = x$.

12. Solve the system $5x + 2y - 8 = 0$ by algebraic methods.
$$3x - 7y - 13 = 0$$

13. Simplify: $3\sqrt{3} - \sqrt{12} + \sqrt{75}$.

14. Simplify: $\dfrac{\sqrt{18} - \sqrt{27}}{3}$.

15. Where does the graph of $y = x^2 + 4x + 3$ intersect the x-axis?

16. Find two consecutive integers such that twice the second less one half the first is 14.

17. A man gave one third of his money to one son, one fourth to a daughter, and had $250 left. How much did he have to start with?

18. A furniture dealer sold a desk and a chair for $640. If the desk sold for $40 more than four times the chair, what was the price of each?

19. In a right triangle, if two thirds of one acute angle is added to one half of the second acute angle, the result is 50°. Find each angle.

20. The hypotenuse of a right triangle is 20 meters long. If one of the remaining sides is 16 meters long, how long is the third side?

APPENDIX
SET NOTATION; FUNCTION NOTATION

SET NOTATION

A **set** is a collection of objects. The objects belonging to a set are called the **members** or **elements** of the set. In mathematics, these members are usually numbers. Braces, { }, are commonly used to enclose the members of the set; the order in which we list the members is not important. For example, {0, 1, 2}, {1, 0, 2}, {1, 2, 0}, {2, 0, 1}, and {2, 1, 0} represent the same set because they contain the same members.

Capital letters, such as A, B, and C, are often used to name sets. For example, if we let

$$A = \{3, 4, 5\}, \qquad B = \{5, 6, 7\}, \qquad \text{and} \qquad C = \{4, 3, 5\},$$

then we can write $A = C$, which indicates that A and C name the same set; and we can write $A \neq B$, which indicates that A and B do not name the same set.

It is often convenient to describe the members of a set by a special notation called **set-builder notation.** In set-builder notation, a short vertical bar is used for the phrase "such that." For example,

$$\{x \mid x > -2, \quad x \text{ is an integer}\} \tag{1}$$

is read "the set of all x *such that* x is greater than -2 and x is an integer."

The symbol $\in$ is often used for the phrase "is an element of." Thus, if we use the letter J to represent the set of integers, we can rewrite (1) as

$$\{x \mid x > -2, \quad x \in J\} \tag{2}$$

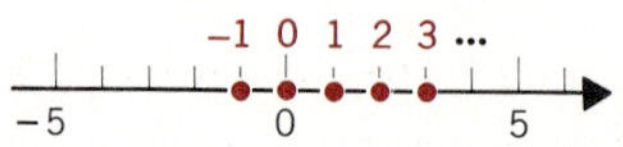

Figure 1

The graph of (2) is shown in Figure 1.

428

As another example, consider

$$\{(x, y) \mid y = x + 2, \quad x \in \{1, 3, 5\}\}, \tag{3}$$

which is read "the set of all ordered pairs (x, y) *such that* y equals $x + 2$ and x is an element of $\{1, 3, 5\}$." In this case,

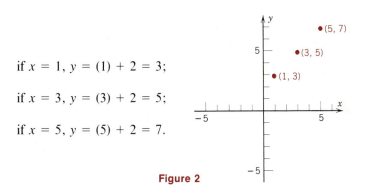

if $x = 1$, $y = (1) + 2 = 3$;

if $x = 3$, $y = (3) + 2 = 5$;

if $x = 5$, $y = (5) + 2 = 7$.

Figure 2

Thus, the set consists of three ordered pairs: $\{(1, 3), (3, 5), (5, 7)\}$. The graph of this set is shown in Figure 2.

In the examples above, the members of set (2) are integers, and the members of set (3) are ordered pairs of integers. But what if the variables represent real numbers rather than integers? Using the letter $\boldsymbol{R}$ for the set of real numbers, we can write

$$\{x \mid x > -2, \quad x \in R\}$$

and

$$\{(x, y) \mid y = x + 2, \quad x, y \in R\}$$

to specify the members of the sets whose graphs are shown in Figure 3.

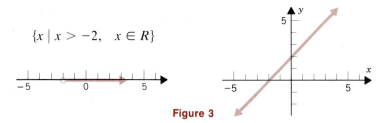

Figure 3

The solid lines in both figures represent an infinite set of points. Note that we use an open dot at -2 in part (a) because -2 does not satisfy the inequality $x > -2$ and therefore is not a member of the set. We would use a solid dot at -2 for $x \geq -2$.

One useful operation on two sets is defined as follows:

> The *union* of two sets A and B is the set of all elements that belong either to A or to B or to both. We use the symbol $\cup$ to designate the operation.

The expression $A \cup B$ is read "the union of set A and set B." For example, if $A = \{0, 1, 3\}$ and $B = \{3, 5, 7\}$,   then

$$A \cup B = \{0, 1, 3, 5, 7\}.$$

Note that $A \cup B$ includes all the members in both sets. We write the element 3 only once in the union even though it appears in both A and B.

The operation is also applicable to sets containing ordered pairs. Thus, if $A = \{(2, 3), (4, 5)\}$ and $B = \{(4, 5), (6, 7)\}$,   then

$$A \cup B = \{(2, 3), (4, 5), (6, 7)\}.$$

A second useful operation on two sets is defined as follows:

> The *intersection* of two sets A and B is the set of all elements that belong to both A and B. We use the symbol $\cap$ to designate the operation.

The expression $A \cap B$ is read "the intersection of A and B." For example, if $A = \{2, 4, 6, 8\}$ and $B = \{5, 6, 7, 8\}$,   then

$$A \cap B = \{6, 8\}.$$

Also, if $A = \{(2, 3), (4, 5), (6, 7)\}$ and $B = \{(3, 6), (4, 5)\}$,   then

$$A \cap B = \{(4, 5)\}.$$

A set that does not contain any members is called the **empty set,** or **null set,** and is designated by the symbol $\varnothing$. Thus, if $A = \{1, 3, 5\}$ and $B = \{2, 4\}$, then

$$A \cap B = \varnothing.$$

If $A = \{(2, 3), (4, 5)\}$ and $B = \{(3, 2)\}$,   then

$$A \cap B = \varnothing.$$

The sets A and B are said to be **disjoint** if they have no members in common.

FUNCTION NOTATION

Sometimes we use a special notation to indicate a relationship between two variables. For example, in the expression

$$f(x) = -2x + 4, \tag{1}$$

the symbol $f(x)$ *plays the same role as* y in the equation

$$y = -2x + 4.$$

Thus, $f(x)$, (read "f of x"), is an alternate way of denoting the dependent variable, y. It can also be used to denote the value of the expression (1) for specific values of x. For example, $f(1)$ represents the value of the expression $-2x + 4$ when x is replaced by 1:

$$f(1) = -2(1) + 4 = 2, \quad \text{or} \quad f(1) = 2.$$

Similarly,

$$f(0) = -2(0) + 4 = 4,$$

and

$$f(2) = -2(2) + 4 = 0.$$

The symbol $f(x)$ is commonly referred to as **function notation.**

EXERCISES

■ *Graph the members of each set. In exercises 1–8, assume that all variables represent integers.*

Sample Problems

a. $\{x \mid x < 2\}$

b. $\{(x, y) \mid y = 2x + 4, \quad x \in \{1, 2, 3\}\}$

a. Note that 2 is not included in the set.

Ans.

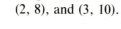

b. First obtain the values of y associated with each value of x.

$\quad$ If $\quad x = 1, \quad y = 6;$

$\quad$ if $\quad x = 2, \quad y = 8;$

$\quad$ if $\quad x = 3, \quad y = 10.$

Graph the ordered pairs $(1, 6)$, $(2, 8)$, and $(3, 10)$.

Ans.

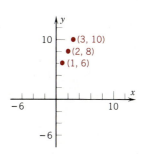

1. $\{x \mid x > 2\}$

2. $\{x \mid x > -3\}$

3. $\{x \mid x \leq -3\}$

4. $\{x \mid x \leq 2\}$

5. $\{x \mid x < 5\}$

6. $\{x \mid x > -5\}$

7. $\{x \mid x \text{ is positive}\}$

8. $\{x \mid x \text{ is negative}\}$

9. $\{(x, y) \mid y = x + 3, \quad x \in \{0, 2, 4\}\}$

10. $\{(x, y) \mid y = 2x - 3, \quad x \in \{-4, -2, 0\}\}$

11. $\{(x, y) \mid y = 3x + 2, \quad x \in \{-2, 0, 2\}\}$

12. $\{(x, y) \mid y = 4x + 4, \quad x \in \{-2, 0, 2\}\}$

■ *Graph each set. Assume that all variables represent real numbers* $(x, y \in R)$.

a. $\{x \mid x < 2\}$

b. $\{(x, y) \mid y = 2x + 4\}$

a. Note that 2 is not included in the set. Thus, an open dot is used at 2.

Ans.

b. Graph $y = 2x + 4$ by the inter-   *Ans.*
cept method.
$$\text{If } x = 0, \quad y = 4;$$
$$\text{if } y = 0, \quad x = -2.$$
First graph $(0, 4)$ and $(-2, 0)$. Then
draw a straight line containing
these intercepts.

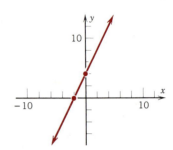

13. $\{x \mid x > 4\}$

14. $\{x \mid x > -2\}$

15. $\{x \mid x \le -2\}$

16. $\{x \mid x \le 4\}$

17. $\{x \mid x \text{ is positive}\}$

18. $\{x \mid x \text{ is negative}\}$

19. $\{(x, y) \mid y = x + 2\}$

20. $\{(x, y) \mid y = x - 2\}$

21. $\{(x, y) \mid y = 2x + 2\}$

22. $\{(x, y) \mid y = 2x - 2\}$

23. $\{(x, y) \mid y = 2x\}$

24. $\{(x, y) \mid y = -2x\}$

■ *Write the union and intersection of each pair of sets by listing the members.*

a. $A = \{-5, 2, 3, 4\}$
$B = \{2, 4, 8\}$

b. $A = \{(-3, 2), (0, 4), (2, 3)\}$
$B = \{(-2, 1), (2, 3)\}$

The union of two sets contains all the members in both sets. The intersection
contains the members that are *common* to both sets.

Ans. $A \cup B$
$= \{-5, 2, 3, 4, 8\};$
$A \cap B = \{2, 4\}$

Ans. $A \cup B$
$= \{(-3, 2), (0, 4), (2, 3), (-2, 1)\}$
$A \cap B = \{(2, 3)\}$

25. $A = \{2, 4, 6\}$
$B = \{4, 6, 8\}$

26. $A = \{1, 3, 5\}$
$B = \{3, 5, 7\}$

27. $A = \{-3, -1, 1, 3\}$
$B = \{-1, 1\}$

28. $A = \{-4, -2, 2, 4\}$
$B = \{-4, 4\}$

29. $A = \{2, 4, 6\}$
$B = \{3, 5\}$

30. $A = \{-2, 0, 2\}$
$B = \{-1, 1\}$

31. $A = \{(0, 1), (1, 2)\}$
$B = \{(0, 2), (1, 2)\}$

32. $A = \{(2, 3), (2, 4)\}$
$B = \{(3, 3), (2, 4)\}$

33. $A = \{(0, 0), (-1, -1), (-2, -2)\}$
$B = \{(-4, -4), (-2, -2), (0, 0)\}$

34. $A = \{(1, 0), (2, 0), (3, 0)\}$
$B = \{(0, 1), (0, 2), (3, 0)\}$

■ *Graph each intersection. Assume that all variables represent real numbers* $(x \in R)$.

Sample Problem

$\{x \mid x > -2\} \cap \{x \mid x \le 3\}$

Graph each inequality on a separate number line. Show the intersection (common interval) on a third line.

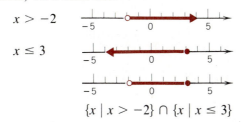

$x > -2$

$x \le 3$

Ans.

$\{x \mid x > -2\} \cap \{x \mid x \le 3\}$

35. $\{x \mid x > 1\} \cap \{x \mid x < 5\}$

36. $\{x \mid x \ge 2\} \cap \{x \mid x \le 6\}$

37. $\{x \mid x > -4\} \cap \{x \mid x \le 2\}$

38. $\{x \mid x \ge -1\} \cap \{x \mid x < 5\}$

39. $\{x \mid x \le 2\} \cap \{x \mid x \ge 2\}$

40. $\{x \mid x \le 3\} \cap \{x \mid x \ge 3\}$

■ *Find each intersection by algebraic methods.*

(*Hint.* Solve the system of equations that determines each set.)

41. $\{(x, y) \mid x + y = 5\} \cap \{(x, y) \mid 2x - y = 4\}$

42. $\{(x, y) \mid 2x - 3y = 3\} \cap \{(x, y) \mid x + y = 4\}$

43. $\{(x, y) \mid 3x + y = -1\} \cap \{(x, y) \mid x - 3y = -7\}$

44. $\{(x, y) \mid y = 3x - 6\} \cap \{(x, y) \mid y = 2x - 4\}$

■ *Evaluate each expression for the given values of the variable.*

Sample Problem

If $f(x) = 2x - 4$,   find $f(-3)$ and $f(0)$.

$$f(-3) = 2(-3) - 4$$
$$= -6 - 4 = -10$$
$$f(0) = 2(0) - 4$$
$$= 0 - 4 = -4$$

Ans. $f(-3) = -10$; $f(0) = -4$

45. If $f(x) = x + 5$,   find $f(-2)$ and $f(2)$.

46. If $f(x) = x - 3$, find $f(-3)$ and $f(3)$.

47. If $f(x) = 2x - 7$, find $f(-5)$ and $f(-1)$.

48. If $f(x) = 3x + 2$, find $f(-4)$ and $f(2)$.

49. If $f(x) = 4 - 3x$, find $f(-1)$ and $f(0)$.

50. If $f(x) = 3 - 2x$, find $f(-4)$ and $f(-2)$.

51. If $f(x) = \dfrac{3x - 4}{2}$, find $f(2)$ and $f(6)$.

52. If $f(x) = \dfrac{5x + 2}{2}$, find $f(-6)$ and $f(-2)$.

53. If $f(x) = \dfrac{3x - 5}{5}$, find $f(-5)$ and $f(0)$.

54. If $f(x) = \dfrac{3x + 4}{4}$, find $f(4)$ and $f(0)$.

55. If $f(x) = \dfrac{2x - 3}{5}$, find $f(-1)$ and $f\left(\dfrac{5}{2}\right)$.

56. If $f(x) = \dfrac{3x - 1}{4}$, find $f(2)$ and $f\left(\dfrac{4}{3}\right)$.

GLOSSARY

Abscissa
The first component in an ordered pair. The distance to a point from the vertical axis in rectangular coordinates; to the right of the axis if the component is positive; to the left of the axis if the component is negative.

Absolute Value
$$|a| = \begin{cases} a, \text{ if } a \text{ is greater than or equal to 0.} \\ -a, \text{ if } a \text{ is less than 0.} \end{cases}$$

Algebraic Expression

Any variable or number or meaningful combination thereof. $3xy - xy^2$, $\dfrac{x + y}{3}$, xyz, and so on, are algebraic expressions.

Associative Law
a. Addition: $a + (b + c) = (a + b) + c$
b. Multiplication: $(ab)c = a(bc)$

Axis
A straight line used as a visual representation of the relative order of the real numbers. A number line.

Base (of a Power)
The number to which an exponent is attached. In the term x^4, x is the base to which the exponent 4 is attached.

Binomial
A polynomial consisting of two terms, $3a + 4b$ is a binomial.

Cartesian Coordinates
See Rectangular Coordinates.

Coefficient
Any factor or group of factors in a product is the coefficient of the remaining factors. In $32ab$, 32 is the coefficient of ab; in ax, a is the coefficient of x.

Commutative Law

a. Addition: $a + b = b + a$

b. Multiplication: $ab = ba$

Complex Fraction

A fraction that contains other fractions in its numerator or denominator or both.

Component (of an Ordered Pair)

Either number of an ordered pair.

Conjugate

The conjugate of the binomial $a + b$ is the binomial $a - b$. (Usually refers to binomials which contain square roots.)

Consecutive Integers

Integers that differ by 1. The numbers $-3, -2, -1, 0, 1, 2$, and 3 are consecutive integers.

Constant

A symbol representing a single number during a particular discussion.

Coordinate(s) of a Point

Numbers giving the position of a point with respect to an origin.

Degree of a Term (with One Variable)

The exponent of the variable. The term $2x^4$ is of degree 4.

Degree of a Polynomial or an Equation

The degree of the term of highest degree.

Dependent Equations

A system of equations where every set of values which satisfies one of the equations satisfies them all.

Dependent Variable

A variable whose values are determined by the values of another variable.

Descending Powers

The arrangement of an expression so that each term is of higher degree in one of the variables than the next succeeding term. Thus, $x^4 + x^3 - 2x^2 + x - 1$ is arranged in descending powers of x.

Difference

The result of subtracting one number from another.

Direct Variation

A relationship determined by the equation $y = kx$, where k is a constant.

Disjoint Sets

Sets with no members in common.

Distributive Law

$a(b + c) = ab + ac$

Equation
An assertion that two expressions are names for the same number.

Equivalent Equations
Equations which have the same solutions. The equations $2x + 2 = 6$ and $2x = 4$ are equivalent, because 2 is the only solution of each.

Equivalent Expressions
Expressions that represent the same number for all values of any variables involved. The expressions, $2c + 3b + c$, $3c + 3b$, and $3(c + b)$ are equivalent, because for all values of b and c they represent the same number.

Even Integers
. . . $-4, -2, 0, 2, 4, 6, . . .$

Exponent
A number placed to the right and above a symbol to indicate how many times the symbol occurs as a factor in a product. In a^3, 3 is the exponent.

Factor
Any of a group of numbers that are multiplied together. To write an algebraic expression as a product of factors.

First-Degree Equation
An equation of degree 1; a linear equation.

Formula
A relationship between quantities expressed in symbols; an equation.

Fraction
A symbol that represents the quotient of two algebraic expressions. $\dfrac{2}{3}, \dfrac{x}{x + 2},$
and $\dfrac{y - 1}{y^2 + y + 3}$ are fractions.

Function Notation
A symbol such as $f(x)$, used to name an expression in x.

Graph
A geometric representation of a numerical relationship.

Incomplete Quadratic Equation
A quadratic equation where either the first-degree term or the constant term is missing. $x^2 - 2 = 0$ and $2x^2 + 3x = 0$ are incomplete quadratic equations.

Inconsistent Equations
Equations that have no common solution. Graphically, inconsistent linear equations (in two variables) appear as parallel lines.

Independent Variable
A variable considered free to assume any one of a given set of values.

Inequality
A relationship between quantities that are unequal.

Integer
The integers include the natural numbers, the negatives of the natural numbers, and 0. One of the numbers . . . , $-3, -2, -1, 0, 1, 2, 3, ...$

Irrational Number
A number that can be associated with a point on the number line but cannot be expressed as a quotient of integers.

Like Terms
Terms whose variable factors are identical. $22xy$ and $5xy$ are like terms.

Linear Equation
An equation of the first degree. The graph of a linear equation in two variables is a straight line.

Lowest Common Denominator (L.C.D.)
The smallest natural number or the polynomial of least degree into which each of the denominators of a given set of fractions divide exactly.

Member of an Equation
The expression to the right of an equal sign constitutes the right-hand member of an equation, and that to the left of an equal sign constitutes the left-hand member. In $2x + 3 = x + 9$, the left-hand member is $2x + 3$ and the right-hand member is $x + 9$.

Monomial
A polynomial consisting of one term. The expression $2xy$ is a monomial.

Natural Number
A positive "counting" number, such as 1, 2, 3, 4, . . .

Negative Number
A number less than 0.

Number Line
A straight line used to represent the relative order of a set of numbers.

Numerical Evaluation
The act of finding the value of an expression. To evaluate $x + 4$ for $x = 3$ means to replace x with 3 and simplify the results (giving 7).

Odd Integers
. . . $-5, -3, -1, 1, 3, 5, . . .$

Ordered Pair
A pair of numbers in which the order of the numbers is important. An ordered pair is usually represented (x, y).

Ordinate
The second component in an ordered pair. The distance to a point from the horizontal axis in rectangular coordinates. The point is above the axis if the second component is positive and below the axis if the second component is negative.

Origin
The point on a line graph corresponding to 0. The point of intersection of coordinate axes.

Parabola
The graph of the equation $y = ax^2 + bx + c$.

Parentheses
Symbols, ( ), used to group factors or terms.

Polynomial
A special kind of algebraic expression. In this book, any term or sum of terms.

Positive Number
A number greater than 0.

Prime Factor
A factor that is a prime number.

Prime Number
A natural number greater than 1 that is divisable only by itself and one.

Principal Square Root
The positive square root of a positive number.

Product
The result of the multiplication of two or more numbers.

Quadrant
One of the four regions into which a set of rectangular axes divides the plane.

Quadratic Equation
An equation of degree 2. $x^2 = 2$, $x^2 - 3x = 0$, and $x^2 + 2x - 1 = 0$ are quadratic equations.

Quadratic Formula

The formula $x = \dfrac{-b \pm \sqrt{b^2 - 4ac}}{2a}$ is used to solve quadratic equations of the form $ax^2 + bx + c = 0$, $a \neq 0$.

Quotient
The result of dividing one number by another.

Radical
A symbol ($\sqrt{}$) indicating the positive square root of a number. Any expression under a radical sign is called a radicand.

Rational Number

The quotient of two integers, $\dfrac{a}{b}$, where b does not equal zero.

Real Number
Any number that is either a rational number or an irrational number.

Reciprocal
The reciprocal of a number is the quotient obtained by dividing the given number into 1. The reciprocal of 3 is $\frac{1}{3}$; the reciprocal of $\frac{3}{4}$ is $\frac{4}{3}$; and so on. The number 0 has no reciprocal.

Rectangular Coordinates
Numbers specifying the distances of points from two perpendicular number lines.

Set
A collection of objects, which are called the members or elements of the set.

Simplify
To find an equivalent expression that is simpler than the original.

Slope
A number (m) associated with a nonvertical line that specifies its direction. For two points (x_1, y_1) and (x_2, y_2) on a line, $m = \dfrac{y_2 - y_1}{x_2 - x_1}$ $(x_2 \neq x_1)$.

Solution of an Equation
A value for the variable that satisfies the equation—that is, for which the equation is a true statement.

Square Root
One of two equal factors of a number. Since $3 \cdot 3 = 9$, the number 3 is a square root of 9. Also, since $(-3)(-3) = 9$, the number -3 is a square root of 9.

Standard Form (for an Equation)
An equation with the left-hand member arranged in descending powers of the variable and the right-hand member 0, such as $3x^2 - 2x + 1 = 0$.

Standard Form (for a Fraction)
A positive fraction with positive denominator, such as, $\dfrac{-a}{b}$ or $\dfrac{a}{b}$.

Standard Form (for a System of Equations)
The equations arranged with like terms in order in the left-hand members. For example, $2x + 3y = 7$
$-x + 5y = 3.$

Sum
The result of the addition of two or more numbers.

Symmetric Property of Equality
If $a = b$, then $b = a$. Thus, if $2x = y + 3$, then $y + 3 = 2x$.

System of Equations
A set of two or more equations considered together.

Term

Any part of an algebraic expression separated from other parts by plus or minus signs.

Trinomial

A polynomial consisting of 3 terms. $2x + 3y + 2$ is a trinomial.

Unlike Terms

Terms that differ in their variable factors. $23xy$ and $4x$ are unlike terms; $3x^2$ and $3x^3$ are unlike terms.

Variable

A symbol representing any one of a given set of numbers.

Variation

See "Direct Variation."

Vertex

The maximum or minimum point of a parabola.

Whole Number

A natural number or 0. One of the numbers 0, 1, 2, 3, ...

ODD-NUMBERED ANSWERS

Note to students:

Partial solutions to all word problems include word phrases that define the variables, equations for the conditions on the variables, and answers to the problems. You may use different equations that are also appropriate, but your answer should be the same as the given answer.

EXERCISES 1.1 (Page 5)

1. 7, 11   **3.** 23, 29   **5.** 2, 3, 5, 7, 11, 13

7. 29, 31   **9.** 47, 53, 59, 61   **11.** $3 \cdot 3 \cdot 5$

13. $2 \cdot 2 \cdot 7$ **15.** $3 \cdot 3 \cdot 3$ **17.** $2 \cdot 2 \cdot 2 \cdot 2 \cdot 2 \cdot 2$

19. $2 \cdot 3 \cdot 5 \cdot 7$ **21.** $2 \cdot 2 \cdot 3 \cdot 13$

23. Addition: increased by, add, more than, exceeded by, sum
Multiplication: times, multiply, product, multiplied by

25. $\underline{3} + (6 + 9)$ **27.** $8 \cdot \underline{6}$ **29.** $3 \cdot (x \cdot y)$ **31.** $(9 \cdot \underline{8}) \cdot 3$

33. $5 + 3$ **35.** $4y$ **37.** $x + 6$ **39.** $2x$

41. $l + w$ **43.** $5c$ **45.** $2(5 + y)$ **47.** $5 + (4x)$

49. $2 + (4 + t)$ **51.** $b(c + 6)$ **53.** $5x(3 + z)$ **55.** $0.50(68)$

57. $0.37c$ **59.** $0.08s$

EXERCISES 1.2 (Page 10)

1. 6   **3.** Not a whole number

5. 9   **7.** 0

9. Not a whole number   **11.** 10

13. 3   **15.** Undefined   **17.** 0   **19.** Undefined

21. 8   **23.** 5

25. Subtraction: take away, less than, difference, subtract, decreased by,
                  diminished by, subtracted from, less
      Division: divide, quotient, divided by

27. $\dfrac{6}{x}$ **29.** $8 - y$ **31.** $a - b$

33. $\dfrac{a}{b}$ **35.** $\dfrac{d}{t}$ **37.** $\dfrac{10}{x + 4}$

39. $y - (4 + x)$ **41.** $\dfrac{x + y}{5}$ **43.** $\dfrac{r + s}{6z}$

45. $2x - 6$ **47.** $\dfrac{2x}{5}$ **49.** $2x - 9$

51. $4x - 2$ **53.** $\dfrac{7}{x} + 5$ **55.** $x(x + 4)$

57. $\dfrac{3x}{x + 7}$

EXERCISES 1.3 (Page 15)

1. 5	**3.** 14	**5.** 5	**7.** 32
9. 27	**11.** 20	**13.** 13	**15.** 4
17. 2	**19.** 4	**21.** 2	**23.** 0
25. 9	**27.** 3	**29.** 3	**31.** 24
33. 4	**35.** 34	**37.** 18	**39.** 12
41. 6	**43.** 7	**45.** 4	**47.** 14
49. 7	**51.** 6	**53.** 6	**55.** 12
57. 3	**59.** 25	**61.** 2	**63.** 3
65. 9	**67.** 6	**69.** 12,000	**71.** 76
73. 50,000		**75.** 30	

EXERCISES 1.4 (Page 21)

1. Yes	**3.** No	**5.** Yes	**7.** Yes
9. No	**11.** 11	**13.** 14	**15.** 3
17. 4	**19.** 175	**21.** 0	**23.** $x = 7$
25. $y = 6$	**27.** $z = 4$	**29.** $x = 16$	**31.** $z = 30$
33. $y = 21$	**35.** $x = 2$	**37.** $z = 6$	**39.** $y = 4$
41. 5	**43.** 11	**45.** 5	**47.** 3

49. 3 **51.** 4 **53.** 20 **55.** 18
57. 12 **59.** 18 **61.** 10 **63.** 0

EXERCISES 1.5 (Page 25)

1. 3 **3.** 1 **5.** 1 **7.** 4
9. 4 **11.** 2 **13.** 3 **15.** 0
17. 1 **19.** 20 **21.** 12 **23.** 20
25. 14 **27.** 25 **29.** 0 **31.** 18
33. 14 **35.** 4 **37.** 60 **39.** 45
41. 8 **43.** 66 **45.** 0 **47.** 144
49. 18 **51.** 10 **53.** 10 **55.** 0
57. 28 **59.** 40

EXERCISES 1.6 (Page 29)

1. $x + 21 = 59$ **3.** $8 = 2x - 6$ **5.** $15 + x = 53$
7. $3x - 12 = 9$ **9.** $2x - 7 = 33$

11. a. Amount Ruth has: x
   b. $x + 21 = 150$
   c. $x = 129$;   Ruth has \$129.

13. a. Price of the used car: x
   b. $x + 3400 = 9200$
   c. $x = 5800$;   the used car costs \$5800.

15. a. Amount Sam earned: x
   b. $3x = 861$
   c. $x = 287$;   Sam earned \$287.

17. a. Number of games the Dodgers played: x
   b. $0.60x = 96$
   c. $x = 160$;   the Dodgers played 160 games.

19. a. Number of students enrolled: x
   b. $0.65x = 377$
   c. $x = 580$;   the enrollment is 580 students.

21. a. Number of hours Neil works per week: x
   b. $2x + 4 = 36$
   c. $x = 16$;   Neil works 16 hours per week.

23. a. Rent on Grace's old apartment: x
   b. $2x - 60 = 540$
   c. $x = 300$;   the rent on Grace's old apartment was \$300.

25. a. Daily rental fee: x
  b. $40 + 14x = 82$
  c. $x = 3$; the daily rental fee is \$3.

27. a. Monthly installment: x
  b. $1200 + 36x = 10{,}200$
  c. $x = 250$; the monthly installment is \$250.

29. a. Number of hours the plumber worked: x
  b. $40 + 18x = 130$
  c. $x = 5$; the plumber worked 5 hours.

31. a. Speed at which Agnes drove: x
  b. $7x = 364$
  c. $x = 52$; Agnes drove at 52 miles per hour.

33. a. Number of hours the trip took: x
  b. $560x = 2800$
  c. $x = 5$; the trip took 5 hours.

35. a. Length of the garden: x
  b. $12x = 180$
  c. $x = 15$; the garden should be 15 feet long.

37. a. Length of the pasture: x
  b. $2x + 2(75) = 500$
  c. $x = 175$; the pasture will be 175 yards long.

39. a. Height of the sail: x
  b. $\dfrac{4x}{2} = 12$
  c. $x = 6$; the sail is 6 meters tall.

CHAPTER 1 REVIEW (Page 34)

1. a. $2 \cdot 2 \cdot 3 \cdot 3 \cdot 3$ b. $2 \cdot 2 \cdot 5 \cdot 29$ c. $7 \cdot 13 \cdot 13$

2. a. $6x$ b. $\dfrac{4 + y}{6}$ c. $y(3 + x)$

3. a. $\dfrac{3x}{6}$ b. $4x + 5$ c. $8x - 5$

4. a. 8   b. undefined   c. 8

5. a. 13   b. 10   c. 7

6. a. 1   b. 16   c. 14

7. a. 2   b. 2   c. 3

8. a. 6   b. 4   c. 4

9. terms

10. a. $3 + x = 27$ b. $2x - 5 = 17$ c. $\dfrac{3x}{4} = 6$

11. *a, c*

12. a. 5   b. 72   c. 16

13. a. 8   b. 4   c. 10

14. a. 49   b. 45   c. 15

15. a. 3   b. 0   c. 0

16. a. Al's price for the turntable: x
   b. $x + 38 = 123$
   c. $x = 85$;   Al's price is $85.

17. a. Price of the house: x
   b. $0.15x = 16,800$
   c. $x = 112,000$;   the house cost $112,000

18. a. Number of channels Stella takes: x
   b. $1.35x + 7.95 = 16.05$
   c. $x = 6$;   Stella takes 6 movie channels.

19. a. Number of hours the trip took: x
   b. $47x = 423$
   c. $x - 9$;   the trip took 9 hours.

20. a. Length of the property: x
   b. $2x + 2(60) = 332$
   c. $x = 106$;   the property is 106 feet long.

EXERCISES 2.1 (Page 40)

1.

3.

5. <   **7.** <   **9.** >   **11.** >

13. <   **15.** <   **17.** >

19.

21.

23.

25. 6   **27.** 10   **29.** -9 **31.** -6
33. $<$ **35.** $=$ **37.** $>$ **39.** $>$
41. $<$ **43.** $>$ **45.** 4   **47.** 7
49. 18   **51.** 17   **53.** 2   **55.** 0
57. Positive   **59.** Negative

EXERCISES 2.2 (Page 43)

1. 7   **3.** -9 **5.** 4   **7.** -6
9. 8   **11.** 0   **13.** 10   **15.** -10
17. 5   **19.** -4 **21.** 5   **23.** 0
25. 8   **27.** -8 **29.** 5   **31.** 6
33. -3 **35.** -3 **37.** 3   **39.** 3
41. -4 **43.** -7 **45.** 0   **47.** 0
49. -15

EXERCISES 2.3 (Page 48)

1. 6   **3.** 8   **5.** 3   **7.** -3
9. 5   **11.** -5 **13.** 0   **15.** 5
17. -3 **19.** -6 **21.** -6 **23.** 4
25. -12 **27.** 2   **29.** 8   **31.** 0
33. 0   **35.** 4   **37.** -3 **39.** 12
41. -14 **43.** -4 **45.** 9   **47.** -2
49. 9

EXERCISES 2.4 (Page 53)

1. -15 **3.** 30   **5.** 0   **7.** -8
9. -15 **11.** -14 **13.** 0   **15.** -24
17. 18   **19.** 0   **21.** 120   **23.** 8
25. 5   **27.** -3 **29.** 0   **31.** Undefined
33. -6 **35.** -5 **37.** -2 **39.** 23
41. -22 **43.** 16   **45.** -36 **47.** -4
49. -6 **51.** 7   **53.** -26 **55.** -28
57. -2 **59.** -45

EXERCISES 2.5 (Page 57)

1. -5	**3.** 1	**5.** 13	**7.** 4
9. -6	**11.** 5	**13.** -175	**15.** 0
17. -4	**19.** -2	**21.** -8	**23.** -4
25. -28	**27.** -15	**29.** -60	**31.** -2
33. 4	**35.** 1	**37.** -7	**39.** -4

41. 18

43. a. Yesterday's temperature: x
b. $x - 13 = -7$
c. $x = 6$; yesterday's temperature was 6°.

45. a. Temperature four days ago: x
b. $x + 4(8) = 26$
c. $x = -6$; four days ago the temperature was $-6°$.

47. a. Number of weeks to diet: x
b. $186 - 4x = 162$
c. $x = 6$; Eric must diet for 6 weeks.

49. a. Store's expenses: x
b. $85,000 - x = -11,500$
c. $x = 96,500$; the store's expenses were $96,500.

51. a. Loss on each suit: x
b. $367,000 - 2600x = 315,000$
c. $x = 20$;   the loss on each suit was $20.

53. a. Revenue per flight: x
b. $2000x - 26,480,000 = -1,630,000$
c. $x = 12,425$; the revenue per flight was $12,425.

CHAPTER 2 REVIEW (Page 61)

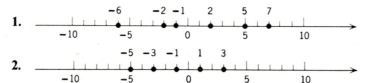

3. $-4, -3, 0, 2, 5$

4. a. $<$	b. $=$	c. $>$
5. a. -2	b. -6	c. 1
6. a. -3	b. -3	c. -3
7. a. -2	b. -4	c. 2
8. a. -5	b. 16	c. -2
9. a. 15	b. -24	c. 0

10. a. −4 b. 8 c. −8

11. a. −5 b. −10 c. −55

12. a. −1 b. −11 c. 3

13. a. 14 b. −2 c. 1

14. a. −3 b. 5 c. −3

15. a. −6 b. −5 c. 12

16. a. 4 b. −8 c. −10

17. a. Yesterday's temperature: x
b. $x - 17 = -23$
c. $x = -6$; yesterday's temperature was −6°.

18. a. Number of weeks Robert dieted: x
b. $231 - 4x = 187$
c. $x = 11$; Robert dieted for 11 weeks.

19. a. Number of boxes sold at a loss: x
b. $13,000 - 2x = 7800$
c. $x = 2600$; they sold 2600 boxes at a loss.

20. a. Bus fare from Columbus to Pasadena: x
b. $25,963 - 127x = 17,835$
c. $x = 64$; the fare is $64.

CHAPTER 2 CUMULATIVE REVIEW (Page 62)

1. Terms; factors

2. Absolute value

3. Multiplication and division; addition and subtraction

4. Equivalent

5. Integers

6.

7.

8. a. 6 b. −6 c. 0
d. Undefined e. 0 f. Indeterminate

9. a. 6 b. 6 c. −6 d. −6

10. $|-5|$

11. a. $-5x$ b. $\dfrac{12}{a - 3}$ c. $(8 + b) - 6$

12. a. $3x + 12 = -18$ b. $\dfrac{x + 7}{-4} = 5$ c. $4x - 6 = 22$

13. a. -15 **b.** 2 **c.** -1

14. a. -8 **b.** -14 **c.** 59

15. a. 28 **b.** -26 **c.** 6

16. a. -6 **b.** 9 **c.** -2

17. a. 4 **b.** -56 **c.** 11

18. a. -24 **b.** 14 **c.** -60

19. a. -5 **b.** 0 **c.** -2

20. a. -3 **b.** 12 **c.** -7

21. a. Average weight loss per member: x
  b. $37x = 666$
  c. $x = 18$; the average weight loss per member was 18 pounds.

22. a. Number of weeks until average temperature is 62°: x
  b. $-16 + 3x = 62$
  c. $x = 26$; the average temperature will be 62° in 26 weeks.

23. a. Caterer's hourly rate: x
  b. $125 + 6x = 209$
  c. $x = 14$; the caterer charges $14 per hour.

24. a. Jorge's monthly payments: x
  b. $568 - 12x = 184$
  c. $x = 32$; Jorge's monthly payments were $32 apiece.

25. a. Number of eggbeaters: x
  b. $75,000 - 12.95x = 10,250$
  c. $x = 5000$; there were 5000 eggbeaters in the first shipment.

EXERCISES 3.1 (Page 70)

1. $6y$ **3.** $-4x$ **5.** $2a$ **7.** $15x$

9. $-2z$ **11.** 0 **13.** $8x$ **15.** $3a$

17. $-3y$ **19.** $6x$ **21.** $-4g$ **23.** $-2s$

25. $2t + 3$ **27.** $12y$ **29.** $x - 2$ **31.** $6 - 9a$

33. $6a$ **35.** $3b$ **37.** $3x$ **39.** $9x + 2$

41. -1 **43.** $2r - 1$ **45.** $3x$ **47.** $y + 7$

49. $2a + 10$ **51.** $x - 5$ **53.** $2x + 3$ **55.** $-2y + (y + 1)$

57. $y - (4 + y)$ **59.** $12 - (3a - 4)$ **61.** $3x - 12$ **63.** $10y - 10$

65. $-2x - 16$ **67.** $25a - 20$ **69.** $3 - 5b$ **71.** $36 - 12t$

73. $-4x - 6$ **75.** $4y - 20$ **77.** $-14z - 15$ **79.** $3x - 12$

81. $5 - 9x$ **83.** $6 - 2a$

85. Let $x = 2$. Then
  $2 + 7x = 2 + 7(2) = 16$, and
  $9x = 9(2) = 18$.

87. Let $a = 2$. Then
  $-(a - 3) = -(2 - 3) = 1$, and
  $-a - 3 = -2 - 3 = -5$.

Thus $2 + 7x$ and $9x$ are not equivalent

Thus $-(a - 3)$ and $-a - 3$ are not equivalent.

89. Let $x = 5$. Then
$2(3x) = 2(3 \cdot 5) = 30$, and
$2 \cdot 3 \cdot 2 \cdot x = 2 \cdot 3 \cdot 2 \cdot 5 = 60$.
Thus $2(3x)$ and $2 \cdot 3 \cdot 2 \cdot x$
are not equivalent.

EXERCISES 3.2 (Page 75)

1. 2	**3.** 0	**5.** 4	**7.** -5
9. 0	**11.** 8	**13.** 5	**15.** 1
17. 3	**19.** 2	**21.** -2	**23.** 2
25. 0	**27.** 10	**29.** 4	**31.** 6
33. 15	**35.** 3	**37.** 2	**39.** 6
41. -2	**43.** -27	**45.** 10	**47.** -3
49. -2	**51.** 7	**53.** -1	**55.** 3
57. -1	**59.** -2	**61.** 3	**63.** -5
65. 4	**67.** -5	**69.** 0	**71.** 2
73. -4	**75.** 1	**77.** 2	**79.** -4
81. 100	**83.** 50	**85.** 3	**87.** 9
89. 5000			

EXERCISES 3.3 (Page 79)

1. Number of professors: x
Number of students: $10x$

3. Number of goats: x
Number of sheep: $x + 50$

5. Delbert's salary: x
Francine's salary: $25,000 - x$

7. Price of the cheaper camera: x
Price of the expensive camera: $x + 300$

9. Number of quarters: x
Number of dimes: $2x$

11. Number of 22¢ stamps: x
Number of 14¢ stamps: $50 - x$

13. Amount invested at 9%: x
Amount invested at 7%: $x + 1000$

15. List price of the TV: x
Discount: $0.20x$

17. Length of first piece: x
Length of second piece: $5x - 2$

19. Steve's speed: x
Ron's speed: $x + 20$

21. Distance Wayne traveled: x
Distance Harry traveled: $3x$

23. Distance Jason drives in one hour: x
Distance Melissa drives in one hour: $90 - x$

25. Number of gallons of oil: x
Number of gallons of vinegar: $10 - x$

27. Number of pounds of caramels: x
Number of pounds of creams: $5 - x$

29. First odd integer: x
Second odd integer: $x + 2$
Third odd integer: $x + 4$

31. a. $x + 4$ b. $5x$ c. $5(x + 4)$

33. $3(n + 6)$

35. a. $27 - n$ b. $3n$ c. $3(27 - n)$

37. a. $n + 16$ b. $5n$ c. $2(n + 16)$

39. a. $10 - x$ b. $2x$ c. $5(10 - x)$

41. a. $x + 4$ b. $25x$ c. $10(x + 4)$

43. a. $n - 3$ b. $n + 2$ c. $25n$
d. $10(n - 3)$

45. $5n + 10(n + 3)$ or $15n + 30$

47. a. $x - 4$ b. $7x$ c. $6(x - 4)$

EXERCISES 3.4 (Page 87)

1. *Steps 1, 2*   Smaller integer: x
Larger integer: $x + 2$
Step 4 $(x + 2) + 4x = 17$
Step 6 $x = 3$; the smaller integer is 3 and the larger is 5.

3. *Steps 1, 2*   Capacity of one tank: x
Capacity of other tank: $x + 4$
Step 4 $3x + 3(x + 4) = 144$
Step 6 $x = 22$; the smaller tank holds 22 gallons; the larger 26.

5. *Steps 1, 2*   First even integer: x
Next even integer: $x + 2$
Step 4 $x + (x + 2) = 26$
Step 6 $x = 12$; the integers are 12 and 14.

7. *Steps 1, 2*   First odd integer: x
Next odd integer: $x + 2$
Step 4 $x + (x + 2) = 32$
Step 6 $x = 15$; the integers are 15 and 17.

9. *Steps 1, 2*   First integer: x
Second integer: $x + 1$
Third integer: $x + 2$

Step 4 $x + (x + 1) + (x + 2) = -33$
Step 6 $x = -12$; the integers are -12, -11, and -10.

11. *Steps 1, 2*   First odd integer: x
Second odd integer: $x + 2$
Third odd integer: $x + 4$

Step 4 $x + (x + 2) + (x + 4) = -21$
Step 6 $x = -9$; the integers are -9, -7 and -5.

13. *Steps 1, 2*   First even integer: x
Second even integer: $x + 2$
Third even integer: $x + 4$

Step 4 $x + (x + 2) + (x + 4) = 4x$
Step 6 $x = 6$; the integers are 6, 8, and 10.

15. *Steps 1, 2*   Votes for losing candidate: x
Votes for winning candidate: $x + 50$

Step 4 $x + (x + 50) = 4376$
Step 6 $x = 2163$; the loser received 2163 votes; the winner 2213.

17. *Steps 1, 2*   Length of shorter piece: x
Length of longer piece: $x + 24$

Step 4 $x + (x + 24) = 144$
Step 6 $x = 60$; the shorter piece is 60 centimeters long; the longer piece is 84 centimeters long.

19. *Steps 1, 2*   Length of longer piece: x
Length of shorter piece: $x - 4$

Step 4 $x + (x - 4) = 20$
Step 6 $x = 12$; one piece is 12 feet long; the other 8 feet.

21. *Steps 1, 2*   Length of first piece: x
Length of second piece: $3x$
Length of third piece: $x + 4$

Step 4 $x + 3x + (x + 4) = 24$
Step 6 $x = 4$; the first piece is 4 feet long; the second piece 12 feet, and the third piece 8 feet.

23. *Steps 1, 2*   Length of first piece: x
Length of second piece: $x + 3$
Length of third piece: $2x$

Step 4 $x + (x + 3) + 2x = 51$
Step 6 $x = 12$; the first piece is 12 feet long; the second 15 feet, and the third 24 feet.

25. *Steps 1, 2*   Length of each of the four pieces: x
   Step 4 $4x + 24 = 340$
   Step 6 $x = 79$;   each of the four pieces is 79 centimeters long.

27. *Steps 1, 2*   Distance from town A to town B: x
   Step 4 $2x + 5 = 19$
   Step 6 $x = 7$;   it is 7 kilometers from town A to town B.

29. *Steps 1, 2*   Distance slower car traveled: x
   Step 4 $x + 54 = 4x$
   Step 6 $x = 18$;   the slower car traveled 18 miles.

31. *Steps 1, 2*   Numbers of nickels: x
      Number of dimes: $x + 3$
   Step 4 $0.05x + 0.10(x + 3) = 1.80$
   Step 6 $x = 10$;   she had 10 nickels and 13 dimes.

33. *Steps 1, 2*   Number of pennies: x
      Number of nickels: $x + 6$
      Number of dimes: $x - 6$
   Step 4 $0.01x + 0.05(x + 6) + 0.10(x - 6) = 1.14$
   Step 6 $x = 9$;   she had 9 pennies, 15 nickels and 3 dimes.

35. *Steps 1, 2*   Number of adult tickets: x
      Number of children's tickets: $1000 - x$
   Step 4 $4.25x + 1.50(1000 - x) = 3150$
   Step 6 $x = 600$;   600 adult tickets were sold, and 400 children's.

37. *Steps 1, 2*   Capacity of smaller truck: x
      Capacity of larger truck: $x + 4$
   Step 4 $5x + 7(x + 4) = 112$
   Step 6 $x = 7$;   the smaller truck holds 7 tons; the larger 11 tons.

39. *Steps 1, 2*   Delbert's salary: x
      Francine's salary: $25,000 - x$
   Step 4 $2x + (25,000 - x) = 34,000$
   Step 6 $x = 9000$;   Delbert's salary is $9000 and Francine's is $16,000.

EXERCISES 3.5 (Page 95)

1. *Steps 1, 2*   Amount invested at 10%: x
      Amount invested at 11%: $x + 6200$
   Step 4 $0.10x + 0.11(x + 6200) = 1060$
   Step 6 $x = 1800$; $1800 is invested at 10%, and $8000 at 11%.

3. *Steps 1, 2*   Amount invested at 8%: x
                       Amount invested at 9%: $34,000 - x$
Step 4 $0.08x = 0.09(34,000 - x)$
Step 6 $x = 18,000$; $18,000 is invested at 8%; $16,000 at 9%.

5. *Steps 1, 2*   Amount invested in stocks: x
                       Amount invested in bonds: $x + 100$
Step 4 $0.08x = 0.06(x + 100) + 8$
Step 6 $x = 700$;   Rhoda invested $700 in stocks and $800 in bonds.

7. *Steps 1, 2*   Amount borrowed at 12%: x
                       Amount borrowed at 15%: $30,000 - x$
Step 4 $0.12x + 0.15(30,000 - x) = 3750$
Step 6 $x = 25,000$;   Mario borrowed $25,000 at 12% and $5000 at 15%.

9. *Steps 1, 2*   Number of quarts of 20% solution: x
Step 4 $0.20x + 0.50(30) = 0.40(x + 30)$
Step 6 $x = 15$;   15 quarts of 20% solution should be added.

11. *Steps 1, 2*   Number of ounces of 50% alloy: x
Step 4 $0.50x + 0.70(40 - x) = 0.55(40)$
Step 6 $x = 30$;   30 ounces of 50% alloy must be used.

13. *Steps 1, 2*   Number of grams of 50¢ metal: x
Step 4 $50x + 32(20) = 40(x + 20)$
Step 6 $x = 16$;   16 grams of 50¢ metal is needed.

15. *Steps 1, 2*   Number of kilograms of fine powder: x
Step 4 $30x + 12(50) = 20(x + 50)$
Step 6 $x = 40$;   40 kilograms of fine powder is needed.

17. *Steps 1, 2*   Hours elapsed when the cyclists are 224 miles apart: x
Step 4 $14x + 18x = 224$
Step 6 $x = 7$;   they will be 224 miles apart after 7 hours.

19. *Steps 1, 2*   Hours Francine drives to catch Delbert: x
Step 4 $70x = 40(x + 3)$
Step 6 $x = 4$;   it takes Francine 4 hours to catch Delbert.

21. *Steps 1, 2*   Speed of paddleboat in still water: x
Step 4 $5(x + 8) = 9(x - 8)$
Step 6 $x = 28$;   the speed of the paddleboat is 28 miles per hour.

23. *Steps 1, 2*   Speed of the balloon in still air: x
Step 4 $6(x + 8) = 8(x - 2)$
Step 6 $x = 32$;   the distance from Fresno to Fullerton is $6(32 + 8) = 240$ miles.

EXERCISES 3.6 (Page 102)

1. $<$ **3.** $>$ **5.** $<$ **7.** $x > 4$

9. $y < -15$ **11.** $x \geq 3$ **13.** $x > 25$ **15.** $y > -5$

17. $x \leq 12$ **19.** $y > 3$ **21.** $x < -21$ **23.** $x \geq 12$

25. $t > 40$

27. $x > 2$;

29. $x > -3$;

31. $x \leq -6$;

33. $x > 3$;

35. $y > -5$;

37. $y < -10$;

39. $x < 3$;

41. $x < 3$;

43. $x \geq 6$;

45. $t < 3$;

47. $y \leq -4$;

49. $x \geq -2$;

51. *Steps 1, 2* Smallest integer: x

Step 4 $x + (x + 1) + (x + 2) > 93$

Step 6 $x > 30$; the smallest integer could be 31, 32, 33, and so on.

53. *Steps 1, 2* Integer: x

Step 4 $4x \geq 10 + 3x$

Step 6 $x \geq 10$; the integer could be 10, 11, 12, and so on.

55. *Steps 1, 2*   Length of shortest piece: x
Step 4 $40 - (x + 3x) \geq 4$
Step 6 $x \leq 9$; the shortest piece can be 9 feet or less.

57. *Steps 1, 2*   Miles driven: x
Step 4 $25 + 0.05\,x \leq 50$
Step 6 $x \leq 500$; he can drive 500 miles or less.

CHAPTER 3 REVIEW (Page 106)

1. a. $8x + 4$ b. $-13y - 1$ **2.** a. 13 b. $12z - 2$
3. a. $12x - 1$ b. $6x$ **4.** a. $-21y - 8$ b. $56 - 14y$
5. a. 5 b. 0 c. -1
6. a. -18 b. 9 c. -3
7. a. -2 b. 6 c. 3
8. $24 - x$ **9.** $10x$ **10.** $25(x + 3)$ **11.** $185(x + 4)$
12. *Steps 1, 2*   First odd integer: x
Second odd integer: $x + 2$
Third odd integer: $x + 4$
Step 4 $x + (x + 2) + (x + 4) = 5x$
Step 6 $x = 3$; the integers are 3, 5, and 7.

13. *Steps 1, 2*   Length of first piece: $x + 3$
Length of second piece: x
Length of third piece: $x + 5$
Step 4 $x + (x + 3) + (x + 5) = 32$
Step 6 $x = 8$; the first piece is 11 feet long; the second is 8 feet, and the third is 13 feet.

14. *Steps 1, 2*   Home team's score: x
Visiting team's score: $x - 17$
Step 4 $x + (x - 17) = 105$
Step 6 $x = 61$; the final score was 61 for the home team and 44 for the visitors.

15. *Steps 1, 2*   Number of electric sharpeners: x
Number of mechanical sharpeners: $86 - x$
Step 4 $8x + 3(86 - x) = 443$
Step 6 $x = 37$; they sold 37 electric sharpeners and 49 mechanical.

16. *Steps 1, 2*   Amount invested at 8%: x
Amount invested at 6%: $x + 300$
Step 4 $0.08x + 0.06(x + 300) = 242$
Step 6 $x = 1600$; he invested $1600 at 8% and $1900 at 6%.

17. *Steps 1, 2*   Number of quarts of pure antifreeze: x

     Step 4 $1.00x + 0.30(5) = 0.50(x + 5)$

     Step 6 $x = 2$; 2 quarts of pure antifreeze must be added.

18. *Steps 1, 2*   Numbers of pounds of cashews: x

                   Number of pounds of peanuts: $20 - x$

     Step 4 $1.20x + 0.80(20 - x) = 0.90(20)$

     Step 6 $x = 5$; 5 pounds of cashews and 15 pounds of peanuts should be used.

19. *Steps 1, 2*   Jake's rate for the Pony Express: x

                   His rate returning home: $x - 8$

     Step 4 $6x = 9(x - 8)$

     Step 6 $x = 24$; Jake rides $6(24) = 144$ miles.

20. a. $y < -5$

b. $x > -6$

c. $x < -8$

CHAPTER 3 CUMULATIVE REVIEW (Page 107)

1. Numerical coefficients

2. Two expressions are equivalent if they name the same number for all replacements of the variable.

3. Equivalent equations are equations that have identical solutions.

4. Distributive

5. Multiply or divide; negative number

6. a. -2 b. 8   c. 48

7. a. -5 b. 0   c. 9

8. a. -2 b. 2   c. 18

9. a. $7x - 3$ b. $-2x - 10$ c. $11x - 11$

10. a. $18x - 16$ b. $2 - 14x$ c. -9

11. a. 7   b. 9   c. -4

12. a. $x + 2$ b. $x + 2$ **13.** $3x$

15. a. $y > -4$

b. $y \geq -14$

16. a. 1   b. 2

17. a. 20   b. −6

18. a. 18   b. 4

19. a. 7   b. −2

20. a. −15   b. 5

21. *Steps 1, 2*   Number of hours worked: x
    Step 4 $1.60x + 0.75 = 7.15$
    Step 6 $x = 4$; he worked 4 hours.

22. *Steps 1, 2*   Amount invested at 10%: x
                     Amount invested at 12%: $x + 500$
    Step 4 $0.10x + 0.12(x + 500) = 324$
    Step 6 $x = 1200$; $1200 is invested at 10% and $1700 at 12%.

23. *Steps 1, 2*   Smaller number: x
                     Larger number: $x + 6$
    Step 4 $10x - 4(x + 6) = 6$
    Step 6 $x + 5$; the numbers are 5 and 11.

24. *Steps 1, 2*   Smaller number: x
                     Larger number: $x + 10$
    Step 4 $8x + 3(x + 10) = 129$
    Step 6 $x = 9$; the numbers are 9 and 19.

25. *Steps 1, 2*   Number of dimes: x
                     Number of nickels: $x + 8$
    Step 4 $0.10x + 0.05(x + 8) = 2.65$
    Step 6 $x = 15$; she had 15 dimes and 23 nickels.

EXERCISES 4.1 (Page 111)

1. 6^2 **3.** x^3 **5.** $(-3)^2 y^2$ **7.** $2x^2 y^3$

9. $2a^2 bc^3$ **11.** $(x - 3)^2$ **13.** $3^2 + 2^3$ **15.** $3x^2 + 5y^3$

17. $(-3)^2 - b^2$ **19.** $x^3 + x^2 y^2$ **21.** $3a^2 - (-3c)^2$ **23.** $(x - y)^2 + y^2$

25. $5xxyyy$ **27.** $-2aabbc$ **29.** $yy(3x)(3x)$

31. $(5a)(5a)(-2b)(-2b)(-2b)$ **33.** $3xx$

35. $(3x)(3x)$ **37.** $(a - 4)(a - 4)(a - 4)$

39. $-yyy(3y + 4)(3y + 4)$ **41.** $xxyyy(x - y)(x - y)$

43. Let y be 2. Then
$$3y^2 = 3 \cdot 2^2$$
$$= 3 \cdot 4 = 12, \text{ and}$$

45. Let a be 2. Then
$$a^3 = 2^3 = 8, \text{ and}$$
$$3a = 3 \cdot 2 = 6.$$
Thus a^3 and $3a$ are not equivalent.

$(3y)^2 = (3 \cdot 2)^2$
$ = 6^2 = 36.$

Thus $3y^2$ and $(3y)^2$ are not equivalent.

47. Let x be 3. Then

$-x^2 = -3^2 = -9,$ and
$(-x)^2 = (-3)^2 = 9.$

Thus $-x^2$ and $(-x)^2$ are not equivalent.

49. -9 **51.** -27 **53.** -12 **55.** 2

57. -4 **59.** -16 **61.** $2 \cdot 3^3$ **63.** $2^2 \cdot 3^2 \cdot 5$

65. $2 \cdot 3 \cdot 5 \cdot 7$ **67.** 5^4

EXERCISES 4.2 (Page 114)

1. -10 **3.** 5   **5.** 256   **7.** -31
9. -200 **11.** -12 **13.** -14 **15.** -32
17. -79 **19.** 0   **21.** -1 **23.** -7
25. 14   **27.** 3   **29.** 90   **31.** -512
33. 12   **35.** -4 **37.** -15 **39.** -48
41. 16   **43.** 10   **45.** -14 **47.** -6
49. 3   **51.** -4 **53.** -12 **55.** 1
57. 3   **59.** -3 **61.** 7   **63.** 0
65. -24 **67.** 5   **69.** 0   **71.** -9
73. -12

EXERCISES 4.3 (Page 119)

1. a. Monomial   b. $2y^2$, 2
3. a. Binomial   b. x, 1; $-y^3$, -1
5. a. Trinomial   b. $-3x^2$, -3; $3y$, 3; $4z$, 4
7. a. Binomial   b. $-x^2y$, -1; yz^2, 1
9. a. Binomial   b. $7x^2y$, 7; x^2y^2, 1
11. a. Monomial   b. $7x^2yz^3$, 7
13. $-2x^2$, 2; x, 1; $-4x$, 1
15. $-y^4$, 4; $-2y^3$, 3; $-y$, 1
17. z^5, 5; $-2z^2$, 2; z, 1
19. 4   **21.** 5   **23.** 4
25. $3x^2$ **27.** $3y^2$ **29.** $3b^4$

31. $3x$ **33.** $-4y$ **35.** $6y^2$

37. $6x$ **39.** $3ab^2$ **41.** $7x - 3y$

43. $4ab^2 + a^2b$ **45.** $9a^2b - ab^2$ **47.** $5x^2y + xy^2$

49. $x^3yz + 5x^2yz + 2xyz$ **51.** $22xz^2 + 17x^2z$ **53.** $3x$

55. $x + 7$ **57.** $2a + 10b$ **59.** $-4y^2 - 3y$

61. $-z^2 - 5z$ **63.** $2xy^2 + 4xy - 2x$ **65.** $x + y + 2z$

67. $-g + 3h - 3k$ **69.** $6x^2 + 2x$ **71.** $-7y^2 + y + 3$

73. $x - 11$ **75.** $-7x^2 + 7x - 1$ **77.** $5y^2 + 4y - 2$

79. Let x be 2. Then

$$7x^2 + 5x^2 = 7(2)^2 + 5(2)^2 = 48, \quad \text{and}$$
$$12x^4 = 12(2)^4 = 192.$$

Thus $7x^2 + 5x^2$ and $12x^4$ are not equivalent.

81. Let x be 1 and y be 2. Then

$$4x - 2y = 4(1) - 2(2) = 0, \quad \text{and}$$
$$2xy = 2(1)(2) = 4.$$

Thus $4x - 2y$ and $2xy$ are not equivalent.

EXERCISES 4.4 (Page 126)

1. x^6 **3.** y^{10} **5.** b^9

7. a^{14} **9.** x^{13} **11.** t^{60}

13. t^{15} **15.** x^{36} **17.** y^{21}

19. b^{70} **21.** x^{200} **23.** c^{540}

25. $32x^{15}$ **27.** $256x^8y^{16}$ **29.** $243a^{15}b^5c^{30}$

31. $-27x^9y^3z^9$ **33.** $4x^5$ **35.** a^3b^5c

37. $10a^2b^5c^4$ **39.** $12x^2y$ **41.** $-5x^3y$

43. $8x^5y^3$ **45.** $8a^4b^4$ **47.** $-36a^3b^7c^4$

49. $162x^6y^{10}z^7$ **51.** $(a + b)^6$ **53.** $(a + b)^8$

55. $(x^2 + 2x)^{24}$ **57.** $(x^2 + 2x)^{11}$ **59.** x^{13}

61. x^7y^6 **63.** $4x^{10}y^{14}$ **65.** $-2a^4b^8c^4$

67. $20a^{13}b^{11}c^{11}$ **69.** $-x^3$ **71.** $-a^3$

73. $-x^3y^4$ **75.** $9x^6y^4$ **77.** $-4x^5y^{10}$

79. $405xy^4z^{11}$ **81.** $6a^2b^6 - b^2$ **83.** $-52y^7$

85. $2a^9 + 2a^8$ **87.** $80ab^4 - 54ab^3 + 3b^3$

89. $4a^5b^3 + a^7b^3 + 3a^3b$

91. Let x be 2. Then

$$x^2 \cdot x^4 = 2^2 \cdot 2^4 = 64, \quad \text{and}$$
$$x^8 = 2^8 = 256.$$

Thus $x^2 \cdot x^4$ and x^8 are not equivalent.

93. Let x be 3. Then

$$x^2 \cdot x^4 = 3^2 + 3^2 = 18, \quad \text{and}$$
$$x^4 = 3^4 = 81.$$

Thus $x^2 + x^2$ and x^4 are not equivalent.

95. Let a be 1 and b be 2. Then

$$a^2 \cdot b^3 = 1^2 \cdot 2^3 = 8, \quad \text{and}$$
$$(ab)^5 = (1 \cdot 2)^5 = 32.$$

Thus $a^2 \cdot b^3$ and $(ab)^5$ are not equivalent.

97. Let x be 2. Then

$$(x^2)^3 = (2^2)^3 = 64, \quad \text{and}$$
$$x^5 = 2^5 = 32.$$

Thus $(x^2)^3$ and x^5 are not equivalent.

99. Let x be 2. Then

$$(x^2)^3 = (2^2)^3 = 64, \quad \text{and}$$
$$x^8 = 2^8 = 256.$$

Thus $(x^2)^3$ and x^8 are not equivalent.

101. Let x be 3 and y be 2. Then

$$(xy)^2 = (3 \cdot 2)^2 = 36, \quad \text{and}$$
$$xy^2 = 3 \cdot 2^2 = 12.$$

Thus $(xy)^2$ and xy^2 are not equivalent.

103. Let x be 5. Then

$$-x^2 = -5^2 = -25, \quad \text{and}$$
$$(-x)^2 = (-5)^2 = 25.$$

Thus $-x^2$ and $(-x)^2$ are not equivalent.

105. Let x be 3. Then

$$-2x^2 = -2 \cdot 3^2 = -18, \quad \text{and}$$
$$(-2x)^2 = (-2 \cdot 3)^2 = 36.$$

Thus $-2x^2$ and $(-2x)^2$ are not equivalent.

107. Let y be -1. Then

$$-2y^3 = -2(-1)^3 = 2, \quad \text{and}$$
$$-(2y)^3 = -[2(-1)]^3 = 8.$$

Thus $-2y^3$ and $-(2y)^3$ are not equivalent.

EXERCISES 4.5 (Page 130)

1. -9 **3.** 5   **5.** 2   **7.** 2

9. -32 **11.** 3   **13.** 2   **15.** 92

17. 120   **19.** $t = \dfrac{d}{r}$ **21.** $l = \dfrac{v}{wh}$ **23.** $d = \dfrac{c}{\pi}$

25. $r = \dfrac{d}{t}$ **27.** $h = \dfrac{v}{lw}$ **29.** $r = \dfrac{I}{pt}$ **31.** $g = \dfrac{v - k}{t}$

33. $m = \dfrac{Fd^2}{kM}$ **35.** $l = \dfrac{P - 2w}{2}$ **37.** $t = \dfrac{A - P}{Pr}$ **39.** $A = s^2$

41. $A = l \cdot w$ **43.** $V = lwh$ **45.** $I = Pr$ **47.** $I = \dfrac{E}{R}$

49. $V = \pi r^2 h$ **51.** $A = \dfrac{h(B + b)}{2}$

EXERCISES 4.6 (Page 134)

1. 18.84   **3.** 12   **5.** 16   **7.** 3

9. 8   **11.** 4   **13.** 9 square centimeters

15. 120 square meters   **17.** 104 square inches

19. 144 square centimeters   **21.** 93.76 square centimeters

23. 624.76 square inches

25. *Steps 1, 2*   Length of the rectangle: l
Step 4 $2l + 2(42) = 210$
Step 6 $l = 63$;   the length is 63 feet.

27. *Steps 1, 2*   Altitude of the triangle: h
Step 4 $144 = \dfrac{16h}{2}$
Step 6 $h = 18$;   the altitude is 18 centimeters.

29. *Steps 1, 2*   Width of the rectangle: w
                Length of the rectangle: $w + 10$
Step 4 $2w + 2(w + 10) = 164$
Step 6 $w = 36$;   the width is 36 meters, and the length is 46 meters.

31. *Steps 1, 2*   Length of the base: b
                Length of equal sides: $b + 4$
Step 4 $b + (b + 4) + (b + 4) = 56$
Step 6 $b = 16$;   the base is 16 meters long; the equal sides are each
                20 meters long.

33. *Steps 1, 2*   Smallest angle: x
                Largest angle: $2x$
                Third angle: $2x - 10$
Step 4 $x + 2x + (2x - 10) = 180$
Step 6 $x = 38$;   the angles are 38°, 76°, and 66°.

35. *Steps 1, 2*   Width of the rectangle: x
                Length of the rectangle: $3x$
Step 4 $2x + 2(3x) = 56$
Step 6 $x = 7$;   the width is 7 feet; the length is 21 feet.

37. *Steps 1, 2*   Length of the side: s
Step 4 $4s = 24$
Step 6 $s = 6$;   the side is 6 feet long.

39. *Steps 1, 2*   Third angle: x
Step 4 $40 + 70 + x = 180$
Step 6 $x = 70$;   the third angle is 70°.

41. *Steps 1, 2*   Smallest angle: x
Second angle: $x + 10$
Third angle: $x + 20$
Step 4 $x + (x + 10) + (x + 20) = 180$
Step 6 $x = 50$;   the angles are 50°, 60°, and 70°.

43. *Steps 1, 2*   Smaller acute angle: x
Larger acute angle: $2x$
Step 4 $x + 2x + 90 = 180$
Step 6 $x = 30$;   the smaller angle is 30°; the larger is 60°.

CHAPTER 4 REVIEW (Page 139)

1. a. $2^2a^2b^3$ b. xy^2z^3 c. $(-3)^2c^2d$

2. a. $2 \cdot 3xyy$ b. $aaabb$ c. $-3 \cdot 3 \cdot 3cdd$

3. a. 5   b. 9   c. -35

4. a. 50   b. 26   c. 52

5. a. -2 b. 0   c. 3

6. Binomial   **7.** Numerical coefficient

8. 1   **9.** 4

10. a. $2x$ b. $8x^2 - xy - 3y^2$ c. $x^2y + 3xy + 3xy^2$

11. a. $x^2 - x - 1$ b. $x^2 - 5x + 13$ c. $-x^2 + 3y^2 - 2z^2$

12. a. $3y + z$ b. $a + 2b - 5c$ c. $-x$

13. a. $-x^3y^3$ b. $12ab^4$ c. r^4s^3

14. a. $72a^9b^8$ b. $-864x^{14}y^9$ c. $63s^8t^{23}$

15. a. $(4x - 1)^9$ b. $(4x - 1)^{20}$ c. $(4x - 1)^{13}$

16. a. $4x^4$ b. $2ab^4 - a^4b^2 + 6a^3b^2$ c. $r^4s^6 - 9r^4s^4$

17. a. $a = \dfrac{f}{m}$ b. $g = \dfrac{v - k}{t}$ c. $b = 2M - a$

18. $V = \dfrac{4}{3}\pi r^3$

19. *Steps 1, 2*   Length of the garden: x
Step 4 $2x + 2(34) = 144$
Step 6 $x = 38$;   the garden is 38 meters long.

20. *Steps 1, 2*   First angle: $x + 15$
Second angle: x
Third angle: $x + 30$
Step 4 $(x + 15) + x + (x + 30) = 180$
Step 6 $x = 45$; the angles are 60°, 45°, and 75°.

CHAPTER 4 CUMULATIVE REVIEW (Page 141)

1. 3^2, $\dfrac{4^2}{2}$, $\dfrac{4 + 2^3}{12}$, $5^2 - 3^2$

2. a. Numerical coefficient   **b.** Exponent

3. a. $2y^3$ **b.** $2 \cdot 2 \cdot 3 \cdot 3xyyy$

4. Positive   **5.** 16

6. a. x^5 **b.** x^6

7. a. 81   **b.** -81

8. a. $-a + b - c$ **b.** $-a - b + c$

9. a. 5   **b.** 0

10. a. $-3x^4$ **b.** $27m^4n^4$

11. a. $2a - 4b + c$ **b.** $-10x^2 - 3x + 15$

12. a. $3x^3 - 2y^3$ **b.** 0

13. a. $p^7q^6 - q^6$ **b.** $27p^6q^4 - 2p^6q^6$

14. 520 feet   **15.** $m = \dfrac{E - h}{kv^2}$

16. a. $y \geq -6$

b. $y < 16$

17. a. 1   **b.** 3

18. *Steps 1, 2*   First even integer: x
Second even integer: $x + 2$
Third even integer: $x + 4$

Step 4 $x + (x + 2) + (x + 4) = -48$

Step 6 $x = -18$; the integers are -18, -16, and -14.

19. *Steps 1, 2*   Time for first leg of the journey: x

Step 4 $40x + 60(16 - x) = 880$

Step 6 $x = 4$; the first leg takes 4 hours.

20. *Steps 1, 2*   Number of quarts of eau de cologne: x
Number of quarts of dish soap: $20 - x$

Step 4 $12x + 2(20 - x) = 4(20)$

Step 6 $x = 4$; she uses 4 quarts of eau de cologne and 16 quarts of dish soap.

21. True   **22.** False   **23.** False

24. True   **25.** False

EXERCISES 5.1 (Page 143)

1. $10a^2 + 6a$ **3.** $-b^2 + 2b$ **5.** $x^2y + xy^2$

7. $-2x^3 - 3x^2y$ **9.** $x^3 - 2x^2 + x$ **11.** $-y^3 + y^2 - 2y$

13. $4x^5 - 12x^4 + 16x^3$ **15.** $-y^6 + y^4 - y^3$

17. $-x^3y - x^2y^2 - xy^3$ **19.** $6x^5 - 12x^3 - 10x^2$

21. $-15a^4 + 10a^3b - 5a^2b^2$ **23.** $-8p^5q^2 + 12p^4q^3 - 4p^3q^4$

25. $-3x^2y^3 - 2x^7y^3 + 7x^2y^7$ **27.** $18x^8 - 30x^7 + 6x^5 - 12x^4 + 6x^3$

29. $-6a^5b^5 + 36a^4b^7 - 18a^6b^4 + 48a^3b^{11}$

31. $-a$ **33.** $2ax + x + a$

35. $-ax + ay - 2y$ **37.** $x^2 + 4x + 1$

39. $3x - 12y$ **41.** $3y^2 - 25y$

43. $ax^2 + 3ax - a$ **45.** $-4abx - aby + 2ab + b$

47. $6ab^2 + a^2b^2$ **49.** $3p^5q^3 - 4p^5q^4 + 11p^3q^4$

EXERCISES 5.2 (Page 146)

1. $3x^2$ **3.** $2x$ **5.** $2a^2b$ **7.** $-4b^2$

9. $3(x + 2)$ **11.** $2(x - 3y)$ **13.** $2y(y - 1)$

15. $y(ay + 1)$ **17.** $3y(3ay + 2)$ **19.** $7x^2(4x + 5)$

21. $4x^3(1 + 5x)$ **23.** $3xy(8y - 9x)$ **25.** $8t^4(9t^2 + 5)$

27. $4a^2b^2(7b - 11a^2)$ **29.** $3(y^2 - y + 1)$ **31.** $a(x + y - z)$

33. $x(x - 3 + y)$ **35.** $2y(2y^2 - y + 1)$ **37.** $6axy(x - 3y + 4)$

39. $7xy(2x^2 - 5xy + 3y^2)$ **41.** $5ab(5ab^2 - 1 - 6a^2b)$

43. $3t^5(16t^2 + 9 - 15t^6)$ **45.** $p^2q(16q^2 - 1 + 11pq - 4p^2q)$

47. $3abc^2(ac - 4b + 5a^4b^2)$ **49.** $-a(a + b)$

51. $-x(1 + x)$ **53.** $-b(ac + a + c)$ **55.** $-3y(2y^2 + y + 1)$

57. $-x(1 - x + x^2)$ **59.** $-xy^2(y^3 + y^2 - 1)$ **61.** $(x - 4)(6 + a)$

63. $(x + 6)(2x - 3)$ **65.** $(3x - 2)(4 - 3x)$ **67.** $(2x + 3)(3x^2 - 1)$

69. $(4 - x)(x^2 + 3x - 2)$ **71.** $d = k(1 + at)$ **73.** $S = kr^2(h + 1)$

75. $V = 2ga^2(D - d)$ **77.** $A = r^2(a + b + c)$

EXERCISES 5.3 (Page 150)

1. $x^2 + 7x + 12$ **3.** $y^2 - 2y - 3$ **5.** $a^2 + 7a + 10$

7. $b^2 - 2b - 8$ **9.** $x^2 + 9x + 8$ **11.** $y^2 - 8y + 7$

13. $a^2 + 8a + 16$ **15.** $b^2 - 25$ **17.** $x^2 + 2x + 1$

19. $y^2 - 1$ **21.** $4 - x^2$ **23.** $36 - y^2$

25. $x^2 + 8x + 16$

27. $x^2 - 14x + 49$

29. $x^2 - 2x + 1$

31. $x^2 + 4x + 4$

33. $x^2 - 4bx + 3b^2$

35. $x^2 + xy - 2y^2$

37. $x^2 + 4ax + 4a^2$

39. $a^2 - 2ab + b^2$

41. $y^2 - 36a^2$

43. $x^2 - t^2$

45. $2x^2 + 6x + 4$

47. $6y^2 + 60y + 150$

49. $6x^2 - 12x + 6$

51. $a^3 + 4a^2 - 5a$

53. $a^3 - 4a$

55. $xy^2 - 6xy + 9x$

57. $x^3 + 4x^2 + 5x + 2$

59. $x^3 - 1$

61. $x^3 - 5x^2 + 8x - 4$

63. $x^3 + 6x^2 + 11x + 6$

65. $x^3 - 6x^2 + 12x - 8$

67. $x^3 - 2x^2 - 5x + 6$

69. Let x be 1. Then

$$(x + 3)^2 = (1 + 3)^2 \qquad x^2 + 3^2 = 1^2 + 3^2$$
$$= 4^2 = 16; \qquad\qquad\quad = 1 + 9 = 10$$

Thus $(x + 3)^2$ and $x^2 + 3^2$ are not equivalent.

EXERCISES 5.4 (Page 155)

1. $(x + 2)(x + 3)$

3. $(x + 6)(x + 5)$

5. $(x + 9)(x + 5)$

7. $(y - 1)(y - 2)$

9. $(y - 7)(y - 9)$

11. $(x - 4)(x + 3)$

13. $(y + 5)(y - 4)$

15. $(a + 7)(a - 5)$

17. $(b - 20)(b + 1)$

19. $(a - 10)(a + 5)$

21. $(b - 9)(b + 5)$

23. $(y - 45)(y + 1)$

25. $2(x + 3)(x + 2)$

27. $y(y - 3)(y + 1)$

29. $5(c - 2)(c - 3)$

31. $4b(a + 6)(a - 3)$

33. $(x + 2a)(x + 2a)$

35. $(a - b)(a - 2b)$

37. $(s + 3a)(s + 2a)$

39. $(y + 5)(y + 2)$

41. $(x - 8)(x - 1)$

43. $(z - 8)(z - 4)$

45. $-(x - 3)(x + 7)$

47. $-(z - 12)(z + 2)$

49. $-(y - 9)(y + 2)$

51. Not factorable

53. Not factorable

55. Not factorable

57. $(x + 4y)(x + y)$

59. Not factorable

EXERCISES 5.5 (Page 158)

1. $2x^2 + 7x + 3$

3. $3y^2 + y - 2$

5. $6y^2 + 11y + 3$

7. $20x^2 + 7x - 6$

9. $4x^2 - 4x - 15$

11. $8y^2 + 6y - 9$

13. $4x^2 + 4x + 1$

15. $25x^2 + 20x + 4$

17. $16y^2 + 40y + 25$

19. $x^2 - 4xy + 4y^2$

21. $9x^2 - 6xy + y^2$

23. $64x^2 + 48xy + 9y^2$

25. $4x^2 + 12xy + 9y^2$

27. $4x^2 - 9$

29. $36y^2 - 25$

31. $4x^2 - a^2$

33. $9x^2 - 4y^2$

35. $16x^2 - 49y^2$

37. $6x^2 - 16x - 6$

39. $12y^2 + 3$

41. $12x^2 - 60x + 75$

43. $2x^3 + x^2 - 10x$

45. $4x^3 + 4x^2 - x$

47. $9r^3 - r$

49. $2x^4y - 5x^3y - 3x^2y$

51. $-18x^5 - 24x^4 - 8x^3$

53. $3x^3y + 5x^2y^2 - 2xy^3$

55. $-36p^4q^2 + 3p^3q^3 + 3p^2q^4$

57. $28x^3y - 7xy^3$

59. $108x^4y^3 - 72x^3y^4 + 12x^2y^5$

61. $6x^3 - 13x^2 + 4x + 3$

63. $-3x^3 + 14x^2 + 18x - 8$

65. $8x^3 - 27$

67. $24x^3 - 28x^2 - 14x + 3$

69. $30x^3 + 13x^2 - 30x + 8$

71. $8x^3 - 12x^2 + 6x - 1$

EXERCISES 5.6 (Page 164)

1. $(3a + 1)(a + 1)$

3. $(2x - 1)(x - 1)$

5. $(3b - 1)(3b - 1)$

7. $(2x - 1)(x - 3)$

9. $(4y - 1)(y - 1)$

11. $(8x + 5)(8x + 3)$

13. $(4y + 1)(y - 1)$

15. $(4a + 5)(a - 1)$

17. $(8x + 1)(2x - 5)$

19. $(16x - 1)(x + 5)$

21. $(2x + 3)(x - 1)$

23. $(2x - 3)(x + 1)$

25. $(3a + 1)(2a + 1)$

27. $(4x - 5)(4x + 1)$

29. $2(3x + 1)(x + 1)$

31. $2(4y + 1)(y - 1)$

33. $9(2x - 3)(x + 1)$

35. $3y(3y + 1)(3y - 2)$

37. $2x^3(3x + 5)(2x - 1)$

39. $2a^5(3a - 2)(6a - 5)$

41. $(2t + s)(t - 3s)$

43. $(3x - a)(x - 2a)$

45. $(4y + b)(y + b)$

47. $(2a + 5b)(2a + 3b)$

49. $3a(4b + a)(b + a)$

51. $2xy(5y + 4x)(5y - 2x)$

53. $3x^3y^2(6x - 1)(2x + 3)$

55. $6x^2y^2(2x - y)(x - 2y)$

57. Not factorable

59. Not factorable

61. Not factorable

63. $(2x - y)(3x + 2y)$

65. Not factorable

67. $(3a + 2b)(3a + 5b)$

EXERCISES 5.7 (Page 167)

1. $(x + 3)(x - 3)$

3. $(x + 1)(x - 1)$

5. $(x + z)(x - z)$

7. $(xy + 4)(xy - 4)$

9. $(ax + 7b)(ax - 7b)$

11. $(6 + x)(6 - x)$

13. $(2b + 3)(2b - 3)$

15. $(5x + 4)(5x - 4)$

17. $(3 + 2x)(3 - 2x)$

19. $(9 + 2x)(9 - 2x)$

21. $(2a + 11b)(2a - 11b)$

23. $(5y + 7x)(5y - 7x)$

25. $(7ax + 12by)(7ax - 12by)$

27. $(2xy + 9)(2xy - 9)$

29. $(6ab + 1)(6ab - 1)$

31. $(x^2 + 3)(x^2 - 3)$

33. $(x^2 + 9)(x + 3)(x - 3)$

35. $(6y^4 + 7)(6y^4 - 7)$

37. $(4x^3 + 3y^2)(4x^3 - 3y^2)$

39. $(9x^4 + y^2)(3x^2 + y)(3x^2 - y)$

41. $(11a^2b^3 + 6)(11a^2b^3 - 6)$

43. $5(x + 1)(x - 1)$

45. $3x(x + 1)(x - 1)$

47. $2(x + 2y)(x - 2y)$

49. $3(ab + 2cd)(ab - 2cd)$

51. $4x^2(y + 2)(y - 2)$

53. $3x^2(x^2 + 2)(x^2 - 2)$

55. $4x^2y^2(x + 2y)(x - 2y)$

57. $2a^2b^2(4a^2 + b^2)(2a + b)(2a - b)$

59. $3xy(x^2 + 4y^2)(x + 2y)(x - 2y)$

CHAPTER 5 REVIEW (Page 169)

1. a. $3x^3 + 3x^2$ b. $2xy^2 - 2x^2y$ c. $-x^2 + y - 1$

2. a. $2a - a^2$ b. $-ab + b^2$ c. $3ab + 3b^2 + 3bc$

3. a. $3a^2(1 - 2b)$ b. $2x(x^2 + 2x + 3)$ c. $-y^2(1 + y)$

4. a. $a^2(1 + b)$ b. $4(b - 1)$ c. $b(1 - b - b^2)$

5. a. $x^2 + x - 6$ b. $6a^2 - 17a + 12$ c. $4a^2 - 12a + 9$

6. a. $2x^3 - x^2 - 3x$ b. $6x^4 + 3x^3 - 30x^2$ c. $16t^4 - 16t^3 + 4t^2$

7. a. $x^3 + 3x^2 - 19x + 3$
  b. $6x^3 - 11x^2 + 10x - 8$
  c. $6x^3 - 5x^2 - 29x + 10$

8. a. $(x - 7)(x + 3)$ b. $(5a + 1)(2a + 3)$ c. Not factorable

9. a. Not factorable   b. $(3b + 1)(b + 1)$ c. $(2b - 1)(b + 2)$

10. a. $2(x + 3)(x + 4)$ b. $3(y + 10)(y - 2)$ c. $4x(x + 1)(x - 1)$

11. a. $(x - 2a)(x - a)$ b. $(x^2 + 5)(x^2 - 5)$ c. $2(2b - c)(b + 2c)$

12. a. Not factorable   b. $2(x + 1)(x + 1)$ c. $(2x + 1)^2$

13. 4   **14.** -5 **15.** $-4y^2$

16. $A = 2kr(h + r)$; 747.32   **17.** $A = P(1 + rt)$; 10,000

18. $x + 3$ **19.** $x - 2$ **20.** $2x(x - 3)$

CHAPTER 5 CUMULATIVE REVIEW (Page 170)

1.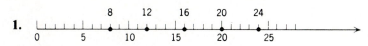

2. $3 \cdot 3 \cdot 3 \cdot 3xxxyyy$ **3.** 112   **4.** $-x^2 - 6x + 5$

5. $-x^2 - x - 3$ **6.** $2x^2 - 11x$ **7.** $x^2 - 10x + 6$

8. $x = -4b$ **9.** $x \le -3$ **10.** nc

11. $(80 - y)°$ **12.** $(y - 6)°$ **13.** $(x - y)°$

14. $47 - x$

15. *Steps 1, 2*   First odd integer: x
                      Next odd integer: $x + 2$
      Step 4 $x + (x + 2) = 56$
      Step 6 $x = 27$; the integers are 27 and 29.

16. *Steps 1, 2*   First integer: x
                      Second integer: $x + 1$
                      Third integer: $x + 2$
      Step 4 $x + (x + 1) + (x + 2) = 111$
      Step 6 $x = 36$; the integers are 36, 37, and 38.

17. *Steps 1, 2*   Larger integer: x
Smaller integer: $x - 5$
  Step 4 $x + 4(x - 5) = 40$
  Step 6 $x = 12$;   the integers are 12 and 7.

18. *Steps 1, 2*   Number of quarters: x
Number of dimes: $x + 1$
Number of nickels: $x + 3$
  Step 4 $0.25x + 0.10(x + 1) + 0.05(x + 3) = 1.45$
  Step 6 $x = 3$;   he had 3 quarters, 4 dimes, and 6 nickels.

19. *Steps 1, 2*   Length of each end piece: x
Length of the middle piece: $3x$
  Step 4 $x + 3x + x = 35$
  Step 6 $x = 7$;   the end pieces are 7 centimeters long; the middle piece is 21 centimeters long.

20. *Steps 1, 2*   Number of hours the sister worked: x
Number of hours the brother worked: $x + 10$
  Step 4 $4x + 5(x + 10) = 185$
  Step 6 $x = 15$;   the sister worked 15 hours; the brother worked 25 hours.

21. True   **22.** False   **23.** True

24. False   **25.** False

EXERCISES 6.1 (Page 175)

1. $\dfrac{4}{7}$ **3.** $\dfrac{3x}{y}$ **5.** $\dfrac{7}{x - y}$ **7.** $\dfrac{x - 3}{4x + 1}$

9. $4 \cdot \dfrac{1}{7}$ **11.** $9 \cdot \dfrac{1}{5}$ **13.** $(x - 3) \cdot \dfrac{1}{4}$ **15.** $2 \cdot \dfrac{1}{x + 3}$

17.

19.

21.

23.

25.

27.

29. $\dfrac{3}{5}$ **31.** $\dfrac{2}{7}$ **33.** $\dfrac{-2}{5}$ **35.** $\dfrac{-a}{b}$

37. $\dfrac{-a}{b}$ **39.** $\dfrac{-x}{y}$ **41.** $\dfrac{7x}{8y}$ **43.** $-c$

45. $\dfrac{-(x+2)}{4}$ **47.** $\dfrac{x+5}{3}$ **49.** $\dfrac{-(2x-1)}{x+2}$ **51.** $\dfrac{x-3}{2}$

53. 0 **55.** 3

57. Let x be 1. Then

$$-\frac{x-1}{2} = -\frac{1-1}{2} \qquad \frac{-x-1}{2} = \frac{-1-1}{2}$$

$$= -\frac{0}{2} = 0; \qquad = \frac{-2}{2} = -1.$$

Thus $-\dfrac{x-1}{2}$ and $\dfrac{-x-1}{2}$ are not equivalent.

EXERCISES 6.2 (Page 181)

1. $\dfrac{2}{3}$ **3.** $\dfrac{-6}{7}$ **5.** $\dfrac{8}{5}$ **7.** $\dfrac{1}{4x}$

9. $\dfrac{-2}{3y^4}$ **11.** $\dfrac{-1}{2x}$ **13.** x^2y **15.** $\dfrac{-1}{xy^2}$

17. $2x^2$ **19.** $\dfrac{3x^2}{2}$ **21.** $\dfrac{5bc^2}{4}$ **23.** $\dfrac{13a^2}{3c^2}$

25. $\dfrac{3}{4}$ **27.** $-4(x-y)$ **29.** 1 **31.** -2

33. $\dfrac{2}{a-x}$ **35.** $\dfrac{-1}{x+4}$ **37.** $\dfrac{1}{x+1}$ **39.** $\dfrac{1}{a-b}$

41. $\dfrac{a-b}{a+b}$ **43.** $\dfrac{-a}{a+1}$ **45.** $\dfrac{x-2}{x-3}$ **47.** $\dfrac{a+3}{a-1}$

49. $\dfrac{2x-3}{2x+1}$ **51.** $\dfrac{3-2a}{a-5}$ **53.** $\dfrac{3x-1}{2x+7}$ **55.** $\dfrac{a-1}{a+2}$

57. $\dfrac{x}{x+1}$ **59.** $\dfrac{x+3y}{2x+3y}$ **61.** b **63.** a

65. b **67.** a **69.** $4x^2y$ **71.** $\dfrac{-5y^2}{x}$

73. $\dfrac{x}{15y^2}$

75. Let x be 1, y be 1. Then

$$\frac{x + y}{x} = \frac{1 + 1}{1} = \frac{2}{1} = 2; \quad \text{but } y = 1.$$

Thus $\dfrac{x + y}{x}$ is not equivalent to y.

EXERCISES 6.3 (Page 186)

1. $2x - 1$ **3.** $y + 2$ **5.** $x + 3$

7. $3y^2 - 2y + 1$ **9.** $2y^2 - y + 3$ **11.** $4y - x + 1$

13. $3x^2 + 2x + \dfrac{-1}{3}$ **15.** $y + 2 + \dfrac{-1}{y}$ **17.** $3x^2 - 2 + \dfrac{-2}{3x^2}$

19. $y^2 - 3y + 2 + \dfrac{-1}{y}$ **21.** $y + 1 + \dfrac{1}{y}$ **23.** $x - 2 + \dfrac{3}{y}$

25. $x + 6$ **27.** $x + 1$ **29.** $2x + 1$

31. $x + 3$ **33.** $2x + 3$ **35.** $2x^2 - x + 1$

37. $x + 1 + \dfrac{-1}{x + 2}$ **39.** $x - 2 + \dfrac{1}{x + 5}$ **41.** $2x - 1 + \dfrac{-1}{x + 1}$

43. $2x - 3 + \dfrac{-2}{2x + 1}$ **45.** $x + 7$ **47.** $x - 6 + \dfrac{29}{x + 6}$

49. $2x^2 + x + 1 + \dfrac{2}{x - 1}$

EXERCISES 6.4 (Page 190)

1. $\dfrac{4}{10}$ **3.** $\dfrac{12}{18}$ **5.** $\dfrac{66}{36}$ **7.** $\dfrac{98}{42}$

9. $\dfrac{10}{6x}$ **11.** $\dfrac{-12ab^2}{12b^3}$ **13.** $\dfrac{-3x^2y}{3y^3}$ **15.** $\dfrac{72}{36}$

17. $\dfrac{xy^2}{xy}$ **19.** $\dfrac{3x^4y}{3x^2y}$ **21.** $\dfrac{x + y}{2(x + y)}$ **23.** $\dfrac{-2a(a + 4)}{5(a + 4)}$

25. $\dfrac{2a(a + 3)}{a + 3}$ **27.** $\dfrac{3(x + y)}{(x - y)(x + y)}$ **29.** $\dfrac{-3(x + 1)}{(2x - 1)(x + 1)}$

31. $\dfrac{7a(b - 3)}{(b + 2)(b - 3)}$ **33.** $\dfrac{a^2}{a(a - 3)}$ **35.** $\dfrac{-3(x - y)}{(x + y)(x - y)}$

37. $\dfrac{y(y + 2)}{(y - 1)(y + 2)}$ **39.** $\dfrac{3x(x - 4)}{(2x - 1)(x - 4)}$ **41.** $\dfrac{(x - 2)(2x - 1)}{(2x + 3)(2x - 1)}$

43. $\dfrac{(2a - b)(3a + 2b)}{(a - b)(3a + 2b)}$ **45.** $\dfrac{(6x + 1)(3x - 4)}{(x - 1)(3x - 4)}$ **47.** $\dfrac{(x + 1)(x - 1)}{x(x - 1)^2}$

49. $\dfrac{3x^2(x - 2)}{2x(2x + 1)(x - 2)}$

EXERCISES 6.5 (Page 196)

1. $\dfrac{1}{25}$ **3.** $\dfrac{1}{x^6}$ **5.** 1 **7.** $\dfrac{3}{64}$

9. $\dfrac{3}{1000}$ **11.** $\dfrac{4}{x^2}$ **13.** 2^{-3} **15.** 5^{-2}

17. 2^{-2} or 4^{-1} **19.** $x \cdot 5^{-2}$ or $x \cdot 25^{-1}$

21. $2 \cdot 10^{-2}$ **23.** $x \cdot 10^{-3}$ **25.** x^5 **27.** $15x^{-6}$

29. 10^{-5} **31.** 8^{-10} **33.** w^{18} **35.** $3^{-2}x^{-2}$

37. $a^{-3}b^{-6}$ **39.** x^{-5} **41.** b^4 **43.** 6^8

45. 10^3 **47.** 10^{-3} **49.** 10^2 **51.** 10^{-2}

53. 4.83×10^2 **55.** 7.2×10^{-2} **57.** 4×10^3 **59.** 6.3×10^{-4}

61. $43{,}000$ **63.** 0.00057 **65.** $82{,}340{,}000$ **67.** 0.000008

69. 5.98×10^{27} **71.** 3.0×10^8 **73.** 1.5×10^{-5} **75.** 400

77. 120 **79.** $150{,}000$

CHAPTER 6 REVIEW (Page 201)

1.

2. a. $4 \cdot \dfrac{1}{9}$ b. $(x + 6) \cdot \dfrac{1}{3}$ c. $2y \cdot \dfrac{1}{x + y^2}$

3. a. $\dfrac{2(x - 3)}{3}$ b. $\dfrac{-(x^2 + 1)}{3}$ c. $\dfrac{-3(2x + y)}{4}$

4. a. $\dfrac{-3}{x + y}$ b. $\dfrac{a}{x}$ c. $\dfrac{-(b - 2)}{4}$

5. a. 0 b. -3

6. a. $\dfrac{8x^2z^2}{7y}$ b. $\dfrac{-3(a - 3b)^2}{2a^2}$

7. a. $\dfrac{1}{b + 1}$ b. $\dfrac{1}{1 - a}$

8. a. $\dfrac{x + 2}{x}$ b. $x - 1$

9. a. $x^2 + 2x - 3$ b. $4x^2 - 2 + \dfrac{3}{2x^2}$

10. a. $2x - 3 + \dfrac{-2}{x - 1}$ b. $2x^2 - 4x + 9 + \dfrac{-21}{x + 2}$

11. a. $\dfrac{8a^3b}{2a^2b^3c}$ b. $\dfrac{6}{2(x - y)}$

12. a. $\dfrac{x(x - 1)}{(x - 2)(x - 1)}$ b. $\dfrac{x(x - 1)(x + 2)}{x^2(x - 2)(x + 2)}$

13. a. $\dfrac{-5}{3 - x}$ b. $\dfrac{-3x(3 + x)}{(3 - x)(3 + x)}$

14. a. $\dfrac{2}{25}$ b. $\dfrac{1}{x^2}$ c. $\dfrac{4}{y^3}$

15. a. $3x^{-3}$ b. $-6x^{-1}$ c. $(5x)^4(2y)^{-4}$

16. a. $10x^{-7}$ b. $16a^{-12}$ c. $x^{-5}y^{15}$

17. a. x^{-3} b. $2x^{-3}$ c. 10^3

18. a. 3.47×10^7 b. 8.73×10^{-4}

19. a. 48,300   b. 0.000381

20. a. 0.00000075   b. 3

CHAPTER 6 CUMULATIVE REVIEW (Page 203)

1. Commutative   **2.** 13   **3.** 0

4. 375   **5.** -28 **6.** $x + 4$

7. $x(x - 1)(x + 1)$ **8.** $5b(b + 1)(b + 1)$ **9.** $x - 4$

10. *Steps 1, 2*   Number of nickels: x
Number of dimes: $50 - x$

Step 4 $0.05x + 0.10 (50 - x) = 3.50$

Step 6 $x = 30$;   there are 30 nickels and 20 dimes.

11. *Steps 1, 2*   First even integer: x
Second even integer: $x + 2$
Third even integer: $x + 4$

Step 4 $x + (x + 2) + (x + 4) = 16 + (x + 6)$

Step 6 $x = 8$;   the integers are 8, 10, and 12.

12. $6x^2 - 8x - 8$ **13.** $(4x + 1)(2x - 3)$ **14.** $x + 2 + \dfrac{-5}{x + 2}$

15. $2x^2 - 1 + \dfrac{1}{x^2}$ **16.** $\dfrac{1}{27}$ **17.** 4.81×10^6

18. 4   **19.** 6   **20.** $x < -2$

21. False   **22.** False   **23.** False

24. False   **25.** False

EXERCISES 7.1 (Page 206)

1. $\dfrac{3}{8}$ **3.** $\dfrac{9}{55}$ **5.** $\dfrac{1}{2}$ **7.** y

9. $\dfrac{4x^2}{5}$ **11.** $4y$ **13.** $\dfrac{-x}{y}$ **15.** $\dfrac{49rt}{4}$

17. $\dfrac{-b}{az}$ **19.** $\dfrac{3}{4}$ **21.** 5 **23.** $\dfrac{-1}{8}$

25. 1 **27.** $\dfrac{x-3}{x+7}$ **29.** $\dfrac{3x-2}{3x+2}$ **31.** $\dfrac{x+1}{x+3}$

33. $\dfrac{2-x}{2x(x+1)}$ **35.** $\dfrac{3x+2}{7x-1}$ **37.** $\dfrac{3x+y}{3y-2x}$ **39.** $\dfrac{y-3}{y-6}$

41. $\dfrac{x^2+2xy}{y(x+y)}$ **43.** $\dfrac{2x}{3}$ **45.** $\dfrac{-2a}{5}$ **47.** $\dfrac{3(a-b)}{4}$

49. $\dfrac{-3(2x-y)}{5}$ **51.** $\dfrac{3}{7}x$ **53.** $\dfrac{-5}{7}a$ **55.** $\dfrac{5}{2}(a-b)$

57. $\dfrac{1}{7}(x+y)$ **59.** $\dfrac{125a^3}{64b^3}$ **61.** $\dfrac{81x^8y^4}{16z^{12}}$ **63.** a^4b^8

65. $\dfrac{625z^{20}}{81x^{24}y^4}$ **67.** Let x be 1, y be 1. Then

$$\frac{1}{x+y}\cdot\frac{y}{3}=\frac{1}{(1)+(1)}\cdot\frac{(1)}{3} \qquad \frac{1}{3x}=\frac{1}{3(1)}$$

$$=\frac{1}{2}\cdot\frac{1}{3}=\frac{1}{6}; \qquad\qquad =\frac{1}{3}.$$

Thus $\dfrac{1}{x+y}\cdot\dfrac{y}{3}$ and $\dfrac{1}{3x}$ are not equivalent.

EXERCISES 7.2 (Page 212)

1. $\dfrac{9}{10}$ **3.** $\dfrac{3}{2}$ **5.** $\dfrac{10}{3}$ **7.** 6

9. $\dfrac{8}{7}$ **11.** $\dfrac{28}{3}$ **13.** 1 **15.** $\dfrac{5}{16b^2}$

17. $\dfrac{-x}{v}$ **19.** $\dfrac{-x}{16y}$ **21.** $\dfrac{b}{6a}$ **23.** $\dfrac{y^4}{3x}$

25. $12y$

27. $\dfrac{2y}{x}$

29. $\dfrac{a+b}{2a}$

31. $\dfrac{a}{2}$

33. $\dfrac{5}{2}$

35. $\dfrac{-5}{6x}$

37. $\dfrac{2x+y}{x+2y}$

39. $\dfrac{y-1}{y+7}$

41. $\dfrac{x-5}{x-2}$

43. $\dfrac{y^2+6y+8}{(y-5)(y-2)}$

45. $\dfrac{x^2-x-12}{(x-1)(2x+1)}$

47. $\dfrac{1}{y-1}$

49. $\dfrac{2x-1}{x+3}$

51. $\dfrac{a^3-a^2-69a+189}{(a-1)(a+6)(a-8)}$

53. $\dfrac{-(x+3)}{2x^2}$

55. $\dfrac{x^3-x^2}{2(x+3)}$

57. 1

59. $\dfrac{x^2-6x+9}{(x+1)(x-2)}$

EXERCISES 7.3 (Page 216)

1. $\dfrac{7}{9}$

3. $\dfrac{5x-3}{11}$

5. $\dfrac{3-x}{5}$

7. $\dfrac{4}{a}$

9. $\dfrac{1}{b}$

11. $\dfrac{4}{x}$

13. $\dfrac{x}{y}$

15. $\dfrac{x+4}{2}$

17. $\dfrac{2x+y}{3x}$

19. $\dfrac{x}{a}$

21. x^2+x

23. $\dfrac{4x}{y}$

25. $\dfrac{x+6}{2}$

27. $\dfrac{a+3b}{a-b}$

29. $\dfrac{-a}{a+b}$

31. $\dfrac{3x+2y}{x+y}$

33. $\dfrac{1}{x+2y}$

35. $\dfrac{6a-b}{3}$

37. $3x+y$

39. $\dfrac{u}{2u-v}$

41. $\dfrac{x-1}{x+2}$

43. $\dfrac{4}{x-1}$

45. $\dfrac{1}{x-1}$

47. $\dfrac{x+5}{x}$

EXERCISES 7.4 (Page 222)

1. 12

3. 18

5. 84

7. 72

9. 24

11. 60

13. x^2y

15. xyz

17. $24x^2y$

19. $(x+y)(x-y)$

21. $x^2(x+2)$

23. $(x+4)(x-1)^2$

25. $-3 \cdot 5(a-2b)$

27. $-x(x-3)(x+3)$

29. $-3 \cdot 4x(x-1)(x+2)$

31. $\dfrac{5}{8}$ **33.** $\dfrac{7}{12}$ **35.** $\dfrac{19}{20}$ **37.** $\dfrac{21x}{8}$

39. $\dfrac{5 + 3a}{ax}$ **41.** $\dfrac{6y + 4x}{3xy}$

43. $\dfrac{yz + xz + xy}{xyz}$ **45.** $\dfrac{8y - 5}{6}$

47. $\dfrac{4x + 7}{6}$ **49.** $\dfrac{5x - y}{6x}$

51. $\dfrac{7x + 3}{(x + 3)(x - 3)}$ **53.** $\dfrac{x^2 + 3xy - 2y^2}{(x - y)(x + y)}$

55. $\dfrac{2x^2 + x - 2}{(x - 2)(x + 2)}$ **57.** $\dfrac{2y^2 - 4y + 14}{(y - 3)(y + 2)}$

59. $\dfrac{y^2 + 4y - 3}{(y + 1)(y + 1)(y - 1)}$ **61.** $\dfrac{3x^2 - 3x}{(x + 1)(x + 2)(x - 2)}$

63. $\dfrac{2x^2 + 2x - 3}{x(x + 3)(x + 3)}$ **65.** $\dfrac{17x^2 - 2x - 4}{(3x + 1)(2x - 3)(4x - 3)}$

67. $\dfrac{x^2 + 2x - 2xy + 2y + y^2}{2(x + 2y)(x - y)}$

69. Let x be 1, y be 1. Then,

$$\frac{1}{x} + \frac{1}{y} = \frac{1}{(1)} + \frac{1}{(1)} \qquad \frac{1}{x + y} = \frac{1}{(1) + (1)}$$

$$= 1 + 1 = 2; \qquad = \frac{1}{2}.$$

Thus $\dfrac{1}{x} + \dfrac{1}{y}$ and $\dfrac{1}{x + y}$ are not equivalent.

EXERCISES 7.5 (Page 229)

1. $\dfrac{3}{10}$ **3.** $\dfrac{-7}{12}$ **5.** $\dfrac{4x - 3y}{12}$ **7.** $\dfrac{1}{3x}$

9. $\dfrac{2y - 3x}{xy}$ **11.** $\dfrac{9y - 5x}{6xy}$ **13.** $\dfrac{-x - 4}{6}$ **15.** $\dfrac{4y - 3}{6}$

17. $\dfrac{-4x - 7}{6}$ **19.** $\dfrac{-4x + 7}{6x}$

21. $\dfrac{x - 5y}{6x}$ **23.** $\dfrac{6a^2 - 5ab + 6b^2}{12ab}$

25. $\dfrac{3}{2(x + y)}$ **27.** $\dfrac{-2}{3(x + 1)}$

29. $\dfrac{9}{4(2a + b)}$ **31.** $\dfrac{19}{6(x - 2)}$

33. $\dfrac{23}{10(3x - 1)}$

35. $\dfrac{11}{3(2x - 3y)}$

37. $\dfrac{-6x}{(x + 3)(x - 3)}$

39. $\dfrac{38}{(3x - 4)(5x + 6)}$

41. $\dfrac{-3a - 9}{(2a + 1)(a - 2)}$

43. $\dfrac{x^2}{(x - 2y)(x + 2y)}$

45. $\dfrac{-8x}{(x + 2)(x - 2)}$

47. $\dfrac{-1}{(x + 2)(x + 3)}$

49. $\dfrac{a^2 - 7ab + 2b^2}{(a + b)(a - b)}$

51. $\dfrac{3}{(x + 1)(x + 1)(x - 2)}$

53. $\dfrac{5x + 12}{(x - 3)(x + 3)(x + 2)}$

55. $\dfrac{x^2 - x}{(x - 2)(x + 3)(x + 5)}$

57. $\dfrac{2x^2 + 13x + 6}{(x + 2)^2(x + 1)}$

59. $\dfrac{5x + 9}{(x + 2)(x + 3)(x - 4)}$

61. $\dfrac{3t^2 + t + 2}{2t(t + 2)(t - 2)}$

63. $\dfrac{-7}{(y - 4)(2y + 1)(3y - 2)}$

65. $\dfrac{6x - 28y}{(x - 2y)(x + 2y)(x - 3y)}$

67. Let x be 2, y be 1. Then,

$$\dfrac{1}{x} - \dfrac{1}{y} = \dfrac{1}{(2)} - \dfrac{1}{(1)} \qquad \dfrac{1}{x - y} = \dfrac{1}{(2) - (1)}$$

$$= \dfrac{1}{2} - 1 = \dfrac{-1}{2}; \qquad\qquad = \dfrac{1}{1} = 1.$$

Thus $\dfrac{1}{x} - \dfrac{1}{y}$ is not equivalent to $\dfrac{1}{x - y}$.

EXERCISES 7.6 (Page 236)

1. $\dfrac{3}{2}$ **3.** $\dfrac{5}{4}$ **5.** $\dfrac{8}{7}$ **7.** 6

9. $\dfrac{b}{3a}$ **11.** $\dfrac{5}{3xy}$ **13.** $\dfrac{5}{3}$ **15.** $\dfrac{5}{2x}$

17. $\dfrac{3}{2}$ **19.** 6   **21.** $\dfrac{1}{4}$ **23.** $\dfrac{1}{4}$

25. $\dfrac{1}{14}$ **27.** 5   **29.** $\dfrac{2ab - a^2}{2ab - b^2}$ **31.** $\dfrac{2x + 1}{x}$

33. $\dfrac{x}{4}$ **35.** $\dfrac{5}{3y^2}$ **37.** xy **39.** $\dfrac{-2(x + z)}{xz}$

41. $3x + y$ **43.** $\dfrac{x^2 + 2xz}{2z + xz^2}$ **45.** $\dfrac{-3b}{2a}$ **47.** $\dfrac{2x}{x - 3}$

49. $\dfrac{ab - a}{2b + 3}$ **51.** $\dfrac{2x - 1}{x}$ **53.** $\dfrac{x + 5}{x + 4}$ **55.** $\dfrac{-a^2 - 6a - 5}{a^2 + 2a - 3}$

57. $\dfrac{x^2 - 2x}{x^2 - 2}$ **59.** $\dfrac{1 + 2y}{1 - 2y}$ **61.** $\dfrac{1 + x^2}{x + x^2}$ **63.** $\dfrac{y + x}{xy}$

65. $\dfrac{2y - x}{y + x}$

EXERCISES 7.7 (Page 243)

1. 11   **3.** 1   **5.** 2   **7.** 1

9. $\dfrac{25}{3}$ **11.** $\dfrac{-27}{8}$ **13.** 6   **15.** 2

17. 15   **19.** 5   **21.** 6   **23.** 4

25. $\dfrac{19}{4}$ **27.** $\dfrac{21}{4}$ **29.** 2   **31.** 1

33. $\dfrac{45}{2}$ **35.** 5   **37.** 4   **39.** 3

41. 3   **43.** 2   **45.** 15   **47.** No solution

49. No solution   **51.** $b = \dfrac{2A}{h} - c$ **53.** $C = \dfrac{5}{9}(F - 32)$

55. $l = \dfrac{2S}{n} - a$ **57.** $A = \dfrac{RB}{B - R}$ **59.** $a = \dfrac{s(1 - r)}{1 - r^3}$

61. $y = \dfrac{nW}{n + 1} + 1$ **63.** $x = \dfrac{2y + 1}{y + 3}$ **65.** $y = m(x - h) + k$

67. $x = a - \dfrac{ay}{b}$ **69.** $e = \dfrac{r}{d + rc}$

EXERCISES 7.8 (Page 249)

1. *Steps 1, 2*   Number: x

Step 4 $\dfrac{1}{2}x + 3x = \dfrac{35}{2}$

Step 6 $x = 5$;   the number is 5.

3. *Steps 1, 2*   First integer: x
Next integer: $x + 1$

Step 4 $\dfrac{1}{2}x + \dfrac{2}{3}(x + 1) = 17$

Step 6 $x = 14$;   the integers are 14 and 15.

5. *Steps 1, 2*   Numerator: x

Step 4 $\dfrac{x}{x + 6} = \dfrac{3}{4}$

Step 6 $x = 18$; the numerator is 18.

7. *Steps 1, 2*   Profit needed: x

Step 4 $\dfrac{2}{3}x = 160$

Step 6 $x = 240$; the business must make \$240.

9. *Steps 1, 2*   Celsius temperature: C

Step 4 $68 = \dfrac{9}{5}C + 32$

Step 6 $C = 20$; the Celsius temperature is 20°.

11. *Steps 1, 2*   Number of kilograms of ore: x

Step 4 $\dfrac{1}{15}(75 + x) = 15$

Step 6 $x = 150$; she needs 150 more kilograms of ore.

13. *Steps 1, 2*   Grade needed on next test: x

Step 4 $\dfrac{72 + 78 + 84 + 94 + x}{5} = 80$

Step 6 $x = 72$; she needs a grade of 72.

15. *Steps 1, 2*   Man's speed: x
Woman's speed: $x + 20$

Step 4 $\dfrac{120}{x + 20} = \dfrac{80}{x}$

Step 6 $x = 40$; the man's speed is 40 miles per hour; the woman's is 60 mile per hour.

17. *Steps 1, 2*   Walking rate: x

Step 4 $\dfrac{15}{x + 2} = \dfrac{7}{x}$

Step 6 $x = \frac{7}{4}$; he walks $1\frac{3}{4}$ miles per hour.

19. *Steps 1, 2*   Rate of slower driver: x
Rate of faster driver: $2x$

Step 4 $\dfrac{300}{2x} + 5 = \dfrac{300}{x}$

Step 6 $x = 30$; the slower driver travels at 30 miles per hour; the faster at 60 miles per hour.

21. *Steps 1, 2*   Rate of the plane: x
Rate of the rocket: $x + 2700$

Step 4 $\dfrac{12{,}600}{x + 2700} = \dfrac{3600}{x}$

Step 6 $x = 1080$;   the rate of the plane is 1080 miles per hour; the rate of the rocket is 3780 miles per hour.

EXERCISES 7.9 (Page 254)

1. $\dfrac{2}{3}$ **3.** $\dfrac{5}{9}$ **5.** $\dfrac{2}{5}$ **7.** $\dfrac{2}{5}$

9. $\dfrac{3}{10}$ **11.** $\dfrac{8}{5}$ **13.** $\dfrac{8}{3} = \dfrac{24}{9}$ **15.** $\dfrac{21}{24} = \dfrac{7}{8}$

17. $\dfrac{15}{x} = \dfrac{10}{4}$ **19.** $\dfrac{6}{2} = \dfrac{x}{x+1}$ **21.** 15   **23.** 15

25. $\dfrac{3}{2}$ **27.** 4   **29.** -2 **31.** 4

33. *Steps 1, 2*   Number of centimeters: x

   Step 4 $\dfrac{22}{1} = \dfrac{x}{2.54}$

   Step 6 $x = 55.88$;   22 inches equals 55.88 centimeters.

35. *Steps 1, 2*   Number of miles: x

   Step 4 $\dfrac{24.6}{1} = \dfrac{x}{0.62}$

   Step 6 $x = 15.25$;   24.6 kilometers equals 15.25 miles.

37. *Steps 1, 2*   Number of kilograms: x

   Step 4 $\dfrac{36}{2.2} = \dfrac{x}{1}$

   Step 6 $x = 16.36$;   36 pounds equals 16.36 kilograms.

39. *Steps 1, 2*   Number of liters: x

   Step 4 $\dfrac{32}{1.06} = \dfrac{x}{1}$

   Step 6 $x = 30.19$;   32 quarts equals 30.19 liters.

41. *Steps 1, 2*   Number of pounds of coffee: x

   Step 4 $\dfrac{x}{3} = \dfrac{3000}{225}$

   Step 6 $x = 40$;   40 pounds of coffee are needed.

43. *Steps 1, 2*   Number of liters of gas: x

   Step 4 $\dfrac{32}{x} = \dfrac{184}{460}$

   Step 6 $x = 80$;   80 liters of gas are needed.

45. *Steps 1, 2* Number of liters of milk: x

Step 4 $\dfrac{3}{x} = \dfrac{2}{90}$

Step 6 $x = 135$; 135 liters of milk will be used.

47. *Steps 1, 2* Number of bricks: x

Step 4 $\dfrac{660}{x} = \dfrac{740}{925}$

Step 6 $x = 825$; 825 bricks will be required.

49. *Steps 1, 2* Cost of 28 pounds of apples: x

Step 4 $\dfrac{20}{28} = \dfrac{1.60}{x}$

Step 6 $x = 2.24$; 28 pounds of apples cost $2.24.

CHAPTER 7 REVIEW (Page 258)

1. a. $\dfrac{x^2}{6}$ b. $\dfrac{5x - 10}{x}$ c. 1

2. a. $\dfrac{7s}{r}$ b. $\dfrac{a^2 - a - ab + b}{a}$ c. x^3

3. a. $\dfrac{4}{5}$ b. $\dfrac{x}{y}$ c. $\dfrac{-5}{3}$

4. a. $\dfrac{7}{6}$ b. $\dfrac{9}{8}$ c. $\dfrac{3}{2y}$

5. a. $\dfrac{7}{3x}$ b. $\dfrac{6s + 5r}{2rs}$ c. $\dfrac{2a - 3b}{a^2 b^2}$

6. a. $\dfrac{4a + 2b}{(a - b)(a + b)}$ b. $\dfrac{a^2 - a + 1}{a(a - 1)(a + 1)}$

  c. $\dfrac{6x + 26}{(x + 1)(x + 5)(x - 5)}$

7. a. $\dfrac{-3}{4}$ b. $\dfrac{2x^2 + x + 2}{2x(2 + x)(2 - x)}$ c. $\dfrac{-1}{(x + 3)(x + 1)}$

8. a. $\dfrac{9}{4}$ b. $\dfrac{5}{7}$ c. $\dfrac{6}{11}$

9. a. $\dfrac{b - a}{b + 2}$ b. $\dfrac{x^2 y - x^2}{y^2(x - 1)}$ c. $\dfrac{1 + 3y}{2y - 3}$

10. a. $\dfrac{x - 2x^2}{1 + 3x^2}$ b. $\dfrac{x^2 - xy^2}{y^2 - yx^2}$ c. $\dfrac{x^2 + xy^2}{y^2 + yx^2}$

11. a. 6   b. $\dfrac{-4}{3}$ c. 7

12. a. $\dfrac{-4}{5}$ b. -9 c. 21

13. a. $\dfrac{-1}{2}$ **b.** 3   **c.** $\dfrac{-2}{3}$

14. a. 3   **b.** 1   **c.** -21

15. a. $\dfrac{216a^6}{b^3c^9}$ **b.** $\dfrac{-8x^9y^3}{125z^{15}}$ **c.** $\dfrac{81x^{20}z^{16}}{16y^{20}}$

16. $p = \dfrac{3qr}{2r - 3q}$

17. *Steps 1, 2*   Number: x

    Step 4 $\dfrac{3x}{x + 10} = \dfrac{1}{2}$

    Step 6 $x = 2$; the number is 2.

18. *Steps 1, 2*   Number of centimeters: x

    Step 4 $\dfrac{12}{1} = \dfrac{x}{2.54}$

    Step 6 $x = 30.48$; 12 inches equals 30.48 centimeters.

19. *Steps 1, 2*   Number of defective parts: x

    Step 4 $\dfrac{92}{3} = \dfrac{276}{x}$

    Step 6 $x = 9$; we would expect 9 defective parts.

20. *Steps 1, 2*   Rate of faster car: x
                      Rate of slower car: $x - 10$

    Step 4 $\dfrac{90}{x} = \dfrac{60}{x - 10}$

    Step 6 $x = 30$; the faster car travels 30 miles per hour; the slower car travels 20 miles per hour.

CHAPTER 7 CUMULATIVE REVIEW (Page 259)

1. $c = \dfrac{5}{9}(F - 32)$

2. $\dfrac{2(4)}{5} + \dfrac{2(4 - 3)}{5} \overset{?}{=} 2$, $\dfrac{8}{5} + \dfrac{2}{5} \overset{?}{=} 2$, yes

3. $\dfrac{42}{13}$ **4.** $\dfrac{x - 3}{x + 2}$

5. $3x(x + 6)(x + 1)(x - 1)$ **6.** $2x^2 - 3x - 1$

7. $\dfrac{5x - 19}{(x - 4)(x - 1)}$ **8.** $2xy^2(2x^2 - y)$

9. $\dfrac{x - 3}{x + 1}$ **10.** $\dfrac{9}{5}$ **11.** 5   **12.** 6

13. $\dfrac{ab}{c}$ **14.** $x < 1$ **15.** $x \leq -1$ **16.** $\dfrac{1}{16}$

17. 0.000325 **18.** $x^2 + 2x + 3$

19. *Steps 1, 2*   Number of kilometers: x

     Step 4 $\dfrac{\frac{1}{4}}{8} = \dfrac{5}{x}$

     Step 6 $x = 160$;   5 centimeters represents 160 kilometers.

20. *Steps 1, 2*   Original amount of money: x

     Step 4 $x - \dfrac{1}{3}x - \dfrac{1}{4}x = 10$

     Step 6 $x = 24$;   he had \$24 originally.

21. False   **22.** True   **23.** False   **24.** False

25. False

EXERCISES 8.1 (Page 264)

1. a. d and t b. d increases   c. t d. d e. 12

3. a. $(2, 1)$ b. $(-1, 4)$ **5.** a. $(2, -5)$ b. $(-2, 3)$

7. a. $(-3, 8)$ b. $(3, -1)$ **9.** a. $\left(-4, \dfrac{-3}{2}\right)$ b. $\left(4, \dfrac{5}{2}\right)$

11. a. $(2, 7)$ b. $(-3, -3)$ **13.** a. $(3, 11)$ b. $(1, 3)$

15. a. $(0, 0)$ b. $(-3, -9)$ **17.** $y = 10 - 3x$

19. $y = \dfrac{-x - 6}{2}$ **21.** $y = \dfrac{2x - 7}{3}$

23. $y = \dfrac{x - 3}{3}$ **25.** a. $(4, 13)$ b. $(1, 7)$

27. a. $(5, -14)$ b. $(3, -10)$ **29.** a. $(-2, -10)$ b. $(0, -2)$

31. a. $\left(-2, \dfrac{7}{2}\right)$ b. $\left(5, \dfrac{-7}{4}\right)$ **33.** a. $\left(10, \dfrac{14}{3}\right)$ b. $\left(5, \dfrac{4}{3}\right)$

35. a. $\left(7, \dfrac{-21}{2}\right)$ b. $\left(5, \dfrac{-13}{2}\right)$ **37.** a. $(-3, -3)$ b. $\left(0, \dfrac{-3}{5}\right)$

39. a. $(-2, 8)$ b. $(1, -10)$

EXERCISES 8.2 (Page 269)

1.

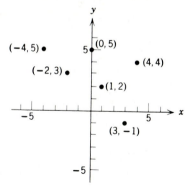

3.

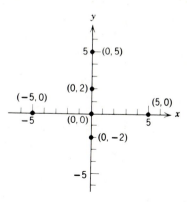

5.

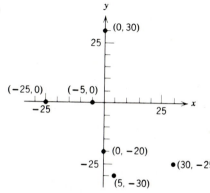

7.

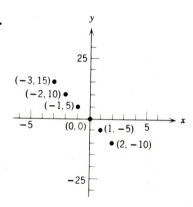

9. Result is a straight line

11. a. Ordinate   b. Abscissa

13. Origin

15. On a straight line bisecting the angles at the origin in the first and third quadrants.

17.

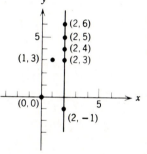

a. Yes   **19.** 2
b. No
c. Yes
d. No

EXERCISES 8.3 (Page 275)

1.

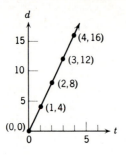

a. (0, 0)
b. (2, 8)
c. (4, 16)
d. On the line through these points
e. (1, 4)(3, 12)
f. Two

3.

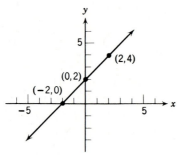

5.

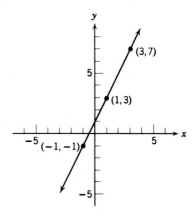

7.

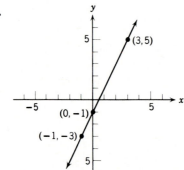

9.

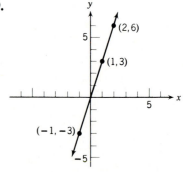

11.

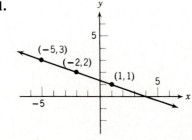

13.

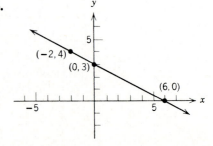

15. a.

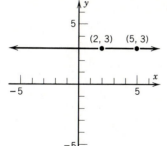

b. Yes
c. Yes
d. No
e. Yes
f. Yes

17.

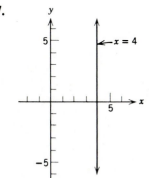

19.

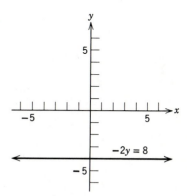

21.

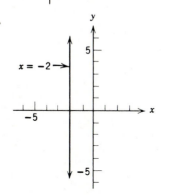

23.

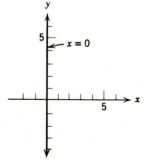

25.

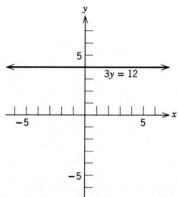

27.

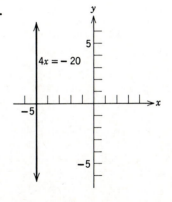

EXERCISES 8.4 (Page 279)

1.

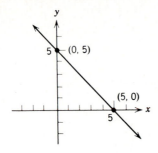

3.

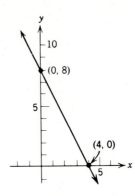

5.

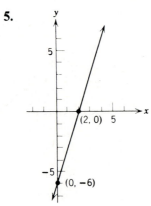

7.

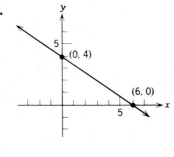

9.

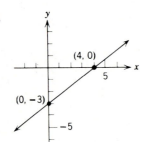

11.

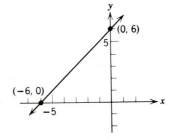

13.

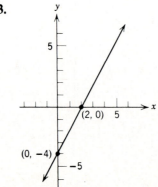

15.

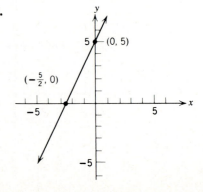

17.

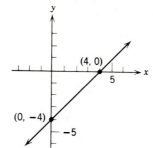

19.

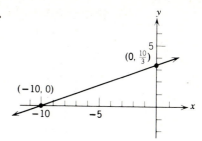

21.

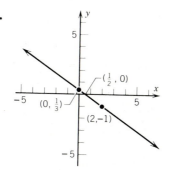

23.

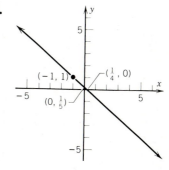

25.

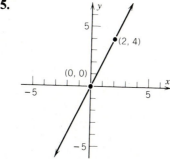

27.

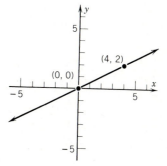

29.

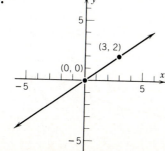

EXERCISES 8.5 (Page 282)

1. $\dfrac{5}{3}$ **3.** $\dfrac{-3}{4}$ **5.** -1 **7.** $\dfrac{-5}{4}$

9. Undefined   **11.** 0

13. a.

b. -2
c. -2
d. Yes

$A\,(-2, 5)\bullet$ 5

$\bullet$
$B\,(1, -1)$

-5

5

-5 $\bullet C\,(3, -5)$

15. a.

b. 3
c. 2
d. No

$\bullet C\,(4, 7)$

5   $\bullet B\,(3, 5)$

$\bullet A\,(2, 2)$

-5 5

5

17. a. 2   b. 2   c.
d. Yes

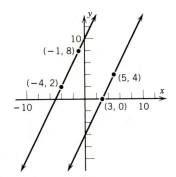

19. a. $\dfrac{1}{2}$ b. $\dfrac{5}{8}$ c.

d. No

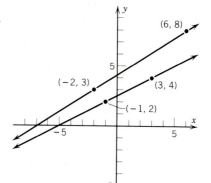

21. a. $\dfrac{1}{4}$ b. -4 c.

d. Yes

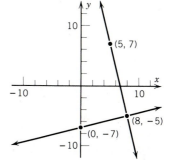

23. a. $\dfrac{5}{4}$ b. 2   c.

d. No

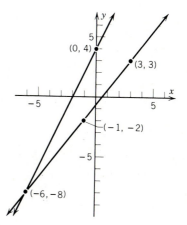

25. $\dfrac{1}{2}$ **27.** $\dfrac{-3}{2}$ **29.** 0

EXERCISES 8.6 (Page 287)

1. $y = 3x + 4$ **3.** $y = -2x - 2$ **5.** $y = 1$

7. $y = \dfrac{1}{2}x - 5$ **9.** $y = -4x - 5$ **11.** $y = \dfrac{2}{3}x + 3$

13. $y = x + 6$ **15.** $y = 4x + 7$ **17.** $y = \dfrac{5}{3}x$

19. $y = 4$ **21.** $y = \dfrac{-1}{8}x + \dfrac{19}{4}$

23. Slope: 5; y-intercept: -2 **25.** Slope: -4; y-intercept: 3

27. Slope: -3; y-intercept: 4   **29.** Slope: -2; y-intercept: $\dfrac{5}{3}$

31. Slope: $\dfrac{-5}{4}$; y-intercept: $\dfrac{-3}{4}$ **33.** Slope: $\dfrac{2}{3}$; y-intercept: -2

35. a. -3 b. -3 c. $y = -3x + 6$

37. a. -2 b. $\dfrac{1}{2}$ c. $y = \dfrac{1}{2}x$

EXERCISES 8.7 (Page 291)

1. $(0, 1)$ is not a solution.   **3.** $(-1, 2)$ is not a solution.

5. $(-2, -1)$ is a solution.   **7.** $(-1, -1)$ is a solution.

9. $(0, -3)$ is not a solution.   **11.** $(-1, 1)$ is a solution.

13.

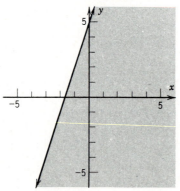

15.

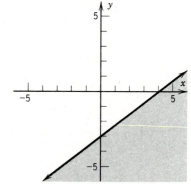

17.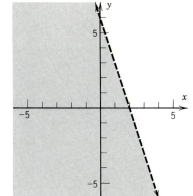

19.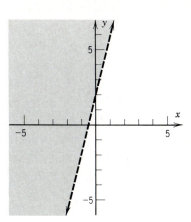

21.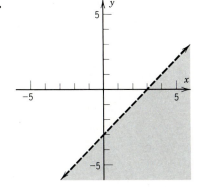

23.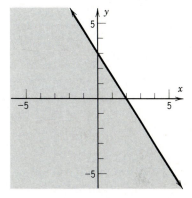

25.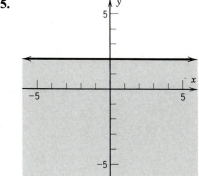

27.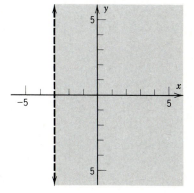

29.

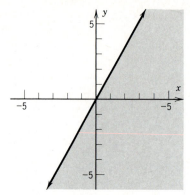

31.

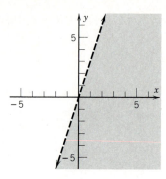

33.

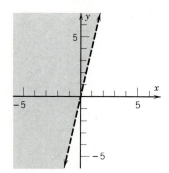

EXERCISES 8.8 (Page 293)

1. $y = kx$

3. $R = kL$

5. $c = kw$

7. $T = ks$

9. 21

11. $\dfrac{1}{2}$

13. $\dfrac{169}{10}$

15. 63

17. 640 foot-pounds

19. 14.4 pounds

21. 1820 pounds

23. 168 miles

25. 94.29 volts

27. 30 pounds per square foot

29. Circumference is doubled

CHAPTER 8 REVIEW (Page 297)

1. a. 0.05

b. I, P

c. Increases

2. $y = 2x - 4$

3. $y = \dfrac{3x + 6}{2}$

4. a. (3, 7)

b. (−2, −3)

c. (0, 1)

d. $\left(-\dfrac{1}{2}, 0\right)$

5. a. $\left(6, \dfrac{4}{3}\right)$ b. $(-5, -6)$

6.

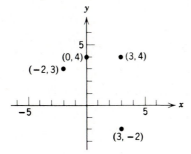

7.

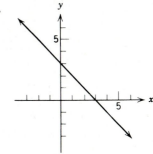

8.

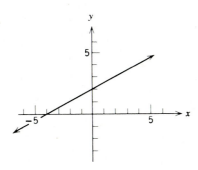

9.

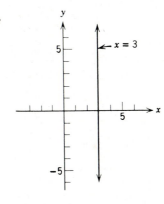

10.

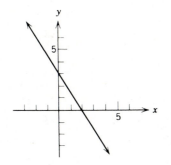

11. x-intercept: 8; y-intercept: -8

12. $\dfrac{-1}{3}$

13. No

14. $y = -2x - 9$

15. $y = \dfrac{2}{3}x + \dfrac{13}{3}$

16. Slope: $\dfrac{5}{2}$; y-intercept: 0

17. 13.2

18.

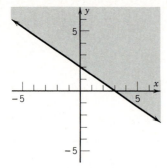

19.

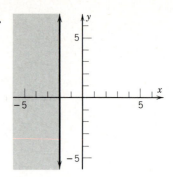

20.

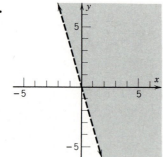

CHAPTER 8 CUMULATIVE REVIEW (Page 298)

1.

2. $\dfrac{1}{2}$

3. 7

4. 2064

5. $\dfrac{3}{2}$

6. $\dfrac{a + b}{a - b}$

7. $\dfrac{24a - 5}{24}$

8. $\dfrac{-x + 8}{(x + 1)(x - 2)}$

9. $\dfrac{3}{5}$

10. $\dfrac{3}{4}$

11. x

12. $x - 4$

13. 12

14. 1

15. 2

16. -6

17.

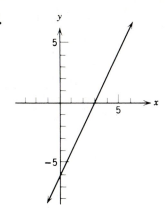

18. Slope: $\dfrac{2}{3}$; y-intercept: -3

19. Slope: $\dfrac{3}{4}$

20.

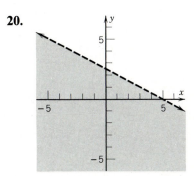

21. False

22. False

23. True

24. False

25. False

EXERCISES 9.1 (Page 301)

1. No   **3.** Yes   **5.** No

7.

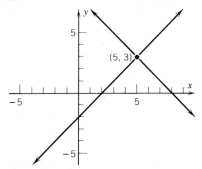

$(5, 3)$

9. (1, 2)

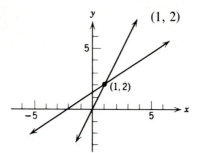

11. (−3, −1)

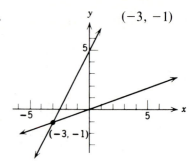

13. (4, 2)

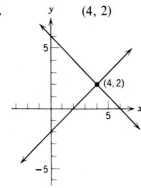

15. (2, 2)

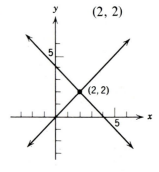

17. (−3, −2)

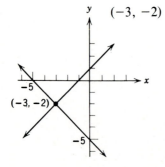

19. (1, 7)

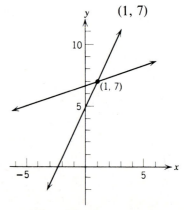

21. 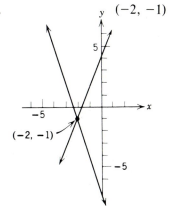 $(-2, -1)$

$(-2, -1)$

23. Dependent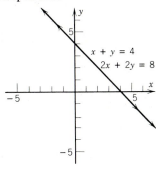

$x + y = 4$
$2x + 2y = 8$

25. Inconsistent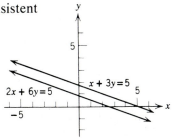

$2x + 6y = 5$ $x + 3y = 5$

27. Dependent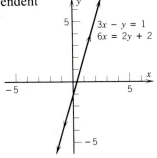

$3x - y = 1$
$6x = 2y + 2$

29. Inconsistent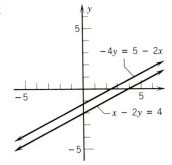

$-4y = 5 - 2x$

$x - 2y = 4$

EXERCISES 9.2 (Page 304)

1. $(3, 2)$ **3.** $(-1, 4)$ **5.** $(6, 2)$ **7.** $(1, 1)$

9. $(7, -2)$ **11.** $(-1, 1)$ **13.** $(1, 1)$ **15.** $(5, 2)$

17. $(1, -1)$ **19.** $(-4, 10)$ **21.** $(5, -1)$ **23.** $(4, 3)$

25. $(-4, 3)$ **27.** $(-2, -1)$ **29.** $(0, 2)$ **31.** $(-4, 2)$

33. $(3, 1)$ **35.** $(4, 3)$

EXERCISES 9.3 (Page 307)

1. (1, 2)   **3.** (2, 2)   **5.** $(-2, -1)$ **7.** $(-3, -2)$

9. $(5, -2)$ **11.** (8, 5)   **13.** (1, 2)   **15.** $(10, -2)$

17. (1, 1)   **19.** $(-1, -3)$ **21.** $(-1, 1)$ **23.** (1, 0)

25. $(1, -1)$ **27.** (4, 3)   **29.** (4, 3)   **31.** $(-3, 2)$

33. (2, 3)   **35.** $(-4, 10)$ **37.** (6, 2)

EXERCISES 9.4 (Page 310)

1. (2, 4)   **3.** (3, 1)   **5.** $(-1, 2)$ **7.** Dependent

9. $(2, -2)$ **11.** $(1, -4)$ **13.** (5, 3)   **15.** (5, 2)

17. $(-3, 1)$ **19.** $(2, -5)$ **21.** (2, 2)   **23.** $(-3, 1)$

25. $\left(\frac{1}{7}, -\frac{9}{7}\right)$ **27.** $\left(\frac{35}{13}, \frac{-3}{13}\right)$ **29.** $(4, -4)$

EXERCISES 9.5 (Page 313)

1. *Steps 1, 2*   Larger number: x
Smaller number: y

Step 4 $x + y = 25$
$x - y = 9$

Step 6 $x = 17, y = 8$;   the numbers are 17 and 8.

3. *Steps 1, 2*   Length of the longer piece: x
Length of the shorter piece: y

Step 4 $x + y = 20$
$x = y + 2$

Step 6 $x = 11, y = 9$;   the pieces are 11 meters and 9 meters.

5. *Steps 1, 2*   Weight of heavier package: x
Weight of lighter package: y

Step 4 $x + y = 28$
$x = 2y - 8$

Step 6 $x = 16, y = 12$;   the packages weigh 16 kilograms and 12 kilograms.

7. *Steps 1, 2*   Value of the car: x
Value of the trailer: y

Step 4 $x + y = 12,000$
$x = y + 5000$

Step 6 $x = 8500, y = 3500$;   the car is worth $8500 and the trailer $3500.

9. *Steps 1, 2* Number of votes for the winner: x
Number of votes for the loser: y

Step 4 $x + y = 7672$
$x = y + 12$

Step 6 $x = 3842, y = 3830$; the winner received 3842 votes; the loser 3830.

11. *Steps 1, 2* Distance to be walked: x
Distance to be driven: y

Step 4 $x + y = 45$
$y = x + 19$

Step 6 $x = 13, y = 32$; 13 kilometers must be walked.

13. *Steps 1, 2* First number: x
Second number: y

Step 4 $x + y = 24$
$\frac{1}{2}x = y + 3$

Step 6 $x = 18, y = 6$; the numbers are 18 and 6.

15. *Steps 1, 2* Number of dimes: x
Number of quarters: y

Step 4 $x + y = 34$
$0.10x + 0.25y = 5.50$

Step 6 $x = 20, y = 14$; there are 20 dimes and 14 quarters.

17. *Steps 1, 2* Number of adults: x
Number of children: y

Step 4 $x + y = 454$
$1.25x + 0.55y = 530.40$

Step 6 $x = 401, y = 53$; there were 401 adults and 53 children.

19. *Steps 1, 2* Amount invested at 8%: x
Amount invested at 12%: y

Step 4 $x + y = 3600$
$0.08x = 0.12y$

Step 6 $x = 2160, y = 1440$; $2160 was invested at 8% and $1400 at 12%.

21. *Steps 1, 2* Amount invested at 8%: x
Amounted invested at 7%: y

Step 4 $x = 2y$
$0.08x + 0.07y = 345$

Step 6 $x = 3000, y = 1500$; he has $3000 invested at 8% and $1500 at 7%.

23. *Steps 1, 2* Number of liters of 12% solution: x
Number of liters of 30% solution: y

Step 4 $x + y = 45$
            $0.12x + 0.30y = 0.24(45)$

Step 6 $x = 15$, $y = 30$;   15 liters of 12% solution and 30 liters of 30% solution should be mixed.

25. *Steps 1, 2*   Number of grams of 50% alloy: x
                        Number of grams of 75% alloy: y

Step 4 $x + y = 87.5$
            $0.50x + 0.75y = 0.60(87.5)$

Step 6 $x = 52.5$, $y = 35$;   52.5 grams of the 50% alloy and 35 grams of the 75% alloy must be mixed.

27. *Steps 1, 2*   Rate of plane from A: x
                        Rate of plane from B: y

Step 4 $4x + 4y = 2400$
            $x = y + 100$

Step 6 $x = 350$, $y = 250$;   rate of plane from A: 350 kilometers per hour; rate of plane from B: 250 kilometers per hour.

29. *Steps 1, 2*   Rate of freight train: x
                        Rate of passenger train: y

Step 4 $y = \dfrac{3}{2}x$

            $\dfrac{180}{x} - 2 = \dfrac{180}{y}$

Step 6 $x = 30$, $y = 45$;   the rate of the freight train is 30 kilometers per hour; the rate of the passenger train is 45 kilometers per hour.

CHAPTER 9 REVIEW (Page 320)

1. $(3, 2)$

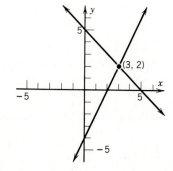

2. $(4, -1)$

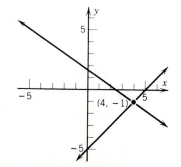

3. Dependent

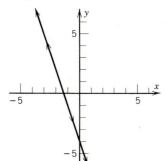

4. Inconsistent

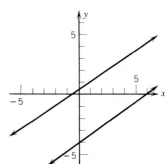

5. (4, 1) **6.** (−3, 0) **7.** (1, 0) **8.** (−1, −6)

9. $\left(2, \dfrac{1}{4}\right)$ **10.** (5, −2) **11.** (−1, 3) **12.** (−1, −5)

13. (2, 1) **14.** (−3, −5) **15.** (8, 2) **16.** $\left(\dfrac{1}{2}, 3\right)$

17. *Steps 1, 2* Weight of heavier package: x
Weight of lighter package: y

Step 4 $x + y = 84$
$x = y + 20$

Step 6 $x = 52, y = 32$; the packages weigh 52 kilograms and 32 kilograms.

18. *Steps 1, 2* Number of dimes: x
Number of quarters: y

Step 4 $x + y = 25$
$0.10x + 0.25y = 3.55$

Step 6 $x = 18, y = 7$; there are 18 dimes and 7 quarters.

19. *Steps 1, 2* First number: x
Second number: y

Step 4 $x + y = -40$
$x - y = -8$

Step 6 $x = -24, y = -16$; the numbers are -24 and -16.

20. *Steps 1, 2* Smaller number: x
Larger number: y

Step 4 $x + y = -2$
$y = 4x + 8$

Step 6 $x = -2, y = 0$; the numbers are -2 and 0.

CHAPTER 9 CUMULATIVE REVIEW (Page 321)

1. -29

2. $\dfrac{2x - 10}{(x + 3)(x - 3)(x - 1)}$

3. $3x^2 - 4x + 4$

4. $\dfrac{9}{8}$

5. $x > \dfrac{7}{2}$ **6.** 1 **7.** $\dfrac{3}{7}$ **8.** 9

9.

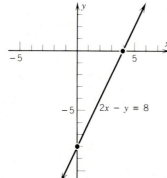

10.

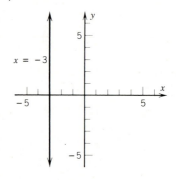

11.

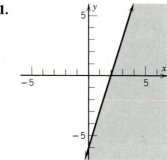

12.

13. Slope: -10

14. Slope: 2; y-intercept: $\dfrac{-3}{2}$

15. $y = -2x - 1$

16. $(-1, 4)$

17. $(1, 7)$

18. $(4, -1)$

19. a. *Steps 1, 2*   Smaller number: x
                     Larger number: $10x + 6$

      Step 4 $x + (10x + 6) = 28$

      Step 6 $x = 2$;   the numbers are 2 and 26.

   b. *Steps 1, 2*   Smaller number: x
                     Larger number: y

Step 4 $x + y = 28$
                $y = 10x + 6$
Step 6 $x = 2, y = 26$; the numbers are 2 and 26.

20. a. *Steps 1, 2*   First integer: x
                    Next integer: $x + 1$
        Step 4 $2x - \frac{1}{2}(x + 1) = 10$
        Step 6 $x = 7$; the integers are 7 and 8.

   b. *Steps 1, 2*   First integer: x
                    Next integer: y
        Step 4 $y = x + 1$
                    $2x - \frac{1}{2}y = 10$
        Step 6 $x = 7, y = 8$; the integers are 7 and 8.

21. False   **22.** False   **23.** False   **24.** False
25. True

REVIEW OF FACTORING (Page 322)

1. $5(x + 2y)$ **2.** $3(x^2 + 2x - 1)$ **3.** $-2(x^2 + 2)$
4. $4x(x - 2)$ **5.** $4(y^2 + 2y - 2)$ **6.** $x(3x - 6 + x^2)$
7. $(x + 3)(x + 2)$ **8.** $(y - 3)^2$ **9.** $(y - 8)(y + 1)$
10. $(y + 7)(y - 5)$ **11.** $2(x + 5)(x - 2)$ **12.** $3(y + 4)(y - 1)$
13. $(2y - 3)(y + 1)$ **14.** $(3y - 1)(2y + 1)$ **15.** $(3x - 2)(2x - 3)$
16. $(3x + 7)(x - 5)$ **17.** $(3x + 4)(2x - 1)$ **18.** $(4y + 1)(2y - 1)$
19. $2(2y + 1)(y + 1)$ **20.** $3(2x + 1)(x + 3)$ **21.** $2(3x + 1)(3x - 8)$
22. $5(8x - 5)(2x + 1)$ **23.** $2(2x + 1)(x - 3)$ **24.** $3(3y - 2)(2y + 1)$
25. $(x + a)^2$ **26.** $(x - a)^2$ **27.** $(y + 3b)^2$
28. $(y - 2b)^2$ **29.** $(x + 4a)^2$ **30.** $(y - 5b)^2$
31. $(x + 4)(x - 4)$ **32.** $(y + 6)(y - 6)$ **33.** $(y + 2)(y - 2)$
34. $(x + 7)(x - 7)$ **35.** $(2x + 5)(2x - 5)$ **36.** $9(y + 3)(y - 3)$
37. $12(y + 2)(y - 2)$ **38.** $2(5x + 4)(5x - 4)$

EXERCISES 10.1 (Page 324)

1. 7   **3.** 2   **5.** $\frac{1}{3}$ **7.** -3

9. 3, 5   **11.** 0, -4 **13.** 3, -3 **15.** $-\frac{1}{3}, \frac{3}{2}$

17. $0, -\dfrac{4}{5}, \dfrac{7}{2}$ **19.** $2, 3$ **21.** $0, 4$ **23.** $0, -3$

25. $2, -3$ **27.** $1, -8$ **29.** $-2, 3$ **31.** $0, 4$

33. $-4, 3$ **35.** $\dfrac{3}{2}, -\dfrac{3}{4}$ **37.** $\dfrac{2}{3}, -\dfrac{2}{3}$ **39.** $0, -\dfrac{3}{2}$

41. $3, 2, 1$ **43.** $0, -2, 1$ **45.** $0, -4, 3$ **47.** $4, -\dfrac{1}{2}, \dfrac{1}{2}$

EXERCISES 10.2 (Page 327)

1. $0, -3$ **3.** $0, \dfrac{5}{2}$ **5.** $0, \dfrac{9}{2}$ **7.** $0, 4$

9. $1, -1$ **11.** $2, -2$ **13.** $4, -4$ **15.** $8, -8$

17. $3, -3$ **19.** $3, -3$ **21.** $1, 2$ **23.** $-2, -2$

25. $-1, 4$ **27.** $3, -7$ **29.** $2, -7$ **31.** $-6, -6$

33. $-1, 6$ **35.** $-1, 4$ **37.** $5, -6$ **39.** $-1, \dfrac{4}{3}$

41. $\dfrac{1}{3}, \dfrac{3}{2}$ **43.** $\dfrac{1}{2}, \dfrac{1}{2}$ **45.** $\dfrac{3}{2}, \dfrac{3}{2}$ **47.** $\dfrac{1}{2}, 1$

49. $-\dfrac{1}{2}, 2$ **51.** $-1, 2$ **53.** $-2, 3$ **55.** $1, 1$

57. $-\dfrac{1}{2}, 2$ **59.** $\dfrac{1}{3}, -\dfrac{3}{2}$

EXERCISES 10.3 (Page 331)

1. $\dfrac{1}{3}, -\dfrac{1}{3}$ **3.** $\dfrac{3}{2}, -\dfrac{3}{2}$ **5.** $4, -4$ **7.** $0, -2$

9. $0, -2$ **11.** $0, 5$ **13.** $\dfrac{3}{2}, -2$ **15.** $-\dfrac{3}{2}, -\dfrac{3}{2}$

17. $-\dfrac{3}{4}, -\dfrac{5}{2}$ **19.** $-5, 3$ **21.** $-1, -2$ **23.** $1, -3$

25. $1, 1$ **27.** $-3, 5$ **29.** $2, 2$ **31.** $10, 20$

33. $2, 13$ **35.** $5, -6$ **37.** $5, -2$ **39.** $-\dfrac{1}{2}$

41. No solution   **43.** $1, 5$ **45.** -2 **47.** -2

49. $1, 6$

EXERCISES 10.4 (Page 336)

1. *Steps 1, 2*   Integer: x

    Step 4 $x^2 = 5x$

    Step 6 $x = 0, 5$;   the integer is either 0 or 5.

3. *Steps 1, 2*   First integer: x
                  Next integer: $x + 1$

    Step 4 $x(x + 1) = 72$

    Step 6 $x = 8, -9$;   the integers are 8 and 9. (-9 is not positive.)

5. *Steps 1, 2*   Integer: x

    Step 4 $x^2 + 2(x + 1)^2 = 66$

    Step 6 $x = 4, \dfrac{-16}{3}$;   the integer is 4. $\left(\dfrac{-16}{3} \text{ is not an integer.}\right)$

7. *Steps 1, 2*   Time when projectile strikes ground: t

    Step 4 $0 = 160 + 48t - 16t^2$

    Step 6 $t = 5, -2$;   the projectile strikes the ground at 5 seconds. (t cannot be -2.)

9. *Steps 1, 2*   Number of hours: t

    Step 4 $80 = 8t^2 - 32t - 16$

    Step 6 $t = 6, -2$;   it would take 6 hours. (t cannot be -2.)

11. *Steps 1, 2*   Number: x

    Step 4 $x + \dfrac{1}{x} = \dfrac{17}{4}$

    Step 6 $x = 4, \dfrac{1}{4}$;   the number is either 4 or $\dfrac{1}{4}$.

13. *Steps 1, 2*   Integer: x
                  Next odd integer: $x + 2$

    Step 4 $\dfrac{1}{x} + \dfrac{1}{x + 2} = \dfrac{8}{15}$

    Step 6 $x = 3, \dfrac{-5}{4}$;   the integers are 3 and 5.

                      $\left(\dfrac{-5}{4} \text{ is not an integer.}\right)$

15. *Steps 1, 2*   Integer: x
                  Next integer: $x + 1$

    Step 4 $\dfrac{1}{x} + \dfrac{2}{x + 1} = \dfrac{5}{6}$

(continued)

Step 6 $x = 3, \dfrac{-2}{5}$;   the integers are 3 and 4.

$\left(\dfrac{-2}{5} \text{ is not an integer.}\right)$

17. *Steps 1, 2*   Speed of boat in still water: x

Step 4 $\dfrac{24}{x + 2} + \dfrac{24}{x - 2} = 5$

Step 6 $x = 10, \dfrac{-2}{5}$;   the speed of the boat is 10 miles per hour. (The speed cannot be negative.)

19. *Steps 1, 2*   Speed on initial trip: x
Speed on return trip: $x + 15$

Step 4 $\dfrac{180}{x} = \dfrac{180}{x + 15} + 1$

Step 6 $x = 45, -60$;   his initial speed was 45 miles per hour; his return speed was 60 miles per hour. (The speed cannot be -60.)

21. *Steps 1, 2*   Width of the rectangle: x
Length of the rectangle: $2x + 2$

Step 4 $x(2x + 2) = 60$

Step 6 $x = 5, -6$;   the width is 5 centimeters and the length is 12 centimeters. (The width cannot be -5).

23. *Steps 1, 2*   Width of the border: x

Step 4 $(40 + 2x)(60 + x) = 2400 + 498$

Step 6 $x = 3, -83$;   the border is 3 feet wide. (The width cannot be -83.)

25. *Steps 1, 2*   Altitude: x
Base: $x + 11$

Step 4 $\tfrac{1}{2}x(x + 11) = 40$

Step 6 $x = 5, -16$;   the altitude is 5 inches and the base is 16 inches. (The altitude cannot be -16.)

CHAPTER 10 REVIEW (Page 341)

1. a. $2, -5$ b. $0, -3$ **2.** a. $0, 2$ b. $0, 2$

3. a. $7, -7$ b. $3, -3$ **4.** a. $-1, 5$ b. $-2, 9$

5. a. $-1, 3$ b. 2 **6.** a. $-4, \dfrac{3}{2}$ b. $-3, \dfrac{1}{2}$

7. a. $-2, \dfrac{1}{3}$ b. $\dfrac{2}{3}, 4$ **8.** a. $-2, 8$ b. $2, -4$

9. a. $-1, 4$ b. $-2, 3$ **10.** a. $6, -6$ b. $\dfrac{3}{5}, \dfrac{-3}{5}$

11. a. $3, -4$ b. 4 **12.** a. $-3, 5$ b. $4, -6$

13. a. -2 b. -2 **14.** a. -3 b. No solution

15. *Steps 1, 2*   Smaller number: x
Larger number: $x + 7$
Step 4 $x(x + 7) = 60$
Step 6 $x = 5, -12$; the numbers are 5 and 12. (-12 is not positive.)

16. *Steps 1, 2*   Smaller number: x
Larger number: $11 - x$
Step 4 $x(11 - x) = 30$
Step 6 $x = 5, 6$; the numbers are 5 and 6.

17. *Steps 1, 2*   First integer: x
Next integer: $x + 1$
Step 4 $\dfrac{1}{x} + \dfrac{1}{x + 1} = \dfrac{11}{30}$
Step 6 $x = 5, \dfrac{-6}{11}$; the integers are 5 and 6.
$\left(\dfrac{-6}{11} \text{ is not an integer.} \right)$

18. *Steps 1, 2*   Width of the rectangle: x
Length of the rectangle: $3x + 6$
Step 4 $x(3x + 6) = 45$
Step 6 $x = 3, -5$; the width is 3 centimeters and the length is 15 centimeters. (The width cannot be -5.)

19. *Steps 1, 2*   Riding rate: x
Walking rate: $x - 2$
Step 4 $\dfrac{6}{x} + \dfrac{4}{x - 2} = 6$
Step 6 $x = 3, \dfrac{2}{3}$; his riding rate was 3 miles per hour and his walking rate was 1 mile per hour. (The walking rate cannot be negative.)

20. *Steps 1, 2*   Average speed: x
Step 4 $\dfrac{360}{x} = \dfrac{360}{x + 5} + 1$
Step 6 $x = 40, -45$; Laura's average speed was 40 miles per hour. (The speed cannot be -45.)

CHAPTER 10 CUMULATIVE REVIEW (Page 342)

1. -1

2. 0

3. $-3, -\dfrac{5}{2}, -\dfrac{5}{3}, 0, \dfrac{3}{8}, \dfrac{3}{7}, 3$

4. $x(x - 1)(x - 2)$

5. $2x^2 + 5x - 12$

6. $\dfrac{x + 2}{3}$

7. $2x - 6$

8. $\dfrac{3 - 2x}{x(x + 1)}$

9. $\dfrac{3(x + 3)}{x^2 + 2x - 3}$

10. $\dfrac{3x + 2}{2x + 3}$

11. $(3, 3)$

12.

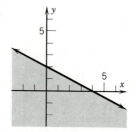

13. 44

14. $y = 2x - 1$

15. *Steps 1, 2*   Denominator: x
Numerator: $x + 5$

Step 4 $\dfrac{x + 5}{x} = \dfrac{4}{3}$

Step 6 $x = 15$;   the fraction is $\dfrac{20}{15}$.

16. *Steps 1, 2*   Weight of lighter package: x
Weight of heavier package: y

Step 4 $x + y = 140$
$y = x + 30$

Step 6 $x = 55, y = 85$;   the packages weigh 55 kilograms and 85 kilograms.

17. $y = 1$ or $y = 0$

18. $x = \dfrac{3}{5}$ or $x = -\dfrac{3}{5}$

19. $x = 3$ or $x = -1$

20. $T = 31.04$

21. False

22. False

23. False

24. False

25. False

EXERCISES 11.1 (Page 346)

1. $9, -9$

3. $11, -11$

5. $\dfrac{4}{5}, \dfrac{-4}{5}$

7. 4

9. -9 **11.** ± 12 **13.** 17 **15.** $\dfrac{1}{6}$

17. $\pm \dfrac{2}{3}$ **19.** $\sqrt{9}$ **21.** $-\sqrt{49}$ **23.** $\sqrt{169}$

25. $-\sqrt{25}$ **27.** $\sqrt{\dfrac{1}{4}}$ **29.** $-\sqrt{\dfrac{49}{64}}$ **31.** x

33. $2x^2$ **35.** $-ac^2$ **37.** $\pm 6a^3$ **39.** $11ab^4$

41. $-13x^3yz^5$ **43.** $8a^{32}b^4$ **45.** $x + y$ **47.** $-2(x + y)$

49. $7(x + 2)^2$ **51.** $\dfrac{a}{b}$ **53.** $\pm \dfrac{3}{xy^2}$ **55.** $\sqrt{x^2}$

57. $\sqrt{x^2y^4}$ **59.** $\sqrt{16y^6}$ **61.** $-\sqrt{64x^8}$ **63.** $\pm \sqrt{81x^4y^6}$

65. $-\sqrt{144x^{24}y^{24}}$ **67.** $\sqrt{(x + y)^2}$ **69.** $\pm \sqrt{(2x + y)^2}$

71. $\sqrt{\dfrac{1}{4}}$ **73.** $\sqrt{\dfrac{9}{16}x^2}$ **75.** $\pm \sqrt{\dfrac{16b^4}{25}}$ **77.** $\sqrt{\dfrac{1}{a^2}}$

79. $-\sqrt{\dfrac{9x^2}{y^2}}$ **81.** $\sqrt{\dfrac{(a + b)^2}{a^2}}$ **83.** 13 **85.** 22

87. $3ab$ **89.** $17x$ **91.** $x \geqslant 3$ **93.** $x \geqslant -7$

95. $x > 6$

EXERCISES 11.2 (Page 350)

1. $\sqrt{10}, -\sqrt{10}$ **3.** $\sqrt{17}, -\sqrt{17}$

5. $\sqrt{22}, -\sqrt{22}$ **7.** Rational; 6

9. Irrational   **11.** Irrational   **13.** Rational; 5   **15.** Rational; 12

17. Irrational   **19.** Rational; $\frac{2}{3}$ **21.** Irrational   **23.** Rational; 3

25. Irrational   **27.** 7.55   **29.** 1.73   **31.** 2.24

33. -5.10 **35.** 3.46   **37.** -8.66 **39.** 1.41

41. -1.73 **43.** 2.73   **45.** 1.59   **47.** 12.94

49. -22.88 **51.** 2.91   **53.** 0.51   **55.** 0.32

57. 0.72

59.

61.

63.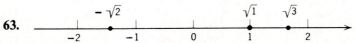

65.

67.

69.

71.   **73.**

75.   **77.**

EXERCISES 11.3 (Page 355)

1. $2\sqrt{2}$ **3.** $3\sqrt{2}$ **5.** $-2\sqrt{5}$ **7.** $-6\sqrt{2}$

9. 8 **11.** $12\sqrt{2}$ **13.** $5\sqrt{5}$ **15.** $-6\sqrt{30}$

17. $\pm 12\sqrt{5}$ **19.** $\pm 18\sqrt{6}$ **21.** $x\sqrt{x}$ **23.** $y^3\sqrt{y}$

25. $-x^5\sqrt{x}$ **27.** x^3 **29.** $\pm x^4$ **31.** x^6

33. $2x$ **35.** $3x\sqrt{x}$ **37.** $-2y^2\sqrt{6y}$ **39.** $8y^2$

41. $7x^3\sqrt{x}$ **43.** $\pm 4x\sqrt{2x}$ **45.** $4x^6\sqrt{5}$ **47.** $-8x^4\sqrt{x}$

49. $\pm y\sqrt{5y}$ **51.** $4x\sqrt{3y}$ **53.** $5xy\sqrt{x}$ **55.** $-3a^2b^3\sqrt{5b}$

57. $\dfrac{3}{4}xy$ **59.** $\pm\dfrac{5}{6}y$ **61.** $2x$ **63.** $6\sqrt{x}$

65. $-49y\sqrt{y}$ **67.** $2x^2\sqrt{y}$ **69.** $-ab^2\sqrt{ab}$ **71.** $\pm 2x^3y\sqrt{y}$

73. $-9a^5b^3\sqrt{b}$ **75.** 10.39 **77.** 16.59 **79.** 14.18

81. -12.32 **83.** -31.18 **85.** 36.59

87. Let x be 16. Then

$$\sqrt{x}+\sqrt{9}=\sqrt{16}+\sqrt{9} \qquad\qquad \sqrt{x+9}=\sqrt{16+9}$$
$$=4+3=7 \qquad\qquad\qquad =\sqrt{25}=5$$

Thus, $\sqrt{x}+\sqrt{9}$ is not equivalent to $\sqrt{x+9}$.

EXERCISES 11.4 (Page 359)

1. $3\sqrt{3}$ **3.** $\sqrt{5}$ **5.** 0

7. $5\sqrt{3}$ **9.** $-4\sqrt{3}$ **11.** $4\sqrt{2}-4\sqrt{3}$

13. $5\sqrt{3}+3\sqrt{2}$ **15.** $8+6\sqrt{6}$ **17.** $-2\sqrt{a}$

19. $12\sqrt{x}-6\sqrt{5x}$ **21.** $2b\sqrt{b}$ **23.** $4\sqrt{3}+4$

25. $-30-5\sqrt{7}$ **27.** $4\sqrt{2}-8\sqrt{3}$ **29.** $2x\sqrt{5}+6x\sqrt{x}$

31. $6x^3y\sqrt{2} - 18xy^2\sqrt{3x}$ **33.** $2(1 + \sqrt{3})$

35. $3(2 - \sqrt{2})$ **37.** $8(1 + 4\sqrt{5})$ **39.** $3(1 + \sqrt{2})$

41. $4(1 - \sqrt{2})$ **43.** $5(3\sqrt{2} - 2\sqrt{3})$ **45.** $3x(\sqrt{x} - 2\sqrt{y})$

47. $3x(\sqrt{y} - 3\sqrt{x})$ **49.** $2y(\sqrt{y} - 3y)$ **51.** $2 + 3\sqrt{3}$

53. $3 - \sqrt{5}$ **55.** $-1 + \sqrt{2}$ **57.** $\dfrac{1 + 3\sqrt{3}}{2}$

59. $\dfrac{x - \sqrt{2}}{x}$ **61.** $2 - \sqrt{y}$ **63.** $\dfrac{2 + \sqrt{2}}{3}$

65. $\dfrac{\sqrt{3} - 1}{5}$ **67.** $\dfrac{2\sqrt{10} - \sqrt{3}}{3}$ **69.** $\dfrac{\sqrt{11} + 1}{a}$

71. $\dfrac{5 + 6\sqrt{2}}{4}$ **73.** $\dfrac{3 + \sqrt{3}}{6a}$ **75.** $\dfrac{6 + 5\sqrt{3}}{15}$

77. $\dfrac{4\sqrt{3} - 3\sqrt{2}}{6}$ **79.** $\dfrac{8 + 3\sqrt{2}}{2}$ **81.** $\dfrac{1 + 2\sqrt{2}}{2x}$

83. $\dfrac{3 - 8\sqrt{y}}{4}$ **85.** $\dfrac{3\sqrt{x} + 2\sqrt{y}}{6}$

EXERCISES 11.5 (Page 363)

1. $\sqrt{15}$ **3.** $\sqrt{30}$ **5.** $3\sqrt{2}$ **7.** 4

9. $x\sqrt{6}$ **11.** $2xy\sqrt{3y}$ **13.** $12a$ **15.** $5x\sqrt{6y}$

17. $\sqrt{30}$ **19.** 10 **21.** $6\sqrt{6}$ **23.** $6x\sqrt{x}$

25. $abc\sqrt{abc}$ **27.** x^4 **29.** $3\sqrt{2} + \sqrt{6}$ **31.** $3\sqrt{2} + 2\sqrt{3}$

33. $4\sqrt{5} + 5\sqrt{2}$ **35.** $\sqrt{6} + 3\sqrt{2}$ **37.** $3 + \sqrt{6}$ **39.** $2\sqrt{5} - 2$

41. $1 - 2\sqrt{2}$ **43.** $2 + 2\sqrt{5}$ **45.** 1 **47.** $7 + 4\sqrt{3}$

49. -11 **51.** $-4 - 3\sqrt{15}$ **53.** $12 + 5\sqrt{35}$ **55.** $37 - 12\sqrt{7}$

57. $93 + 24\sqrt{15}$

EXERCISES 11.6 (Page 367)

1. 3 **3.** $5x$ **5.** $\dfrac{4}{3}$ **7.** $\dfrac{1}{5a}$

9. $2\sqrt{7b}$ **11.** $\dfrac{10ab\sqrt{a}}{7}$ **13.** 2 **15.** $\dfrac{18\sqrt{3}}{y}$

17. $\dfrac{5\sqrt{2}}{2}$ **19.** $\dfrac{2\sqrt{x}}{x}$ **21.** $\dfrac{\sqrt{3}}{3}$ **23.** $\dfrac{\sqrt{3ab}}{b}$

25. $\dfrac{3\sqrt{x}}{x}$ **27.** $\dfrac{5\sqrt{y}}{y}$ **29.** $\sqrt{2a}$ **31.** $\sqrt{6x}$

33. $4\sqrt{a}$ **35.** $2\sqrt{3xy}$ **37.** $\dfrac{2\sqrt{6}}{3}$ **39.** $\dfrac{3\sqrt{2}}{4}$

41. $\dfrac{6\sqrt{10}}{5}$ **43.** $\dfrac{5\sqrt{x}}{x}$ **45.** $\dfrac{2\sqrt{2x}}{x}$ **47.** $\dfrac{x\sqrt{y}}{y}$

49. $\sqrt{2}+1$ **51.** $\dfrac{5\sqrt{3}-7}{2}$ **53.** $\dfrac{\sqrt{21}+3}{4}$ **55.** 0.38

57. 2.12 **59.** 1.34 **61.** 1.73

EXERCISES 11.7 (Page 373)

1. 1 **3.** -1 **5.** 3

7. -5 **9.** -2 **11.** -4

13. 5 **15.** 9 **17.** x^2

19. a^9 **21.** $-2x^2z^5$ **23.** $y^2\sqrt[4]{y^3}$

25. $a^3b\sqrt[6]{a^3b}$ **27.** $5x^6y^6\sqrt[3]{2x^2y}$ **29.** $x^2\sqrt[4]{8xy^3}$

31. $-10y^6z^6\sqrt[3]{3x^2z^2}$ **33.** $6a^8b^7\sqrt[3]{5b^2}$ **35.** -8

37. 9 **39.** 10 **41.** 180

43. 48 **45.** 40

EXERCISES 11.8 (Page 376)

1. $\sqrt[3]{7}$ **3.** $\sqrt[3]{13^4}$ **5.** $\sqrt[4]{(2a)^3}$ **7.** $\sqrt[5]{\left(\dfrac{2x}{3}\right)^2}$

9. $4\sqrt[6]{x^5}$ **11.** $x\sqrt{y^3}$ **13.** $x^{3/5}$ **15.** $8^{3/4}$

17. $(3x)^{4/3}$ **19.** $2^{1/5}x^{3/5}$ **21.** $x^{2/3}y^{4/3}$ **23.** $x^{1/4}y^{9/4}$

25. 4 **27.** 5 **29.** -3 **31.** 4

33. $-3/5$ **35.** $2/3$ **37.** $2x^2$ **39.** $2x^3$

41. $4x^3y^4$ **43.** $\dfrac{-y^2}{3}$ **45.** $\dfrac{xy^4}{z^2}$ **47.** $\dfrac{x^3y^4}{z^2}$

49. $30^{1/3}$; irrational   **51.** $64^{1/3}$; rational   **53.** $18^{1/4}$; irrational

55. $81^{1/4}$; rational   **57.** $(-27)^{1/3}$; rational   **59.** $81^{1/3}$; irrational

61. 9 **63.** $6^{5/3}$ **65.** $x^{1/5}$ **67.** $a^{7/3}$

69. $xy^{6/5}$ **71.** 2^{11} **73.** $\sqrt[3]{5}$ **75.** 2

77. $\sqrt{x}$ **79.** $\sqrt{5}$

CHAPTER 11 REVIEW (Page 379)

1. $\sqrt{6}, -\sqrt{\frac{2}{3}}$ **2.** a. 9.644   b. 13.712   c. 1.414

3. a.   b.

4. a. $6\sqrt{2}$ b. $-3\sqrt{10}$ c. $5\sqrt{7}$

5. a. y^2 b. $x^7\sqrt{x}$ c. $xy^3\sqrt{xy}$

6. a. $3x\sqrt{3}$ b. $3x\sqrt{3}$ c. $x\sqrt{3}$

7. a. $4\sqrt{7}$ b. $2\sqrt{6}$ c. $5\sqrt{2}$

8. a. $-8\sqrt{3}$ b. $\sqrt{x}$ c. $2x\sqrt{y}$

9. a. $3\sqrt{y} - 18$ b. $y\sqrt{xy} - 2y^2$ c. $\sqrt{6} - 3\sqrt{2}$

10. a. $2(1 - \sqrt{2})$ b. $3(\sqrt{3} - \sqrt{5})$ c. $x^2(3 - \sqrt{5})$

11. a. $1 - 2\sqrt{3}$ b. $2 - 2\sqrt{3}$ c. $1 - \sqrt{2}$

12. a. $\dfrac{3 - \sqrt{3}}{2}$ b. $\dfrac{3\sqrt{15}}{5}$ c. $\dfrac{4 - \sqrt{3}}{6}$

13. a. $\dfrac{2\sqrt{5} - 9}{6}$ b. $\dfrac{2\sqrt{3} - 5}{5}$ c. $\dfrac{3\sqrt{6} + 8}{4}$

14. a. $\sqrt{15}$ b. $16\sqrt{2}$ c. $3x\sqrt{2x}$

15. a. $5\sqrt{6}$ b. 12   c. $2x\sqrt{3x}$

16. a. $2\sqrt{3} - \sqrt{6}$ b. $\sqrt{14} - 7\sqrt{2}$ c. $4 - 2\sqrt{6}$

17. a. 1   b. -2 c. $12 - 7\sqrt{3}$

18. a. 3   b. $\dfrac{a\sqrt{6b}}{b}$ c. $a\sqrt{3}$

19. a. $\dfrac{\sqrt{15}}{5}$ b. $\dfrac{2\sqrt{15}}{5}$ c. $\dfrac{\sqrt{6x}}{2x}$

20. a. $3(\sqrt{2} - 1)$ b. $\sqrt{6} + 2$ c. $2\sqrt{5} + 2\sqrt{3} + \sqrt{15} + 3$

21. a. 5   b. -9 c. 3

22. a. $5x\sqrt[3]{3x^2y}$ b. $x^4y^6z^2\sqrt[6]{x^4z}$ c. $2a^4\sqrt[4]{5b^3}$

23. a. $x^5\sqrt[5]{2}$ b. $9a^3b^{10}\sqrt[3]{3}$ c. $2xy^3\sqrt[4]{8y^2}$

24. a. -2 b. 7   c. 11

25. a. $x^{9/4}$ b. $x^{2/3}y^{7/3}$ c. $x^{1/6}y^{5/6}$

26. a. $3\sqrt[3]{x^5}$ b. $\sqrt[5]{x^2y^3}$ c. $\sqrt[3]{(2xy^2)^4}$

27. a. irrational   b. irrational   c. rational

28. a. $5x^5$ b. $-4xy^{27}$ c. $3x^4y^2$

29. a. $8^{7/4}$ b. $x^{2/5}$ c. $4xy^2$

30. a. $\sqrt[3]{5^2}$ b. $\sqrt[3]{x}$ c. $\sqrt[3]{6}$

CHAPTER 11 CUMULATIVE REVIEW (Page 381)

1. Two

2. -4

3. $-6ab^3c$

4. $\dfrac{5a - 9b}{6}$

5. $\dfrac{a - 1}{4x(2a + 5)}$

6. $x^2 + 3x - 1$

7. $(2x - 3)(2x - 3)$

8. -5

9. 6

10. $146 - n$

11. $25x$

12. Dependent

13. a. 3.27×10^6 b. 6.49×10^{-5}

14. $-\frac{1}{2}, 3$

15. *Steps 1, 2*   Smaller number: x
Larger number: $2x - 8$

Step 4 $x + (2x - 8) = 154$

Step 6 $x = 54$;   the numbers are 54 and 100.

16. Slope: $\dfrac{-3}{2}$

17. $8\sqrt{2} - 4\sqrt{5}$

18. 1

19. $\sqrt{6x}$

20. $\sqrt{6}$

21. True

22. False

23. False

24. False

25. False

EXERCISES 12.1 (Page 384)

1. $2, -2$

3. $4, -4$

5. $7, -7$

7. $\sqrt{3}, -\sqrt{3}$

9. $\sqrt{10}, -\sqrt{10}$

11. $\sqrt{6}, -\sqrt{6}$

13. $2\sqrt{3}, -2\sqrt{3}$

15. $3\sqrt{2}, -3\sqrt{2}$

17. $3\sqrt{2}, -3\sqrt{2}$

19. $2, -2$

21. $\sqrt{5}, -\sqrt{5}$

23. $0, 0$

25. $2\sqrt{5}, -2\sqrt{5}$

27. $3, -3$

29. $2\sqrt{2}, -2\sqrt{2}$

31. $25, -25$

33. $8, -8$

35. $5\sqrt{2}, -5\sqrt{2}$

37. $3, -1$

39. $7, -3$

41. $6, 4$

43. $a + 5, a - 5$

45. $a + 3, -a + 3$

47. $a + b, a - b$

49. $-3 + \sqrt{2}, -3 - \sqrt{2}$

51. $-5 + \sqrt{5}, -5 - \sqrt{5}$

53. $-10 + 2\sqrt{2}, -10 - 2\sqrt{2}$

55. $5 + \sqrt{a}, 5 - \sqrt{a}$

57. $-1 + \sqrt{b}, -1 - \sqrt{b}$

59. $b + \sqrt{a}, b - \sqrt{a}$

61. $\dfrac{1}{3}, -3$

63. $\dfrac{13}{5}, \dfrac{-9}{5}$

65. $\dfrac{-1 + 2\sqrt{2}}{2}, \dfrac{-1 - 2\sqrt{2}}{2}$

67. $\dfrac{2 + \sqrt{13}}{6}, \dfrac{2 - \sqrt{13}}{6}$

69. $2, \dfrac{-1}{2}$

71. $4 + \sqrt{5}, 4 - \sqrt{5}$

73. $\sqrt{a}, -\sqrt{a}$

75. $ab\sqrt{3a}, -ab\sqrt{3a}$

77. $\dfrac{\sqrt{30b}}{6}, -\dfrac{\sqrt{30b}}{6}$

79. $t = \dfrac{\sqrt{2gs}}{g}, \quad t = -\dfrac{\sqrt{2gs}}{g}$

81. $r = \dfrac{\sqrt{\pi A}}{2\pi}, \quad r = -\dfrac{\sqrt{\pi A}}{2\pi}$

83. $d = \dfrac{\sqrt{3Ik}}{I}, \quad d = -\dfrac{\sqrt{3Ik}}{I}$

EXERCISES 12.2 (Page 391)

1. $1; (x + 1)^2$

3. $9; (x - 3)^2$

5. $25; (x - 5)^2$

7. $36; (x - 6)^2$

9. $\dfrac{9}{4}; \left(x + \dfrac{3}{2}\right)^2$

11. $\dfrac{121}{4}; \left(x + \dfrac{11}{2}\right)^2$

13. $\dfrac{4}{9}; \left(x - \dfrac{2}{3}\right)^2$

15. $2, -6$

17. $1, 1$

19. $4, -5$

21. $-1, -2$

23. $-1, 4$

25. $-2, 5$

27. $1 + \sqrt{2}, 1 - \sqrt{2}$

29. $\dfrac{3 + \sqrt{21}}{2}, \dfrac{3 - \sqrt{21}}{2}$

31. $\dfrac{-1 + \sqrt{13}}{2}, \dfrac{-1 - \sqrt{13}}{2}$

33. $\dfrac{1}{2}, -\dfrac{3}{2}$

35. $\dfrac{1}{2}, -2$

37. $-\dfrac{5}{2}, 3$

EXERCISES 12.3 (Page 395)

1. $a = 1, \quad b = -3, \quad c = 2$

3. $a = 1, \quad b = -1, \quad c = -30$

5. $a = 1, \quad b = -2, \quad c = 0$

7. $a = 4, \quad b = 0, \quad c = -3$

9. $a = 2, \quad b = -7, \quad c = 6$

11. $a = 6, \quad b = -5, \quad c = 1$

13. $a = 1, \quad b = -8, \quad c = 4$

15. $a = 2, \quad b = -2, \quad c = -1$

17. $a = 6, \quad b = 7, \quad c = -3$

19. $a = 9, \quad b = 6, \quad c = -8$

21. $1, 2$

23. $-2, 6$

25. $3, -5$

27. $\dfrac{-3 + \sqrt{13}}{2}, \dfrac{-3 - \sqrt{13}}{2}$

29. $\dfrac{3 + \sqrt{17}}{2}, \dfrac{3 - \sqrt{17}}{2}$

31. $0, 2$

33. $0, 5$

35. $0, 7$

37. $0, 3$

39. $\dfrac{\sqrt{3}}{2}, -\dfrac{\sqrt{3}}{2}$

41. $2, \dfrac{3}{2}$

43. $\dfrac{1}{3}, -\dfrac{1}{2}$

45. $\dfrac{5}{2}, -\dfrac{1}{3}$

47. $1 + \sqrt{2}, 1 - \sqrt{2}$

49. $2 + \sqrt{6}, 2 - \sqrt{6}$

51. $\dfrac{3 + \sqrt{17}}{3}, \dfrac{3 - \sqrt{17}}{4}$

53. $\dfrac{3}{2}, -\dfrac{5}{2}$

55. $\dfrac{3}{2}, \dfrac{2}{3}$

57. $-\dfrac{3}{2}, 3$

EXERCISES 12.4 (Page 404)

1. $(0, -3)$ **3.** $(-1, 0)$ **5.** $(-2, 5)$

7. $(0, -2)$ **9.** $(1, 0)$ **11.** $(-2, 0)$

13. $(0, 12)$ **15.** $(2, 2)$ **17.** $(4, 0)$

19.

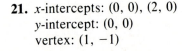

21. x-intercepts: $(0, 0)$, $(2, 0)$
y-intercept: $(0, 0)$
vertex: $(1, -1)$

23. x-intercepts: $(-2, 0)$, $(2, 0)$
y-intercept: $(0, -4)$
vertex: $(0, -4)$

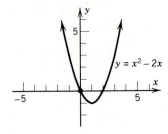

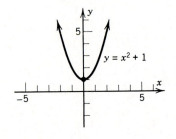

25. x-intercepts: $(-3, 0)$, $(3, 0)$
y-intercept: $(0, 9)$
vertex: $(0, 9)$

27. x-intercepts: none
y-intercept: $(0, 1)$
vertex: $(0, 1)$

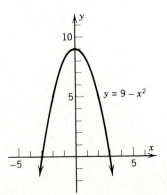

29. *x*-intercept: (0, 0)
   y-intercept: (0, 0)
   vertex: (0, 0)

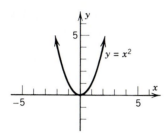

31. *x*-intercept: (0, 0)
   y-intercept: (0, 0)
   vertex: (0, 0)

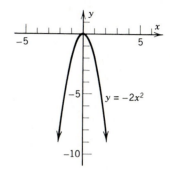

33. *x*-intercept: (−2, 0)
   y-intercept: (0, 4)
   vertex: (−2, 0)

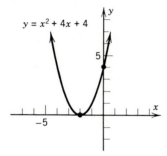

35. *x*-intercepts: (1, 0), (3, 0)
   y-intercept: (0, −3)
   vertex: (2, 1)

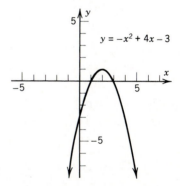

37. *x*-intercepts: none
   y-intercept: (0, 11)
   vertex: (−3, 2)

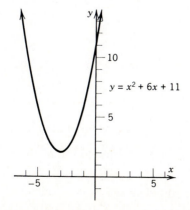

39. *x*-intercepts: (−4, 0), (3,0)
   y-intercept: (0, 12)
   vertex: $(-\frac{1}{2}, \frac{49}{4})$

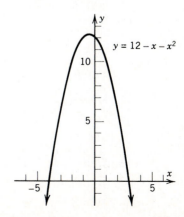

EXERCISES 12.5 (Page 407)

1. *Steps 1, 2*   Length of diagonal: x
   Step 4 $(3)^2 + (4)^2 = x^2$
   Step 6 $x = \pm 5$;   the diameter is 5 centimeters long.

3. *Steps 1, 2*   Length of diagonal: x
   Step 4 $(3)^2 + (3)^2 = x^2$
   Step 6 $x = \pm 3\sqrt{2}$;   the diameter is $3\sqrt{2}$ kilometers long.

5. *Steps 1, 2*   Distance from home-plate to second base: x
   Step 4 $(90)^2 + (90)^2 = x^2$
   Step 6 $x = \pm 127$;   the distance is 127 feet.

7. *Steps 1, 2*   Length of rectangle: x
   Step 4 $x^2 + (5)^2 = (13)^2$
   Step 6 $x = \pm 12$;   the rectangle is 12 inches long.

9. *Steps 1, 2*   Width of rectangle: x
   Step 4 $(9)^2 + x^2 = (11)^2$
   Step 6 $x = \pm 2\sqrt{10}$;   the rectangle is $2\sqrt{10}$ yards wide.

11. *Steps 1, 2*   Width: x
   Length: $x + 3$
   Step 4 $x^2 + (x + 3)^2 = (15)^2$
   Step 6 $x = -12, 9$;   the rectangle is 9 meters wide and 12 meters long.

13. *Steps 1, 2*   Length: x
   Width: $x - 3$
   Step 4 $x^2 + (x - 3)^2 = 29$
   Step 6 $x = -2, 5$;   the rectangle is 5 millimeters long and 2 millimeters wide.

15. *Steps 1, 2*   Width: x
   Length: $2x$
   Step 4 $x^2 + (2x)^2 = (3\sqrt{5})^2$
   Step 6 $x = \pm 3$;   the rectangle is 3 yards wide and 6 yards long.

17. *Steps 1, 2*   Length of wire: x
   Step 4 $(40)^2 + (30)^2 = x^2$
   Step 6 $x = \pm 50$;   the wire must be 50 meters long.

19. *Steps 1, 2*   Distance from tip of shadow to top of tree: x
   Step 4 $(30)^2 + (30)^2 = x^2$
   Step 6 $x = \pm 30\sqrt{2}$;   the distance is $30\sqrt{2}$ meters.

EXERCISES 12.6 (Page 412)

1. 21   **3.** 81   **5.** No solution   **7.** 5

9. 5   **11.** 1   **13.** 5   **15.** −3

17. −¾   **19.** 8   **21.** 16   **23.** ⅑, 1

25. 3, 7   **27.** No solution **29.** 5, 21

CHAPTER 12 REVIEW (Page 415)

1. a. −5, 5   b. −3, 3   **2.** a. $-\sqrt{7}, \sqrt{7}$ b. $-\sqrt{6}, \sqrt{6}$

3. a. −1, 5   b. −4, −2   **4.** a. 3, 11   b. $a + c, a - c$

5. a. $-3 + \sqrt{a}, -3 - \sqrt{a}$ b. $2 - a, -2 - a$

6. a. 1, −4   b. $\dfrac{3 + \sqrt{21}}{2}, \dfrac{3 - \sqrt{21}}{2}$

7. a. −1   b. −1, 2   **8.** −1, 5

9. 3, −5   **10.** $\dfrac{-3 + \sqrt{5}}{2}, \dfrac{-3 - \sqrt{5}}{2}$

11. $\dfrac{4}{3}, 3$ **12.**

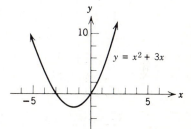

13.   **14.**

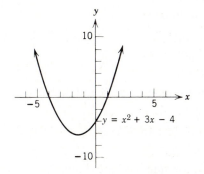   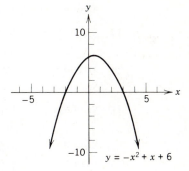

15. ±16   **16.** $\pm 6\sqrt{2}$

17. *Steps 1, 2*   Length of diagonal: x

   Step 4 $(7)^2 + (9)^2 = x^2$

   Step 6 $x = \pm\sqrt{130};$ the diagonal is $\sqrt{130}$ meters long.

18. *Steps 1, 2*   Length of diagonal: x

    Step 4 $(8)^2 + (8)^2 = x^2$

    Step 6 $x = \pm 8\sqrt{2}$;   the diagonal is $8\sqrt{2}$ centimeters long.

19. *Steps 1, 2*   Length of cable: x

    Step 4 $(18)^2 + (24)^2 = x^2$

    Step 6 $x = \pm 30$;   the cable is 30 meters long.

20. *Steps 1, 2*   Width: x

                 Length: $3x$

    Step 4 $x^2 + (3x)^2 = (5\sqrt{10})^2$

    Step 6 $x = \pm 5$;   the rectangle is 5 meters wide and 15 meters long.

21. 7   **22.** No solution   **23.** 12

24. $\dfrac{-6}{25}$, 2   **25.** 5

CHAPTER 12 CUMULATIVE REVIEW (Page 416)

1. $2 \cdot 2 \cdot 2 \cdot 3xxxyy$ **2.** 1

3. 1   **4.** $2a^2$ **5.** 0   **6.** (1, 2)

7. $\dfrac{3x + 4}{x(x + y)}$ **9.** $5x^2\sqrt{y}$ **10.** $a - x$ **11.** Two

12. Either $a = 0$, $b = 0$ or both a and $b = 0$

13. *Steps 1, 2*   Number: x

    Step 4 $x - 9 = 18$

    Step 6 $x = 27$;   the number is 27.

14. *Steps 1, 2*   First number: x

                 Second number: $4x$

    Step 4 $x + 4x = 40$

    Step 6 $x = 8$;   the numbers are 8 and 32.

15. *Steps 1, 2*   First number: x

                 Second number: $16 - x$

    Step 4 $x(16 - x) = 63$

    Step 6 $x = 7, 9$;   the numbers are 7 and 9.

16. *Steps 1, 2*   Second angle: x

                 First angle: $2x$

                 Third angle: $3x$

    Step 4 $x + 2x + 3x = 180$

    Step 6 $x = 30$;   the angles are 30°, 60°, and 90°.

17. *Steps 1, 2*   Length of altitude: x
Length of base: $4x$

Step 4 $\frac{1}{2}x(4x) = 18$

Step 6 $x = \pm3$; the altitude is 3 centimeters and the base is 12 centimeters.

18. *Steps 1, 2*   Length of first side: $\frac{2}{3}x$
Length of second side: $\frac{2}{3}x + 2$
Length of third side: x

Step 4 $x + \frac{2}{3}x + (\frac{2}{3}x + 2) = 30$

Step 6 $x = 12$; the sides are 8 inches, 10 inches, and 12 inches long.

19. *Steps 1, 2*   Width: x
Length: $2x$

Step 4 $x(2x) = 242$

Step 6 $x = \pm11$; the width is 11 centimeters and the length is 22 centimeters.

20. *Steps 1, 2*   Length of altitude: x
Length of base: $3x$

Step 4 $\frac{1}{2}x(3x) = 4(6)$

Step 6 $x = \pm4$; the altitude is 4 centimeters and the base is 12 centimeters.

21. False   **22.** False   **23.** True

24. False   **25.** False

FINAL CUMULATIVE REVIEW I (Page 418)

1. 71, 73, 79, 83, 89, 97   **2.** $a + b = b + a$; $ab = ba$

3. $-4, -2, -1, 3, 5, 6$ **4.** $V = \pi r^2 h$

5. $30 - x$ **6.** $2a^2$

7. $2x + 1 + \dfrac{-1}{3x}$

8. *Steps 1, 2*   First odd integer: x
Next odd integer: $x + 2$

Step 4 $x(x + 2) = 195$

Step 6 $x = -15, 13$; the integers are 13 and 15.

9. $\dfrac{2b}{a + 2b}$ **10.** $\dfrac{3a + 3}{4a}$

11. $\dfrac{3ax^2y}{10}$ **12.** 4

13.

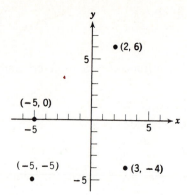

14. (1, 1)

15. (4, 0)   **16.** $\dfrac{\sqrt{y}}{4}$ **17.** 2   **18.** 3.46

19. -3

20. *Steps 1, 2*   Length of side a: x
Length of side b: $x + 2$
Length of side c: $2x$

Step 4 $x + (x + 2) + 2x = 34$

Step 6 $x = 8$; the lengths of the sides are 8 centimeters, 10 centi-meters, and 16 centimeters.

FINAL CUMULATIVE REVIEW II (Page 419)

1. 2   **2.** Zero   **3.** No   **4.** $\dfrac{24}{b}$

5. $5b$ **6.** $a < -5$

7. *Steps 1, 2*   Capacity of smaller truck: x
Capacity of larger truck: $x + 5$

Step 4 $5x + (x + 5) = 47$

Step 6 $x = 7$; the trucks' capacities are 7 tons and 12 tons.

8. $\dfrac{a^2 + b^2}{ab}$ **9.** $\dfrac{3(x - 3)}{x^2 - 9}$ **10.** $\dfrac{2}{3x}$ **11.** Proportion

12.

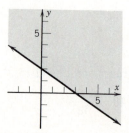

13. $x + y = 1$

14. $(-1, 3)$ **15.** 8, -2

16. $\sqrt{5}, \sqrt{6}, \sqrt{7}, \sqrt{8}, \sqrt{10}$ **17.** $4xy\sqrt{10xyz}$

18. $\dfrac{2 - \sqrt{2}}{2}$ **19.** $\dfrac{-5 \pm \sqrt{17}}{2}$ **20.** $x = \dfrac{25}{3}$

FINAL CUMULATIVE REVIEW III (Page 420)

1. $2 \cdot 2 \cdot 2 \cdot 3 \cdot 3 \cdot 5$ **2.** 30

3. $-a^2 - 2a$ **4.** No

5. 625 square millimeters   **6.** $ab(c - 1)$

7. $3(2x - 3)(x + 1)$ **8.** $5n + 10(n + 6)$ or $15n + 60$

9. $\dfrac{-6b}{5}$ **10.** $\dfrac{5}{6x}$ **11.** $x - 8$

12. $\dfrac{2}{a + b}$

13. *Steps 1, 2*   Numerator: x
                     Denominator: $x + 4$

       Step 4 $\dfrac{x}{x + 4} = \dfrac{5}{6}$

       Step 6 $x = 20$;   the numerator is 20.

14. *Steps 1, 2*   Number of bricks required for 10 meters: x

       Step 4 $\dfrac{228}{4} = \dfrac{x}{10}$

       Step 6 $x = 570$;   570 bricks are required.

15. Linear

16. $y = \dfrac{4x + 2}{3}$ **17.** 1

18. $6a^3b^2c^2\sqrt{c}$ **19.** $3\sqrt{a}$

20. *Steps 1, 2*   Distance from foot of pole: x

       Step 4 $x^2 + (6)^2 = (14)^2$

       Step 6 $x = \pm 4\sqrt{10}$;   the wire will reach $4\sqrt{10}$ meters.

FINAL CUMULATIVE REVIEW IV (Page 421)

1.

2. $r + s$; rs **3.** -3 **4.** 5

5. *Steps 1, 2*   First even integer: x
                     Next even integer: $x + 2$
                     Third even integer: $x + 4$

(continued)

Step 4 $x + (x + 2) + (x + 4) = -78$

Step 6 $x = -28$; the integers are -28, -26, and -24.

6. $-3bx$ **7.** $2(x - 2)(x + 2)$

8. *Steps 1, 2* Number of adult tickets: x

Number of children's tickets: $260 - x$

Step 4 $4x + 2(260 - x) = 820$

Step 6 $x = 110$; 110 adult tickets were sold and 150 children's tickets.

9. $x - 1$ **10.** $\dfrac{3x + 5}{(x + 2)^2}$ **11.** $16x$ **12.** 9

13. *Steps 1, 2* Number of miles: x

Step 4 $\dfrac{\frac{1}{4}}{8} = \dfrac{6}{x}$

Step 6 $x = 192$; 6 inches represents 192 miles.

14. Slope: $\dfrac{4}{3}$, y-intercept: $\dfrac{-5}{3}$ **15.** $\dfrac{4}{3}$

16. $(0, -6)$; $(2, 0)$; $(-3, -15)$ **17.** $\dfrac{a}{4}$

18. $0, \dfrac{1}{4}$ **19.** $(-6, 0)$ and $(5, 0)$

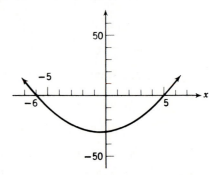

20. *Steps 1, 2* Length of rectangle: x

Width of rectangle: y

Step 4 $2x + 2y = 56$

$xy = 192$

Step 6 $x = 16$, $y = 12$; the rectangle is 16 centimeters long and 12 centimeters wide.

FINAL CUMULATIVE REVIEW V (Page 421)

1. 1 **2.** Integers **3.** 10

4. $(y - 1)(y - 11)$ **5.** $-8b^2$ **6.** $\dfrac{2(w + 8)}{3}$

7. $\dfrac{-(1 - a)}{3}$ or $\dfrac{-1 + a}{3}$

8. $x^2 - 2x + 1$ **9.** $\dfrac{h^2}{6 + h}$ **10.** $\dfrac{875}{t}$ mi.

11. 15   **12.** Means; extremes

13. 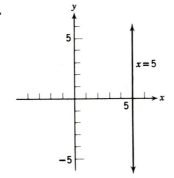   **14.** $(1, 7)$

15. -3 and 7

16.

17. 9   **18.** $\dfrac{-5 \pm \sqrt{17}}{2}$ **19.** $\dfrac{2}{3}$ and -3

20. a. 4.87×10^7 b. 6.32×10^{-5}

FINAL CUMULATIVE REVIEW VI (Page 422)

1. -20 **2.** 11   **3.** x **4.** $-17°$

5. $\dfrac{a + 2}{a - 1}$

6. *Steps 1, 2*   Number of votes for loser: x
Number of votes for winner: $x + 62$

    Step 4 $x + (x + 62) = 3626$
    Step 6 $x = 1782$; the winner received 1844 votes, and the loser 1782 votes.

7. $3y^2$ **8.** $(a - 5b)(a + 5b)$

9. $5(r - 8)$ or $5r - 40$ **10.** 60

11. $\dfrac{11}{2}$ **12.** $\dfrac{-3x - 11}{6}$ **13.** 22   **14.** 2

15. $\dfrac{2E}{v^2}$ **16.** 8.46   **17.** (1, 2)

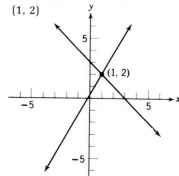

18. 0 and $\dfrac{-3}{2}$ **19.** $-1 + \sqrt{5}$ **20.** ± 12

FINAL CUMULATIVE REVIEW VII (Page 423)

1. -1 **2.** $1 \pm \sqrt{6}$ **3.** $4b - 5c$

4. $x^3 \cdot x^2 = x^5$ **5.** $(2y + 3)(2y + 5)$ **6.** $6(22 - y)$

7. $\dfrac{3}{y}$ **8.** $\dfrac{A}{1 + rt}$ **9.** $\dfrac{b - a}{b + a}$

10. *Steps 1, 2*   Rate of first car: $x + 10$
                Rate of second car: x

  Step 4 $\dfrac{90}{x + 10} = \dfrac{60}{x}$

  Step 6 $x = 20$; the first car travels at 30 miles per hour; the second car at 20 miles per hour.

11. $(3, 13)$; $(-2, -2)$; $(0, 4)$; $(6, 22)$

12. $(2, -3)$ **13.** $(3, -2)$

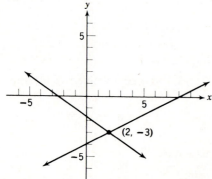

14. 0, 2   **15.** $\frac{1}{2}$, -18 **16.** 16

17. $y = -3x + 5$ **18.** $2\sqrt{3b}$ **19.** $\frac{\pm\sqrt{c}}{b}$

20. *Steps 1, 2*   Rate of the boat: x

   Step 4 $\dfrac{9}{x+2} + \dfrac{9}{x-2} = 6$

   Step 6 $x = -1, 4$;   the rate of the boat was 4 miles per hour.

FINAL CUMULATIVE REVIEW VIII (Page 424)

1. $\dfrac{-8}{3}$ **2.** 5   **3.** $n + 2, n + 4, n + 6$

4. $(a - 4b)(a - 2b)$ **5.** $-4xy$

6. *Steps 1, 2*   Length of longer piece: x
                  Length of shorter piece: y

   Step 4 $x + y = 64$
                  $2x = 5y$

   Step 6 $x = 45\frac{5}{7}$, $y = 18\frac{2}{7}$;   the cable should be cut $18\frac{2}{7}$ meters from
                  one end.

7. $\dfrac{x+6}{x-2}$ **8.** $\dfrac{2}{a+b}$

9. $b - a$ **10.** $x + 8 + \dfrac{33}{x-5}$

11. $(-1, 0)$

12. $(1, 7)$

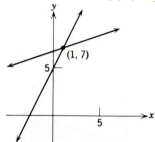

13. 67 kilograms and 79 kilograms   **14.** $-2, 7$

15. $10xy\sqrt{2x}$ **16.** $7\sqrt{5}$

17.

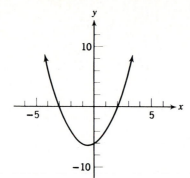

x-intercepts: $(-3, 0)$, $(2, 0)$
y-intercept: $(0, -6)$
vertex: $(-\frac{1}{2}, -6\frac{1}{4})$

18. $\sqrt{10}$

19. $\dfrac{1 \pm \sqrt{15}}{2}$

20. *Steps 1, 2*   Width of rectangle: x
Length of rectangle: $2x - 4$

Step 4 $x(2x - 4) = 240$

Step 6 $x = -10, 12$;   the width is 12 centimeters, and the length is 20 centimeters.

FINAL CUMULATIVE REVIEW IX (Page 425)

1. 4

2. -4

3. $a^2 - 3ab$

4. *Steps 1, 2*   First even integer: x
Second even integer: $x + 2$
Third even integer: $x + 4$

Step 4 $x + (x + 2) + (x + 4) = 78$

Step 6 $x = 24$;   the integers are 24, 26, and 28.

5. $2(x - 11)(x - 1)$

6. $n + 28$

7. $a < 5$

8. $\dfrac{2y - 3x}{xy}$

9. $\dfrac{1}{16b}$

10. $2x - 1 + \dfrac{1}{x + 2}$

11. $\dfrac{AB}{A + B}$

12.

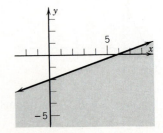

13.

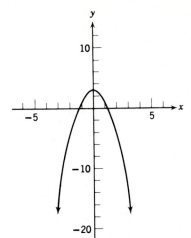

14. $(1, 2)$

15. *Steps 1, 2*   Number of nickels: x
Number of dimes: $x + 3$
Number of quarters: $x + 5$

Step 4 $0.05x + 0.10(x + 3) + 0.25(x + 5) = 3.15$

Step 6 $x = 4$;   there are 4 nickels, 7 dimes, and 9 quarters.

16. $3x \sqrt{10}$ **17.** $1 + 3 \sqrt{3}$ **18.** 4.23

19. *Steps 1, 2*   Distance from starting point: x

Step 4 $(12)^2 + (16)^2 = x^2$

Step 6 $x = \pm 20$;   the plane is 20 kilometers from its starting point.

20. $(0, -1)$

FINAL CUMULATIVE REVIEW X (Page 426)

1.

2. $2 \cdot 2 \cdot 2 \cdot 2 \cdot 3 \cdot 5 \cdot x \cdot x \cdot y$

3. $-5, -2, -1, 1, 3, 5, 7$

4. $-2a - 5b$

5. $2ab - 2b^2$

6. $3(x - 4y)(x - 2y)$

7. $\dfrac{1}{3a + 4}$

8. $\dfrac{b}{7}$

9. $\dfrac{a - 5}{(a - 3)^2}$

10. 2

11.

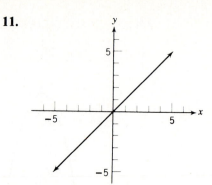

12. $(2, -1)$

13. $6\sqrt{3}$

14. $\sqrt{2} - \sqrt{3}$

15. $(-3, 0)$ and $(-1, 0)$

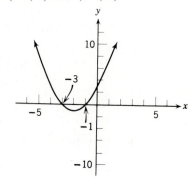

16. *Steps 1, 2*   First integer: x
Next integer: $x + 1$

     Step 4 $2(x + 1) - \frac{1}{2}x = 14$

     Step 6 $x = 8$;   the integers are 8 and 9.

17. *Steps 1, 2*   Original amount: x

     Step 4 $x - \frac{1}{3}x - \frac{1}{4}x = 250$

     Step 6 $x = 600$;   he originally had $600.

18. *Steps 1, 2*   Price of chair: x
Price of desk: $4x + 40$

     Step 4 $x + (4x + 40) = 640$

     Step 6 $x = 120$;   the chair cost $120 and the desk $520.

19. *Steps 1, 2*   First angle: x
Second angle: $90 - x$

     Step 4 $\frac{2}{3}x + \frac{1}{2}(90 - x) = 50$

     Step 6 $x = 30$;   the angles are 30° and 60°.

20. *Steps 1, 2*   Length of third side: x

     Step 4 $(16)^2 + x^2 = (20)^2$

     Step 6 $x = \pm 12$;   the third side is 12 meters long.

APPENDIX EXERCISES (Page 431)

1.

3.

5.

7.

9.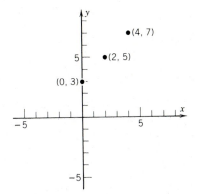

11.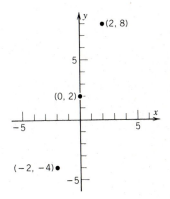

13.

15.

17.

19.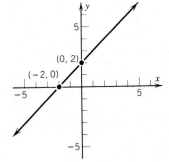

21.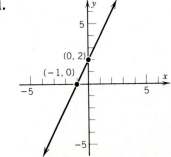

23.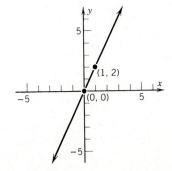

25. $A \cup B = \{2, 4, 6, 8\}$
$A \cap B = \{4, 6\}$

27. $A \cup B = \{-3, -1, 1, 3\}$
$A \cap B = \{-1, 1\}$

29. $A \cup B = \{2, 3, 4, 5, 6\}$
$A \cap B = \varnothing$

31. $A \cup B = \{(0, 1), (1, 2), (0, 2)\}$
$A \cap B = \{(1, 2)\}$

33. $A \cup B = \{(0, 0), (-1, -1), (-2, -2), (-4, -4)\}$
$A \cap B = \{(0, 0), (-2, -2)\}$

35.

37.

39.

41. $\{(3, 2)\}$

43. $\{(-1, 2)\}$

45. $f(-2) = 3,\ f(2) = 7$

47. $f(-5) = -17,\ f(-1) = -9$

49. $f(-1) = 7,\ f(0) = 4$

51. $f(2) = 1,\ f(6) = 7$

53. $f(-5) = -4,\ f(0) = -1$

55. $f(-1) = -1,\ f\left(\frac{5}{2}\right) = \frac{2}{5}$

Table of Metric Conversions
(approximate conversion factors)

Length

1 in. = 2.54 cm (exact)	1 mm = 0.04 in.
1 ft = 0.30 m	1 cm = 0.39 in.
1 yd = 0.91 m	1 m = 39.37 in.
1 mi = 1.61 km	1 km = 0.62 mi

Area

1 sq in. = 6.45 sq cm	1 sq mm = 0.0016 sq in.
1 sq ft = 0.09 sq m	1 sq cm = 0.16 sq in.
1 sq yd = 0.84 sq m	1 sq m = 10.76 sq ft
1 sq mi = 2.59 sq km	1 sq km = 0.39 sq mi

Volume

1 cu in. = 16.39 cc	1 cc = 0.06 cu in.
1 cu ft = 0.028 cu m	1 cu m = 35.31 cu ft
1 cu yd = 0.76 cu m	1 cu m = 1.31 cu yd

Mass and Weight

1 oz = 28.35 g	1 g = 0.035 oz
1 lb = 454 g	1 kg = 2.20 lb
1 ton = 0.907 metric ton	1 metric ton = 2205 lb = 1.10 ton

Capacity

1 pint = 0.47 liter	1 ml = 0.002 pints
1 qt = 0.95 liter	1 liter = 1.06 qt
1 gal = 3.78 liters	1 kl = 265 gal

Table of Squares, Square Roots, and Prime Factors

No.	Square	Square Root	Prime Factors	No.	Square	Square Root	Prime Factors
1	1	1.000		51	2,601	7.141	$3 \cdot 17$
2	4	1.414	2	52	2,704	7.211	$2^2 \cdot 13$
3	9	1.732	3	53	2,809	7.280	53
4	16	2.000	2^2	54	2,916	7.348	$2 \cdot 3^3$
5	25	2.236	5	55	3,025	7.416	$5 \cdot 11$
6	36	2.449	$2 \cdot 3$	56	3,136	7.483	$2^3 \cdot 7$
7	49	2.646	7	57	3,249	7.550	$3 \cdot 19$
8	64	2.828	2^3	58	3,364	7.616	$2 \cdot 29$
9	81	3.000	3^2	59	3,481	7.681	59
10	100	3.162	$2 \cdot 5$	60	3,600	7.746	$2^2 \cdot 3 \cdot 5$
11	121	3.317	11	61	3,721	7.810	61
12	144	3.464	$2^2 \cdot 3$	62	3,844	7.874	$2 \cdot 31$
13	169	3.606	13	63	3,969	7.937	$3^2 \cdot 7$
14	196	3.742	$2 \cdot 7$	64	4,096	8.000	2^6
15	225	3.873	$3 \cdot 5$	65	4,225	8.062	$5 \cdot 13$
16	256	4.000	2^4	66	4,356	8.124	$2 \cdot 3 \cdot 11$
17	289	4.123	17	67	4,489	8.185	67
18	324	4.243	$2 \cdot 3^2$	68	4,624	8.246	$2^2 \cdot 17$
19	361	4.359	19	69	4,761	8.307	$3 \cdot 23$
20	400	4.472	$2^2 \cdot 5$	70	4,900	8.367	$2 \cdot 5 \cdot 7$
21	441	4.583	$3 \cdot 7$	71	5,041	8.426	71
22	484	4.690	$2 \cdot 11$	72	5,184	8.485	$2^3 \cdot 3^2$
23	529	4.796	23	73	5,329	8.544	73
24	576	4.899	$2^3 \cdot 3$	74	5,476	8.602	$2 \cdot 37$
25	625	5.000	5^2	75	5,625	8.660	$3 \cdot 5^2$
26	676	5.099	$2 \cdot 13$	76	5,776	8.718	$2^2 \cdot 19$
27	729	5.196	3^3	77	5,929	8.775	$7 \cdot 11$
28	784	5.292	$2^2 \cdot 7$	78	6,084	8.832	$2 \cdot 3 \cdot 13$
29	841	5.385	29	79	6,241	8.888	79
30	900	5.477	$2 \cdot 3 \cdot 5$	80	6,400	8.944	$2^4 \cdot 5$
31	961	5.568	31	81	6,561	9.000	3^4
32	1,024	5.657	2^5	82	6,724	9.055	$2 \cdot 41$
33	1,089	5.745	$3 \cdot 11$	83	6,889	9.110	83
34	1,156	5.831	$2 \cdot 17$	84	7,056	9.165	$2^2 \cdot 3 \cdot 7$
35	1,225	5.916	$5 \cdot 7$	85	7,225	9.220	$5 \cdot 17$
36	1,296	6.000	$2^2 \cdot 3^2$	86	7,396	9.274	$2 \cdot 43$
37	1,369	6.083	37	87	7,569	9.327	$3 \cdot 29$
38	1,444	6.164	$2 \cdot 19$	88	7,744	9.381	$2^3 \cdot 11$
39	1,521	6.245	$3 \cdot 13$	89	7,921	9.434	89
40	1,600	6.325	$2^3 \cdot 5$	90	8,100	9.487	$2 \cdot 3^2 \cdot 5$
41	1,681	6.403	41	91	8,281	9.539	$7 \cdot 13$
42	1,764	6.481	$2 \cdot 3 \cdot 7$	92	8,464	9.592	$2^2 \cdot 23$
43	1,849	6.557	43	93	8,649	9.644	$3 \cdot 31$
44	1,936	6.633	$2^2 \cdot 11$	94	8,836	9.695	$2 \cdot 47$
45	2,025	6.708	$3^2 \cdot 5$	95	9,025	9.747	$5 \cdot 19$
46	2,116	6.782	$2 \cdot 23$	96	9,216	9.798	$2^5 \cdot 3$
47	2,209	6.856	47	97	9,409	9.849	97
48	2,304	6.928	$2^4 \cdot 3$	98	9,604	9.899	$2 \cdot 7^2$
49	2,401	7.000	7^2	99	9,801	9.950	$3^2 \cdot 11$
50	2,500	7.071	$2 \cdot 5^2$	100	10,000	10.000	$2^2 \cdot 5^2$

INDEX

SUMMARY OF ALGEBRAIC OPERATIONS AND PROPERTIES

CHAPTER 1

commutative properties

$$a + b = b + a$$
$$a \cdot b = b \cdot a$$

associative properties

$$(a + b) + c = a + (b + c)$$
$$(a \cdot b) \cdot c = a \cdot (b \cdot c)$$

If $a = b$, then

$$a + c = b + c, \quad a - c = b - c$$

$$\frac{a}{c} = \frac{b}{c} \quad \text{and} \quad a \cdot c = b \cdot c$$

If $a = b$, then $b = a$.

CHAPTER 2

To add two numbers with

  like signs: add the absolute values of the numbers and prefix the common sign of the numbers to the sum.

  unlike signs: find the nonnegative difference of the absolute values of the numbers and prefix the sign of the number with the larger absolute value to the difference.

To subtract two numbers, use the property

$$a - b = a + (-b).$$

To find the product of two numbers, multiply the absolute values of the numbers. If the factors have like signs, the product is positive; if they have unlike signs, the product is negative.

To find the quotient of two numbers, find the quotient of the absolute values of the numbers. If the dividend and divisor have like signs, the quotient is positive; if they have unlike signs, the quotient is negative.

CHAPTER 3

distributive property

$$a \cdot c + b \cdot c = (a + b) \cdot c$$

$$a \cdot (b + c) = a \cdot b + a \cdot c$$

If $a < b$, then $a + c < b + c$.

If $a < b$ and $c > 0$, then $ac < bc$.

If $a < b$ and $c < 0$, then $ac > bc$.

CHAPTER 4

$$a^m \cdot a^n = a^{m+n}$$

$$(a^m)^n = a^{mn}$$

$$(ab)^n = a^n b^n$$

CHAPTER 5

$$(a + b)(c + d) = ac + ad + bc + bd$$

$$(a - b)(a + b) = a^2 - b^2$$